FOCUS ON SCIENCE AND TECHNOLOGY FROM A GEORGIAN PERSPECTIVE

SCIENTIFIC REVOLUTIONS

Additional books in this series can be found on Nova's website
under the Series tab.

Additional E-books in this series can be found on Nova's website
under the E-books tab.

FOCUS ON SCIENCE AND TECHNOLOGY FROM A GEORGIAN PERSPECTIVE

SERGO GOTSIRIDZE
EDITOR

Nova Science Publishers, Inc.
New York

LIBRARY OF CONGRESS CATALOGING-IN-PUBLICATION DATA

Focus on science and technology from a Georgian perspective / editor, Sergo Gotsiridze.
    p. cm.
  Includes bibliographical references and index.
  ISBN 978-1-61209-970-5 (hardcover)
  1. Research--Georgia (Republic) 2. Science--Georgia (Republic) 3.
Technology--Georgia (Republic) I. Gotsiridze, Sergo.
  Q180.G28F63 2011
  509.4758--dc22
                          2011005300

Published by Nova Science Publishers, Inc. † New York

CONTENTS

PREFACE[*]

This new book presents and discusses current research from Georgian authors in all areas of science and technology. Topics discussed include mathematical simulation of capillary blood flow disorders; infinite versions of some classical results in linear algebra and vector analysis; measurability of unions of plane disks; modelling of elastic waves generated by a point explosion; investment environment in Georgia; uniform measures in banach spaces and Riemann integrability and uniform distribution in infinite-dimensional rectangles. (Imprint: Nova Press)

In Chapter 1, the process of rapprochement and aggregation of erythrocytes and decomposition of the aggregates in the capillary blood motion is simulated. The mathematical model describing capillary blood motion is based on the mechanics of solid deformable bodies.

Blood motion along larger vessels is usually described as a flow of a viscous liquid or suspension along cylindrical vessels. In the case of microvessels, where the erythrocyte size is comparable to the vessel diameter, it does not seem reasonable to regard the blood moving along microvessels either as a suspension or as a viscous liquid. A description involving mechanics of both liquid and solid bodies would result in a too sophisticated model.

We, certainly, realize that neither the erythrocyte nor the plasma have pronounced characteristics of solid deformable bodies, nevertheless our model is based on the mechanics of solid deformable bodies, in particular, on linear elasticity.

Our approach is justified by the following considerations. Apparently, there is no marked boundary between liquids and solid deformable bodies. Therefore we can assume that erythrocytes are formations that possess the properties of solid bodies while the plasma can also be considered solid due to its high viscosity.

We believe that our approach will provide a first approximation to the process of rapprochement of erythrocytes or their getting apart in living blood capillaries.

In Chapter 2, the anti-tumor effects of polyvaccine of Staphylococcus-Proteus-Escherichia-Klebsiella and polychemotherapy (Doxorubicin, Vincristin, and Cyclophos-

[*]Versions of these chapters were also published in Georgian International Journal of Science and Technology, Volume 2, Issues 1-3, edited by Sergo Gotsiridze, published by Nova Science Publishers, Inc. They were submitted for appropriate modifications in an effort to encourage wider dissemination of research.

phanum) at sarcoma-45 growth in non-purebred laboratory rats have been studied. Anti-tumor effects were evaluated according to the following parameters: Index of malignant tumor growth and development (rate of inoculated tumor development; variations in size/volume of tumor tissue (VTT); body mass index of experimental animals (BMI); and prolongation of experimental animals' life-span (LS).

Results of experiments have shown that application of bacterial vaccines being used against hospital infections is well tolerated in rats and does not stimulate, or support malignant tumor growth. Immunization using tetravaccine of Staphylococcus, Proteus, E.colli, Klebsiella has positive anti-tumor influance on experimental animals manifested by delay/inhibition of tumor growth, regression of tumor tissue, prolongation of animals' life-span. Bacterial vaccines in combination with chemical remedies significantly improve effects of chemotherapy. At immunization with tetravaccine tumor did not develop in 15% of cases, shrinking/complete regression of tumor were observed in 35% of cases; and prolongation of LS consisted 43,2%; In case of combined treatment (immunization and poly-chemotherapy) all experimenttal animals survived - tumor did not develop in 20% of cases, and in remaining 80% of cases the shrinking and complete regression of tumor was observed.

In Chapter 3, a generalized mathematical scheme of random processes, with the help of which it is possible to describe a wide range of real queuing systems (data transmitting networks, computer systems, etc.) and to solve some problems of the reliability theory, is offered. The necessity in developing such a scheme has arisen in the connexion with an increasing sophistication of real automated control systems when the available methods do not cover all problem statements that interest a practical worker. So, it is necessary to develop an analytical apparatus capable to consider a great variety of active factors in order to create the models which fit the original adequately, i.e. are maximally approximated to the original.

After the collapse of the Communist regimes and their command economies, the countries of the former Soviet Union found themselves with only a very small amount of goods to supply to the global market. In fact, no markets existed for many types of products. There was no way that they could have existed in that an economy of this type is nothing more than a corpse or a so-called "necroeconomy." The purpose of Chapter 4 is to distinguish the various economic foundations of post-Communist capitalism and to examine the key economic problems of this type of society in the context of the modern financial crisis. As international experience shows, dead firms do exist and "successfully" function in the most developed of economies as well with Japan being the most obvious example. These insolvent and, in fact, bankrupt firms which continue to operate despite their "mortality" are commonly referred to as "zombie-firms." Unlike developed economies, which are exposed to the threat of the zombie-ing of the economy under the conditions of a financial crisis, this threat is even greater for the countries of post-Communist capitalism owing also to their exposure to necroeconomy.

Chapter 5 discusses the task of defining the optimal height of composed welded beams given in the works of different prominent authors. The study points out those inaccuracies and inconsistencies that exist despite preliminary estimates and that make it difficult and often impossible to properly use a mathematical device. The study shows how to define the above-mentioned inconsistencies and namely the coefficient c, how to define the dependence of δ_{well} on h as well as how to avoid them in order to make a correct selection of beams' heights from the perspective of both mathematical and construction mechanics methods.

In Chapter 6, the integration of differential equations of classic theory of elasticity with partial derivatives for concrete task considering the given boundary conditions is the intractable problem. The new direction, elaborated by I. Gudushauri /1/ gives possibility to overcome successfully difficulties of this kind.

In Chapter 7, infinite versions of Cramer's, Ostrogradsky's and Liouville's theorems were considered in some infinite-dimensional topological vector spaces.

Chapter 8 describes the class of all infinite-dimensional diagonal matrices $\mathbb{A}$ and the class of all infinite collections of continuous impulses $\mathbb{F}$ such that Mankiewicz generator $\mathbb{G}_{\mathbb{M}}$ preserves a phase flow (in $\mathbb{R}^{\mathbb{N}}$) defined by a non-homogeneous differential equation $\frac{d\Psi}{dt} = A \times \Psi + f$ for $A \in \mathbb{A}$, $f \in \mathbb{F}$. An analogous question is studied for Preiss-Tišer generators in $\mathbb{R}^{\mathbb{N}}$.

In Chapter 9 we show that if S is a planar set of positive two-dimensional Lebesgue measure ℓ_2, then for arbitrary c, the S contains the vertices of a triangle of area c if and only if it is not bounded by a natural norm. Also, we consider a certain generalized problem of P. Erdösh asking whether there exists a positive constant c such that every planar set with outer measure larger than c contains the vertices of a triangle of area one. We prove that in the system of axioms $ZFC + MA$, the answer to this question is no.

Specialized in Solovay model [9], we prove in Chapter 10 that any Haar null set in $R^n (n > 2)$ can be decomposed into two at most 1-dimensionally Haar null sets. Moreover, we describe an algorithm which gives a partition $\{E_k : 1 \leq k \leq n\}$ of the Lebesgue null set $E \subset \mathbb{R}^n$ such that e_k is transverse to E_k for $1 \leq k \leq n$. This partition answers negatively on H. Shi's question [6]. In addition, we consider J.Mycielski's extension of this result in the theory (ZF) & (DC).

For $n \in \mathbf{N}$, let $G_{P\&T}^n$ be an n-dimensional Preiss-Tišer generator in an infinite-dimensional topological vector space V (cf. [7]). For $D \subseteq V$, we establish in Chapter 11 the validity of the following formula

$$\dim(D) = \inf\{m - 1 : G_{P\&T}^{(m)}(\overline{D}) = 0\},$$

where $\overline{D}$ denotes a closed convex hull of the set D in V.

In Chapter 12, using N.Lusin's arithmetic example of a non-Borel analytic set [4], an example of an uncountable family of polygons is constructed in the Euclidean plane $\mathbb{R}^2$ such that their union is non-Borel. In the system of axioms $ZF + DC$ it is proved that for every uncountable family of plane disks their union is ℓ_2-measurable iff every subset of the real axis $\mathbb{R}$ is ℓ_1-measurable.

In Chapter 13, is studied some differential properties of functions $f(x)$ belonging to the intersection of Lizorkin-Triebel-Morrey type spaces with dominant mixed derivatives $S_{p_\mu,\theta_\mu,a,æ,\tau}^{l^\mu} F(G)$ $(\mu = 1, 2, \ldots, N)$.

As discussed in Chapter 14, before the 60s of the XXI century small business was considered to have no prospects and to be the most weakly developed sector of economy. It started development after the end of the II World War, when developed countries orientated at full employment of the population and mass production of goods. After that small business has become one of the indispensable and dynamically developing parts in the economy of developed countries. Due to its special place it plays an essential role not only in socio-economic, but in public-political life of these countries as well. Through small business these countries achieve common national welibeing, social security and stability of wide circles of the society. Quite important part of GDP and national income is being formed in the sphere of small business. Diverse produce of both production designation and wide consumption is being prepared here in large volumes. However, more important is the situation that a considerable part of the population able to work is engaged just in small business. For example, in 6 countries of EU (Spain, the Netherlands, Belgium, Germany, France, and Italy) a specific share of employed in small and medium enterprises fluctuates within 40-70%. Small and medium enterprises of these countries make up 90% of their general amount and they produce 35-71% of newly formed value of the produce (1, p 8).

As explained in Chapter 15, nowadays chemically active antiwear and antiscoring dopes and dispersive additives (graphite, molybdenum disulfide, powder-like lead, copper, babbit etc.) are used for improving the tribological properties of plastic lubricants.

Currently technical carbon (carbon black), which is characterized by high thermal and chemical stability and also by low cost, is widely used as a lubricant additive. At the sume time, commercial carbon black does not manifest high antiscoring properties. Moreover, in some cases, by catalyzing the oxidotiun processes, it makes worse the thermooxidation stability of dispersion medium and, hence, the rheological and protective properties of lubricants. Consequently, we have modified the commercial carbon black for improving its efficiency.

On the point explosion area boundary, the spectrum of elastic oscillations, which generate volumetric standing waves, and also the index of its discreteness (ratio between individual frequencies) should not differ essentially from the frequency spectrum of volumetric seismic waves recorded at a far distance from the point explosion center. Therefore, in Chapter 16, using the proposed model, after defining the basic frequencies of natural oscillations of the point explosion area boundary by the spectral analysis of seismograms, we can obtain a sufficiently correct analytic solution of the inverse seismological problem which consists in defining the linear characteristics of the point explosion area and estimating the energy released.

As explained in Chapter 17, very low frequency (VLF) electromagnetic radiation (in diapason 1 kHz - 1MHz) in atmosphere, generated during earthquake preparation period, may be connected with linear size, characterizing incoming earthquake source. In order to argue this hypothesis very simple quasi-electrostatic model is used: local VLF radiation may be the manifestation of own electromagnetic oscillations of concrete seismoactive segments of lithosphere-atmosphere system. This model explains qualitatively well-known precursor effects of earthquakes. At the same time, it will be principally possible to forecast expected earthquake with certain precision if we use this model after diagnosing existed data.

On the base this model we consider modeling task according to which simultaneously of origination of the own electromagnetic oscillation of the LAI system some segment or certain delay it is possible to observe periodic perturbations of the geomagnetic fields in the incoming earthquakes epicentral area, frequencies of which will be rather less than characteristic frequencies of the VLF electromagnetic emission.

The aim of Chapter 18 was to determine the percentage of CD34+, CD34-CD133+ and GP-A+ cells in peripheral blood nuclear cells of the study population and its relation to the severity of the condition.

The data obtained indicate that symptoms of HF are associated with the increase in peripheral blood stem cells and progenitors when compared to the control subjects. Increase in GP-A is specific only for relatively mild forms of CHF (I-IIfc, NYHA), while aggravation of the condition (HF III-IV fc, NYHA) is associated with significant drop in GP-A levels in peripheral blood. Stem cell and progenitor CD34+, CD34-CD133+) number increases in parallel with the progression of CHF.

In Chapter 19, investigations were arranged to evaluate the effect of pre-fermentation enzyme maceration on the quantity of phenol compounds, to research the influence of the used enzyme preparation on red wine quality, to determine resveratrol and to enhance its content in Georgian wine, derived from grape variety "Saperavi".

Folin - Ciocalteu spectrophotometric method was applied for determination of total phenols. Resveratrol and anthocyanins were determined by HPLC; Volatile and titriable acids, SO2, sugar and alcohol were measured by conventional methods.

The obtained results reveal, that pre-fermentation enzyme maceration with the above mentioned enzyme preparation increased the content of anthocyanins (by 221mg/l) and phenol compound (by 790 mg/l); Also were increased of alcohol content (by 1, 2 vol. %) and the amount of dry extract (by 3, 2 g/l). Both cis- and trans- forms of resveratrol are presented in Georgian wine prepared from grape variety "Saperavi". It was found, that the enzymatic treatment caused the additional extraction of this compound in wine. There were no big differences between titriable and volatile acids in control and experimental samples. Organoleptic evaluation confirmed that pre-fermentation enzyme maceration with LAFASE ® HE GRAND CRU did not give additional astringency and bitterness.

LAFASE ® HE GRAND CRU can contribute to improve Georgian wine quality.

An accumulated experience in geology shows ineffectiveness of the existing approaches to make a prognosis of the deposits when the available crust characteristics do not have explicit relationships to a given deposit. That takes generally place in cases of more valuable deposits as gold, diamond and so forth. Chapter 20 proposes a use of an Artificial Neuron Network for prognosis of diamond deposits that was successfully carried out about 20 years ago while the worked out method makes it possible its application to any other ores or natural resources.

From the contemplation of economic development, Georgia is now on the transitional level. So, for the improvement of the economic and social state of the country, the investment flow has the great importance, as explained in Chapter 21. The main aim of the country nowadays is integration in the world economic sphere, which can be reached by harmonization of business environment to the world main demands in this field. If we look through the history we will see that the only way to overcome unemployment and poverty is holding the institutional reforms, growing investment activity. From this point of view the possibilities of the local funds are scant. That is why the main priority for the country is to attract the foreign investments, which is quite difficult as the strategic products, like oil and gas, interesting for the foreign investors, are not presented in great volume in Georgia. This is the motivating reason for Georgia to use open economic policy, which is based on the experience of the countries, which developed quickly since 60th of XX century. We are talking about "Asian tigers" - Thailand, Singapore, and Hong Kong, which used the open economic policy and development of attractive business environment, in order to reach success. But these actions are to be based on the important institutional changes. In Georgia such institutional changes must be directed at harmonization with the economic sphere of the world. Besides it is clear that any successful convergence does not happen only as a result of market economy, but it takes place as a result of free market economy. That is why it is very important to take into the consideration the institutional reforms, being executed in the country, as they may become the foundation of the fast economic development of Georgia.

As discussed in Chapter 22, globalization has resulted in increasing levels of trade and investment throughout the world. Reductions in the levels of domestic production, an increase in the number of trade blocs and agreements, and improved technology have facilitated this increase in trade and investment. Strengthening the world's trade system is a determinant factor for democratic processes throughout the world. When a trade system is free and open, the economy has a better chance to increase, which conditions high standards of living. Globalization has significantly boosted economic growth in East Asian economies such as Hong Kong (China), the Republic of Korea, and Singapore. But not all developing countries are equally engaged in globalization or in a position to benefit from it. In fact, except for most countries in East Asia and some in Latin America, developing countries have been rather slow to integrate with the world economy. An actively trading country benefits from the new technologies that "spill over" to it from its trading partners, such as through the knowledge embedded in imported production equipment. These technological spill over's are particularly important for developing countries, because they give them a chance to catch up more quickly with the developed countries in terms of productivity. Former centrally planned economies, which missed out on many of the benefits of global trade because of their politically imposed isolation from market economies, today aspire to tap into these benefits by reintegrating with the global trading system. But active participation in international trade also entails risks, particularly those associated with the strong competition in international markets. The countries, where trade liberation happened, and who have studied competition in global markets, economy growth and development are becoming more advanced, compared with those countries where local markets are protected.[1]

In Chapter 23, it is stated for what α (summability exponent) and for what classes of functions the mean convergence of Cesaro (C, α)-mean Fourier-Laplace series takes place on a sphere.

In chapter 24 we prove that a non-existence of a measurable cardinal implies a non-existence of such a translation-invariant Borel measure μ in an arbitrary infinite-dimensional Banach space for which the closed unit ball has μ measure 1. This answers negatively to the question of D. Fremlin [3]. For an arbitrary infinite parameter set α, we construct a uniform measure in the Banach space ℓ^α and show that this measure has no a uniqueness property.

In Chapter 25 we consider a concept of the uniformly distribution for increasing sequences of finite subsets in an infinite-dimensional rectangle, and by using the technique of the infinite-dimensional Lebesgue measure, introduce a notion of Reimann integrability for functions defined on the entire rectangle. Further, we prove an infinite-dimensional version of Weyl famous result.

---

[1]B. Ohlin. "Interregional and International Trade". CAMB. 1997 P 13.

Our method of construction of measures is a concept of strict standard and strict ordinary products of an infinite family of (no only σ-finite) measures, which allows us to construct Mankiewicz and Preiss-Tišer generators on $\mathbf{R}^{\infty}$. We show in Chapter 26, that if $f : \mathbf{R}^{\infty} \to R$ is a Lipschitz function and $R^{(N)}$ is a group of all eventually zero sequences, then, in Solovay model, f is λ-almost everywhere $R^{(N)}$-Gâteaux differentiable on $\mathbf{R}^{\infty}$, where λ denotes the strict standard Lebesgue measure on $\mathbf{R}^{\infty}$.

It is shown in Chapter 27, how Lebesgue constants for Cesáro (C, α)-means of Fourier series in terms of generalized spherical functions depend on the summation index α.

In: Focus on Science and Technology...
Editor: Sergo Gotsiridze, pp. 1-6

ISBN: 978-1-61209-970-5
© 2011 Nova Science Publishers, Inc.

Chapter 1

MATHEMATICAL SIMULATION OF CAPILLARY BLOOD FLOW DISORDERS

N. Khomasuridze[*]*, Z. Siradze and D. Gorgidze*

Mathematical Simulation of Capillary Blood Flow Disorders
Institute of Applied Mathematics of Tbilisi State University,
2 Universiteti str., Tbilisi

Abstract

The process of rapprochement and aggregation of erythrocytes and decomposition of the aggregates in the capillary blood motion is simulated. The mathematical model describing capillary blood motion is based on the mechanics of solid deformable bodies.

Blood motion along larger vessels is usually described as a flow of a viscous liquid or suspension along cylindrical vessels. In the case of microvessels, where the erythrocyte size is comparable to the vessel diameter, it does not seem reasonable to regard the blood moving along microvessels either as a suspension or as a viscous liquid. A description involving mechanics of both liquid and solid bodies would result in a too sophisticated model.

We, certainly, realize that neither the erythrocyte nor the plasma have pronounced characteristics of solid deformable bodies, nevertheless our model is based on the mechanics of solid deformable bodies, in particular, on linear elasticity.

Our approach is justified by the following considerations. Apparently, there is no marked boundary between liquids and solid deformable bodies. Therefore we can assume that erythrocytes are formations that possess the properties of solid bodies while the plasma can also be considered solid due to its high viscosity.

We believe that our approach will provide a first approximation to the process of rapprochement of erythrocytes or their getting apart in living blood capillaries.

A mathematical model of capillary blood flow is suggested which describes the process of rapproachement of erythrocytes and their subsequent aggregation and which, on the other hand, cane also be used to study the process of dissociation of the aggregated erythrocytes.

---

[*] E-mail address: burjanadze@medinserv.ge

Capillary blood flow is viewed as a slow flow of a highly viscous incompressible liquid (the plasma) containing autonomous formations (erythrocytes) along a cylindrical tube. In other words, a boundary value contact problem of a liquid mixture consisting of plasma with its coefficient of dynamic viscosity μ_p and erythrocytes, which represent a liquid with a another coefficient of dynamic viscosity μ_v.

As we have already mentioned, capillary blood flow (the flow of plasma and erythrocytes along capillaries) is represented as a slow flow of a highly viscous incompressible liquid, consequently in our case Navier-Stokes' equations take the following form:

$$gradP + rotrot\vec{U} = 0,$$
$$div\vec{U} = 0,$$
(1)

where $\vec{U} = u\vec{i} + v\vec{j} + w\vec{k}$ is the vector of displacement velocity in the Cartesian system of coordinates x, y, z; $P = -\dfrac{1}{3\mu}(X_x + Y_y + Z_z)$ X_x, Y_y, Z_z and $X_y = Y_x \, Y_z = Z_y \, X_z = Z_x$ are velocities of variation in normal and tangential stresses; μ is the coefficient of dynamic viscosity.

System (1) is called Stokes' equations. These equation describe blood flows in narrow blood vessels, of a diameter not exceeding 15-20 mcm, with a sufficient degree of precision [1], [2].

Before we go on with description we should note that Equation (1) can also describe elastic equilibrium of incompressible solids In that case $\vec{U} = u\vec{i} + v\vec{j} + w\vec{k}$ is a displacement vector, X_x, Y_y, Z_z are normal and tangential stresses, $P = -\dfrac{1}{3\mu}(X_x + Y_y + Z_z)$ and μ is a shear modulus. This circumstance makes it evident that there is no district borderline between liquid and solid media and shows their intrinsic similarity. Therefore one of our approaches to capillary blood flow simulation is based on the elasticity of a compressible solid with Poisson's ratio of 0.48-0.49.

Consider the equation for the flow of a viscous incompressible liquid (1).

The equation which establish the connection between the velocity of stress variations and the displacement velocities have the following form:

$$X_x = \mu(2\frac{\partial u}{\partial x} - P), X_y = Y_x = \mu(\frac{\partial v}{\partial x} + \frac{\partial u}{\partial y}),$$

$$Y_y = \mu(2\frac{\partial v}{\partial y} - P), Y_z = Z_y = \mu(\frac{\partial w}{\partial y} + \frac{\partial v}{\partial z}),$$
(2)

$$Z_x = \mu(2\frac{\partial w}{\partial z} - P), X_z = Z_y = \mu(\frac{\partial u}{\partial z} + \frac{\partial w}{\partial x}).$$

The capillary itself with blood flowing along it is represented as a non-deformable circular cylindrical shell with a highly viscous incompressible liquid-plasma-flowing along it, the coefficient of the dynamic viscosity of the liquid being μ_p, the liquid contacts the erythrocytes in it which are represented as independent liquid formations with the coefficient of dynamic viscosity μ_r. A two-dimensional case of the of the described flow, when $u = 0, v = v(y,z), P = P(y,z)$, is shown in Figure 1.

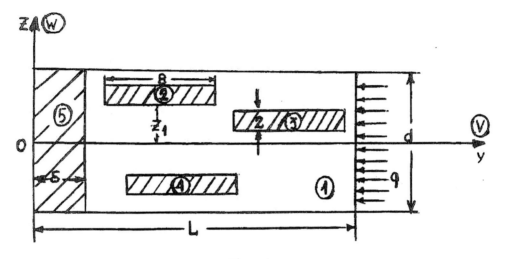

Figure 1.

In this case equations (1) and (2) will take the following form:

$$\frac{\partial P^{(n)}}{\partial y} + \frac{\partial K_x^{(n)}}{\partial z} = 0, \quad \frac{\partial v^{(n)}}{\partial y} + \frac{\partial w^{(n)}}{\partial z} = 0,$$

$$\frac{\partial P^{(n)}}{\partial z} - \frac{\partial K_x^{(n)}}{\partial y} = 0, \quad \frac{\partial w^{(n)}}{\partial y} - \frac{\partial v^{(n)}}{\partial z} = K_x^{(n)}, \tag{3}$$

here $K_x^{(n)} = \mu(2\frac{\partial w}{\partial y} - \frac{\partial v}{\partial z}), \ n = 1,2,3,4,5,$

$$Y_y = \mu(2\frac{\partial v}{\partial y} - P), \ Y_z = \mu(2\frac{\partial w}{\partial y} + \frac{\partial v}{\partial z}),$$

$$Z_z = \mu(2\frac{\partial w}{\partial z} - P), \ X_x = -\mu P, \tag{4}$$

$$P = -\frac{1}{3\mu}(X_x + Y_y + Z_z).$$

In Figure 1 the domains occupied by plasma, erythrocytes and hypothetical level are denoted by 1, 2,3,4 and 5, respectively.

The flow in a capiilary tube of a diameter d is created by the pressure q applied at the cylinder face $y = L$. It should be noted that L is either the entire length of the capillary or part of this length. Finally, δ is the width of the layer of the hypothetical liquid with the coefficient of the dynamic viscosity μ_h.

Since the Fahraeus-Lindqvist effect takes place when blood flows along capillaries with $z = \pm d/2$ and $\delta < y < L$ we consider a sliding contact corresponding to boundary conditions when plasma particles (only plasma particles are assumed to contact capillary walls) cannot move along z, but can easily move along y

$$a) \ z = \pm d/2 : \ Y_z^{(1)} = 0, \ w^{(1)} = 0;$$
$$b) \ y = L : \ Y_y^{(1)} = q, \ Y_z^{(1)} = 0. \tag{5}$$

There are sliding contact conditions between plasma and erythrocytes, e.g., for $z = z_1$ we have

$$w^{(1)} = w^{(3)}, \ Y_z^{(1)} = 0,$$
$$Z_z^{(1)} = Z_z^{(3)}, \ Z_z^{(3)} = 0. \tag{6}$$

The following boundary and contact conditions for the hypothetical layer will appear in the present paper (the hypothetical layer itself is denoted by 5 in Figure 1):

For $z = \pm d/2$ and $0 < y \leq \delta$ we have

$$a) \ w^{(5)} = 0, \ v^{(5)} = 0,$$
$$b) \ Z_z^{(5)} = 0, \ Z_y^{(5)} = 0; \tag{7}$$

For $z = d/2$ and $0 < y \leq \delta$ we have $w^{(5)} = 0$, $v^{(5)} = 0$, or $\tag{7'}$

For $z = -d/2$ and $0 < y \leq \delta$ we have $Z_z^{(5)} = 0$, $Z_y^{(5)} = 0$;

For $y = 0$ we have $\qquad\qquad Y_y^{(5)} = 0, \ w^{(5)} = 0; \tag{8}$

For $y = \delta$ $\qquad$
$$v^{(1)} = v^{(5)}, \ w^{(1)} = w^{(5)},$$
$$Y_y^{(1)} = Y_y^{(5)}, \ Y_z^{(1)} = Y_z^{(5)}. \tag{9}$$

We must say a few words about the role of the hypothetical layer. The mechanism of intravascular aggregation of erythrocytes is determined by a number of factors, i.e. a growing concentration of adhesive molecules, higher viscosity of blood plasma, changes in capillary curvature, etc. Erythrocyte aggregation can also be induced artificially – applying suitable agents at small capillary intervals. We simulate these critical situations by means of a suitable selection of the dynamic viscosity μ_h and boundary conditions on the interval $z = \pm \frac{d}{2}, 0 < y \leq \delta$. In particular, if $\mu_h > \mu_p$ and Condition (7a) is satisfied, the blood flow slows down, which, in its turn, simulates rapproachement of erythrocytes with their subsequent aggregation. On the other hand, if instead of Conditions (7a) Condition (7b) is considered and $\mu_h > \mu_p$ then we can simulate a growing velocity on the interval $0 < y < \delta$ and, hence dissociation of aggregated erythrocytes.

As for Conditions (7'), they simulate a turn in the flow of erythrocytes together with plasma.

The corresponding boundary value and contact problems are supposed to have an approximate solution, in particular, using the method of finite difference. In this case it is convenient to apply a general solution of Papkovitch-Nneuber type and assume

$$\vec{U} = grad(\frac{1}{2}\vec{R} \cdot \vec{\Phi}) - \vec{\Phi},$$

$$P = div \cdot \vec{\Phi},$$

(10)

where

$$\vec{R} = x\vec{i} + y\vec{j} + z\vec{k}, \quad \vec{\Phi} = \varphi_1\vec{i} + \varphi_2\vec{j} + \varphi_3\vec{k}$$

with

$$\Delta\varphi_n = 0, \quad n = 1,2,3; \quad \Delta = \frac{\partial^2}{\partial x^2} + \frac{\partial^2}{\partial y^2} + \frac{\partial^2}{\partial z^2}.$$

In the two-dimensional case under consideration we have

$$v = \frac{1}{2}\frac{\partial}{\partial y}(y\varphi_2 + z\varphi_3) - \varphi_2,$$

$$w = \frac{1}{2}\frac{\partial}{\partial z}(y\varphi_2 + z\varphi_3) - \varphi_3,$$

(11)

$$P = \frac{\partial\varphi_2}{\partial y} + \frac{\partial\varphi_3}{\partial z}.$$

For the domains 1-5 in Figure 1 (all of them are polygonal) the finite difference method is very convenient due to the simplicity of the representation of the boundary and contact conditions given in the paper. This simplicity is also due to representations (10) and (11). In particular, (10) and (11) enable one to write algebraic equations with no pacing of the finite difference table (naturally, the spacing is assumed to be similar for all domains in both directions).

We should note that the method of finite difference can be applied likewise effectively for some other erythrocyte arrangement patterns in plasma.

After solving system (1) with the corresponding boundary and contact conditions we do the following: we write the expression $G(y) = \int_{-d/2}^{d/2} v^2 dz$ then the sexpression

$G_*(y) = \int_{-d/2}^{d/2} v_*^2 dz$ in the case when the hypothetical layer is absent, i.e. when $\mu_p = \mu_h$ and

boundary condition (9) if extended to the entire length of the cylinder L.

The ratio G/G_* will show that the blood flow either slows down or speeds up. When the ratio low, the blood flow slows down, which contributes to the aggregation of erythrocytes and, vice versa, a higher ratio gives the opposite effect.

In conclusion, the authors would like to express their gratitude to Prof. G.Cokelet for his valuable remarks and kind attention.

References

[1]   M.Sugihara-Seky and R.Skalak. Numerical study of Asymmetric Flow of Rod Blood Cells in Capillaries. *Microvascular Resaerch* **36**, 64-74 (1988)

[2]   R.Skalak, P.H. Chen and S.Chien. Effect of Hematocrit and Roulaux on Apparent Viscosity in Capillaries. *Biorheology,* **29**, 67-82 (1972).

In: Focus on Science and Technology…
Editor: Sergo Gotsiridze, pp. 7-13

ISBN: 978-1-61209-970-5
© 2011 Nova Science Publishers, Inc.

Chapter 2

ANTI-TUMOR EFFECTS OF POLYVACCINE OF *STAPHYLOCOCCUS-PROTEUS-ESCHERICHIA-KLEBSIELLA* AND POLYCHEMOTHERAPY

Ketevan Gambashidze[*,1]*, Paata Khorava*[2] *and Nino Bejitashvili*[1]

[1]Tbilisi State Medical University, Department of Pathophysiology
[2]National Cancer Center, Tbilisi, Georgia

Abstract

In the present work the anti-tumor effects of polyvaccine of Staphylococcus-Proteus-Escherichia-Klebsiella and polychemotherapy (Doxorubicin, Vincristin, and Cyclophos-phanum) at sarcoma-45 growth in non-purebred laboratory rats have been studied. Anti-tumor effects were evaluated according to the following parameters: Index of malignant tumor growth and development (rate of inoculated tumor development; variations in size/volume of tumor tissue (VTT); body mass index of experimental animals (BMI); and prolongation of experimental animals' life-span (LS).

Results of experiments have shown that application of bacterial vaccines being used against hospital infections is well tolerated in rats and does not stimulate, or support malignant tumor growth. Immunization using tetravaccine of Staphylococcus, Proteus, E.colli, Klebsiella has positive anti-tumor influance on experimental animals manifested by delay/inhibition of tumor growth, regression of tumor tissue, prolongation of animals' life-span. Bacterial vaccines in combination with chemical remedies significantly improve effects of chemotherapy. At immunization with tetravaccine tumor did not develop in 15% of cases, shrinking/complete regression of tumor were observed in 35% of cases; and prolongation of LS consisted 43,2%; In case of combined treatment (immunization and polychemotherapy) all experimenttal animals survived - tumor did not develop in 20% of cases, and in remaining 80% of cases the shrinking and complete regression of tumor was observed.

---

[*] E-mail address: ketitemo@hotmail.com ketigam@tsmu.edu. Telephone: (995-32)-379-885 (home) 995-91-21-55-92 (mobile), Address, to whom correspondence should be sent: 166/1 Ambrolaury str. Tbilisi, 0160, Georgia.

Introduction

According to the WHO data the number of cases of oncological disease as well as the mortality rate among cancer patients has been increasing worldwide. 6 million people died from cancer in the world during 2007 (American Cancer Society). Contemporary treatment of cancers is mainly through surgical intervention, chemical- and radiotherapy. However, even proficient treatments can fail, and as a consequence, prognosis is poor. The main problems with current treatment, again, even if carried out proficiently, include the side effects of chemo- and radiotherapy such as: myelodepression; cardio-, nephro-, hepato- and neurotoxicity; undercurrent microbial complications due to leucopenia and agranulocytosis; plus additional undesirable effects.

During the last decade, immunological methods have been employed as supplemental anti-tumor therapies. Nowadays, two main directions in anti-tumor immunotherapy involve specific- and nonspecific immunotherapy. Specific immunotherapy with the use of monoclonal antibodies has shown positive treatment effects. However, there are a number of obstacles to successful therapy. Some cells may express tumor antigens while others do not. Tumor blood flow is not always optimal and it is difficult to reliably get the therapy to the site. Very rarely do we see cross-reactivity with normal tissue antigens. At last, creation of a preparation is very expensive. LAK (Lymphokine-activated killers) cell therapy has shown promising results in animal studies, where it shrunk tumors in animals with lung, liver, and other cancers. While clinical trials in humans have not yet been as successful, researchers are constantly improving LAK cell techniques. Although, the early results of immunotherapy through suppression of T regulatory cells are encouraging it might lead to autoimmune diseases. Specific immunotherapy using dendritic cells is associated with technical problems and it is still under investigation [4.13.14].

Recent fundamental investigations in cancer immunotherapy have shown perspectives of administration of bacterial preparations at cancers (nonspecific immunotherapy). It is well known that at cancer growth the tumor microenvironment infiltrating immune cells – macrophages (M2- regulatory macrophages), foam cells, Myeloid-derived suppressor cells (MDSC), Th17 and T-reg lymphocytes are involved in the processes of immunosupresion thereby supporting the malignant tumor growth. Especially should be emphasized the role of T-reg subpopulation. T-regs producing cytokines (IL-10 and TGF- beta) directly, or indirectly results in inhibition of effector immune cells affecting processes of proliferation and activation of T-regulatory (phenotype CD4$^+$, CD25$^+$, FoxP3) and CD8 T-reg/suppressor (CD8+, CD25+, FoxP3) lymphocytes. Quantitative increase and exaggerated functional activity of T-reg cells results in reduction of defensive reactions directed against malignant growth. Thus, increased concentration of T-reg cells could be considered as indicator of unfavorable prognosis for the oncological diseases respectively [2.3.4].

In order to mobilize and activate anti-tumor immune protection, the components of bacterial origin, or microbial "patterns" (Pathogen-associated molecular patterns - PAMP) could be used successfully. PAMPs are small molecular motifs consistently found on pathogens. Bacterial lipopolysaccharide (LPS) is considered to be the prototypical PAMP. The main targets for PAMPs are dendritic cells, monocyte/macrophages, neutrophils, and NK-cells. They activate innate immune responses as a consequence of identifying non-self molecules, protecting the host from infections.

Bacterial LPS have potential to neutralize suppressive activity of T-reg lymphocytes emphasizing perspectives of administration of microbial preparations in anti-tumor immunotherapy [1.9.10.14]. Under the influence of LPSs decreases production of suppressive cytokines (among them TGF-beta), increases secretion of IL-1, TNF-alpha, alpha- and beta-IFN inducing amplification of anti-tumor defensive reactions (e.g. TNF-alpha activates NO synthesis in macrophages, reveals pro-apoptotic influence on endotheliocytes of tumor tissue leading to hemorrhagic necrosis of cancer cells) [3.4.15]

Bacterial components against cancers have been used since XIX century (Coley W., 1893; Nauts, H. Coley 1975-1982; Johnston 1962; Havas H. 1958-1990; Antopol W. 1972; Oberling 1975; Ray P. 1982; Gordon R. 1983; Klausner J. 1985; Zarkua Z., 1986; Jaeckle K., 1990; Kolmel K. 1991; Zhao Y. 1991; Tang Z., 1991; Abe H. 1991; Brightbill 1999; Kiseliovski M., 2001; Okamoto M. 2003; Seya T. 2003; Korosteliov C., 2003; Hobohm U. 2008, et all).

One of the earliest immunotherapy with the use of bacterial preparations against malignant tumors has been carried out by W. Coley. Efficacy of treatment was impressive. Shrinking of tumors and recovery were detected in 45% of cases [5.6.15.17.19.20]. "Coley's toxin" - mixture of thermolysates of *Serracia marcescens* and *Streptococcus pyogenes* for treatment purposes (at uterine and kidney tumors, lymphoma, melanoma, etc.) recently has been used in China, Germany, USA, Guatemala [20]. Bacterial preparations have been used in veterinary as well for treatment of sarcoidosis and carcinomas in horses, oral melanomas in dogs, leukemias in cats, etc. [7.11.16.18].

Although bacterial preparations revealed well-expressed anti-tumor activity they have not been incorporated in clinical practice due to following problems: used bacterial preparations were nonstandard and treatment effects were different respectively; has not been suggested optimal treatment schemes/regimen; treatment is accompanied by undesirable side effects (toxic-pyrogenic symptoms, i.e., fever, pain, infiltrations at the site of injection, and even endotoxic shock at overdoses); lack of theoretical basis and explanations of exact anti-tumor immunological mechanisms of bacterial [5.6.8.10.12].

Thus, an essential problem, though, is to find effective bacterial LPSs/TLR agonists, having the best immunoregulatory properties and minimal side effects or complications at administration.

Proceeding from the aforesaid, we decided to study effects of purified bacterial preparations of new generation with relatively less toxic effects directed towards stimulation of host innate responses enhancing anti-cancer resistance of the organism in combination with chemotherapy. Vaccines of *Staphylococcus, Proteus, E.colli, Klebsiella* have usually been suggested for prophylaxis of hospital infections.

Material and Method

Experiments have been carried out on 3 months 130-135 g. 80 mongrel laboratory rats. Experimental model of malignant tumor was created using transplanted malignant tumor strain of Sarcoma-45 (Sa-45), through subcutaneal inoculation of 25% suspension of cancer tissue. Selected tumor strain successfully inoculates (100%) and develops in non-purebred experimental animals.

Experimental animals were divided into four major groups. The group I involved 20 rats with Sa-45; the group II involved 20 rats with Sa-45 subjected to chemotherapy (intraperitoneal injections of Doxorubicin with the dose of 1,4 mg/kg; Vincristin - 0,04 mg/kg; and Cyclophosphanum - 20 mg/kg) on 17th and 25th days after tumor inoculation; The group III involved 20 rats with Sa-45 and immunized with tetravaccine of *Staphylococcus, Proteus, E.colli, Klebsiella* (days of intraperitoneal immunization were 2nd ,5th , 8th and 11th after tumor inoculation); The group IV involved 20 rats with Sa-45 and immunized with tetra vaccine and thereafter subjected to chemotherapy.

Anti-tumor effects of bacterial preparations were evaluated according to the following parameters:

1. Index of malignant tumor growth and development (rate of inoculated tumor development; variations in size/volume of tumor tissue (VTT);
2. Body mass index of experimental animals (BMI);
3. Prolongation of experimental animals' life-span (LS);

Results

Experiments have shown that at Sa-45 growth the BMI of experimental rats decreases progressively in parallel with malignant tumor growth, while VTT in contrast – increases. The mean LS was 27,3 ± 2,3 days (results of investigations are presented on table N1).

In the group II animals subjected to polychemotherapy there was not statistically significant difference in BMI compared to untreated animals. However, VTT at the end of treatment (22nd and 29th days after Sa-45 inoculation) was reduced by 49,5% and 66,5% respectively (p<0,001), and the LS was prolonged by 29,3%.

In the group III animals – immunized with tetra vaccine *Staphylococcus- Proteus- E.colli- Klebsiella,* after Sa-45 inoculation tumor did not develop in 3 cases. In remaining 17 rats with Sa-45, on 22nd day after tumor inoculation complete regression of tumor tissue was observed in 3 cases, while on 29th day - shrinking and complete regression of tumor tissue was obvious in 4 cases. In remaining 10 rats the BMI on 22nd and 29th days after tumor inoculation was increased by 12,5% and 20,4% compared to animals treated with doxorubicin, vincristin and cyclophosphanum, and by 11,5% and 19,4% compared to untreated the group I animals.

VTT at the beginning of the tumor growth (7th and 17th days) was reduced by 39,4% and 13,6% compared to animals subjected to chemotherapy, and by 37,6% and 16,8% - compared to untreated animals. However on 22nd and 29th days of tumor growth VTT was increased by 45,6% and 56,3 % compared to treated animals, but still reduced by 26,5% and 47,6% compared to untreated the group I animals. The LS was prolonged by 12,7% compared to treated animals and by 45,8% compared to untreated animals.

In the group IV animals – immunized with tetra vaccine and subjected to polychemotherapy after Sa-45 inoculation tumor did not develop in 4 cases. On 22nd day after tumor inoculation complete regression of tumor tissue was observed in 10 cases, while on 29th day - shrinking and complete regression of tumor tissue was obvious in remaining 6 cases.

Thus, by the end of 29th day after tumor inoculation complete disappearance of tumor tissue was observed in all experimental animals from the group IV. The BMI on 22nd and 29th days after Sa-45 inoculation was increased by 16,2% and 26,3%, compared to untreated animals, by 17,2% and 27,3% compared to treated animals, and by 4,2% and 5,8% compared to pure immunized animals.

Table N1. Volume of tumor tissue (VTT), Body mass index of rats (BMI), and duration of animals' life-span (LS) at Sarcoma-45 growth, after polychemotherapy, immunization with tetravaccine, and combination of immunization-polychemotherapy

Groups	N	Indices	Days after Sa-45 inoculation			
			7th	17th	22nd	29th
I Sa-45	20	BMI (g)	135,4 ± 2,3	132,1 ± 2,4	125,0 ± 1,9	117,3 ± 1,7
		VTT (mm³)	13,3 ± 1,3	21,4 ± 1,2	31,3 ± 2,4	37,6 ± 2,3
		LS (day)	27,3 ± 2,3			
II Sa-45 + Chemotherapy 17th-25th day	20	BMI (g)	135,2 ± 1,4	130,0 ± 1,5	123,9 ± 2,3	116,3 ± 1,4
		VTT (mm³)	13,7 ± 0,5	20,6 ± 1,2	15,8 ± 0,4	12,6 ± 1,3
		LS (day)	35,3 ± 0,8			
III Sa-45 + Tetra vaccine 2nd-5th-8th-11th day	20	BMI (g)	136,1 ± 1,3	135,2 ± 1,5	139,4 ± 1,7	140,0 ± 1,4
		VTT (mm³)	8,3 ± 0,1 Tumor did not develop in 3 cases	17,8 ± 0,9	23,0 ± 0,5 Complete regression of tumor in 3 cases	19,7 ± 0,2 Complete regression of tumor in 4 cases
		LS (day)	39,8± 1,3			
IV Sa-45 + Tetra vaccine 2nd-5th-8th-11th day + Chemotherapy 17th-25th day	20	BMI (g)	137,3 ± 1,9	135,2 ± 1,4	145,2 ± 2,1	148,1 ± 2,0
		VTT (mm³)	9,4 ± 0,2 Tumor did not develop in 4 cases.	16,6 ± 0,3	12,1 ± 0,1 Complete regression of tumor in 10 cases.	--- Complete regression of tumor in remaining 6 cases.
		LS (day)	Complete recovery of all investigated animals			

The VTT on 7th, 17th and 22nd days after Sa-45 inoculation was reduced significantly by 29,3%, 22,4%, and 61,3% compared to untreated rats; by 31,4%, 19,4% and 23,4% compared to treated animals; on the 7th day VTT was increased by 13,3%, but reduced on 17th and 22nd days - by 6,7% and 47,4% compared to pure immunized rats. Concerning the LS of experimental animals from the group IV – as it was mentioned above, of 20 rats, 4 were resistant to malignant growth and tumor did not develop, while remaining 16 rats with Sa-45, after shrinking and complete regression of tumor tissue (22th and 29th days of tumor growth) survived.

Conclusion

Thus, according to the obtained results of experiments could be concluded that application of bacterial vaccines being used against hospital infections is well tolerated in rats and does not stimulate, or support malignant tumor growth. Supposedly, increased BMI is good evidence that points on the processes of detoxication developed in immunized animals. Moreover, immunization using tetra vaccine of *Staphylococcus, Proteus, E.colli, Klebsiella* has positive anti-tumor influance on experimental animals manifested by prolongation of animals' life-span.

Bacterial vaccines in combination with chemical remedies significantly improve effects of chemotherapy.

It is worth of note and should be ampassized that at immunization with tetravaccine tumor did not develop in 15% of cases, shrinking and complete regression of tumor - in 35% of cases; prolongation of LS consisted 43,2%;

In case of combination - immunization and polychemotherapy, results were incredibly amazing – all experimenttal animals survived (100%). Tumor did not develop in 20% of cases, and in remaining 80% of cases the shrinking and complete regression of tumor was observed.

References

[1] Abe H. - Antitumor effect of LPS immobilized beads. //*Nippon Geka Gakkai Zasshi*. 1991;92:627-35.

[2] Akhmatova N., Lebedinskaia E., Kuzmenko O. et all - Effect of bacterial vaccine on anti-yumor immunity and functional activity of mononuclears suppressed by cysplatine //*Sibirian Oncology Jurnal* 2009. №1 (31), p. 53-71.

[3] Berezhnaia N. - The role of cells of immune system in microenvironment of the tumor. Cytokine cells – involved in inflammation // *Oncology*, 2009, v. 11, № 1, p. 121-144.

[4] Berezhnaia N., Chernukh V. - Immunology of cancer growth //*Naukova dumka*, 2005, „Kiev".

[5] Coley WB. Further observations upon the treatment of malignant tumors with the toxins of *erysipelas* and *Bacillus prodigiosus* with a report of 160 cases. //*John Hopkins Hospital Bulletin*, 1896 ;7:175-181.

[6] Coley, William B. "A Preliminary Note on the Treatment of Inoperable Sarcoma by the Toxic Product of Erysipelas." //*Post-graduate,* 1893, 8:278-281.

[7]   Gordon B., Matus R., Saatz S. et al. Protein A – Independent Tumorocidal Responses in Dogs After Extracorporeal Perfusion of Plasma Over Staphylococcus aureus. //*J.N.C.I.*,1983, 70, 6, 1127-1132.

[8]   Havas H, Axelrod R, Burns M, et al., Clinical results and immunologic effects of a mixed bacterial vaccine in cancer patients. //*Medical Oncology and Tumor Pharmacotherapy*, 1993;10(4):145-158.

[9]   Hobohm U, Stanford JL, Grange JM.Pathogen-associated molecular pattern in cancer immunotherapy. //*Crit. Rev. Immunol.* 2008;28(2):95-107.

[10]   Johnston B. Clinical effects of Coley's Toxins. 1. Controlled study. 2. A seven-year study. //*Cancer Chemotherapy Reports*, 1962;21:19-68.

[11]   Klausner J.S., Miller W.I., Brien D.O., Branda R.F. Effects of Plasma Treatment with Purified Protein A and Staphylococcus aureus Cowan 1 on Spontaneus Animal Neoplasms. //*Cancer Res.*,1985, 45, 1263-1268.

[12]   Kolmel K, et al. Treatment of advanced malignant melanoma by a pyrogenic bacterial lysate. A pilot study. //*Onkologie.* 1991;14:411-417.

[13]   Korosteliov S. - Anti-cancer vaccines //*Modern oncology*, 2003, v. 5, №4, p. 12-17.

[14]   Murphy K., Travers P., Walport M. - *Janeway's Immunobiology*, 2008, Garland Science.

[15]   Olishevski S., Kozak V. et all. - Immunostimulating CpG-DNA: Perspectives of application in oncology //*Oncology,* 2006, v. 8, № 2, p.43-44.

[16]   Ray P., Raychaudhuri S., Alien P. Mechanism of Regression of Mammary Adenocarcinomas in Rats following of Plasma Adsorption Over Protein A.-conteining Nonviable Staphylococcus aureus // *Cancer*, 1982; 42, 4970-4981.

[17]   Richardson M, Ramirez T, Russell N., Moye L. Coley toxins immunotherapy: a retrospective review. //*Alternative Therapies in Health and Medicine*, 1999 ; 5(3): 42-47.

[18]   Tizard Y., *Veterinary immunology*, 2004, Philadelphia, Pennsylvania.

[19]   Tsung K., Norton JA. "Lessons from Coley's Toxin". //*Surgical oncology* , 2006, . 15 (1): 25–28.

[20]   Zacharski L, Sukhatme V. Coley's toxin revisited: Immunotherapy or plasminogen activator therapy of cancer? //*J. Thromb. Haemost.*, 2005; 3:424-27.

In: Focus on Science and Technology…
Editor: Sergo Gotsiridze, pp. 15-36

ISBN: 978-1-61209-970-5
© 2011 Nova Science Publishers, Inc.

Chapter 3

ON EMBEDDED MARKOVIAN PROCESSES WITH THE SUPPLEMENTARY VARIABLE

I.S. Mikadze[*], *Sh.Sh. Nachkebia and Z.I. Mikadze*
Georgian Technical University, Georgia

Abstract

In this work, a generalized mathematical scheme of random processes, with the help of which it is possible to describe a wide range of real queuing systems (data transmitting networks, computer systems, etc.) and to solve some problems of the reliability theory, is offered. The necessity in developing such a scheme has arisen in the connexion with an increasing sophistication of real automated control systems when the available methods do not cover all problem statements that interest a practical worker. So, it is necessary to develop an analytical apparatus capable to consider a great variety of active factors in order to create the models which fit the original adequately, i.e. are maximally approximated to the original.

Keywords: random processes, queue, waiting time, reliability theory.

Introduction

A general tendency of analytical simulation of complex systems is finding of such a Random process which could be considered as the Markovian process. There exist the method of embedded Markov chains offered by D.Kendall [1] and the method of linear Markovian processes [2,3]. The latter is the development of the supplementary variable Method introduced by D.Cox [4]. The essence of the embedded Markov chains method is as follows. Time moments $\{t_n\}$ $(t_n < t_{n+1})$ are chosen so that the random process values $\{v(t_n)\}$ should form a discrete Markov chain. Then, by the methods usually applied to the Markov chains, the distribution of the random values $v(t_n)$ is studied. According to this distribution, the conclusion about the properties of the initial process $v(t)$ can be made.

---

[*] E-mail address: z_mikadze@mail.ru

The random process $\xi(t)$ is a linear Markovian process [2,3] if a set of its states X consists of two subsets X_0 and X_1. Moreover, when $\xi(t) \in X_0$, at the moment t, operations are not performed in the sets of elements of the (l,j,z) type, where l is the index determining the type of the operation taking place in the system at the moment t (by the random distribution law, on more than one process can take place in the system at the same moment); j is some supplementary discrete parameter; z is the time passed from the beginning of the operation; the probability of going over the time interval $(t,t+\Delta t)$ depends only on the states of the process $\xi(t)$ at the moment t.

Below, a class of the Markovian processes described by the combined method of the embedded Markov chains and the supplementary variable (with the linear Markovian process) are considered. Hereinafter we will call it "the embedded Markovian process with the supplement variable". The present work generalizes works [6-13].

Definition of the Embedded Markovian Process with the Supplementary Variable

Let there exist the system, in which operations of various types take place; at the same time, there cannot be simultaneously more than one operation. The system state is described by the random process $\xi(t)$, the set of states X of which consists of three subsets: X_0, X_1, and X_2. The sets X_0, X_1 and X_2 are finite or countable. If $\xi(t) \in X_0$, the operations are not performed in the system at the moment t and the state of the random process $\xi(t)$ can be characterized by some independent integer parameters $(J_1, J_2, \ldots J_n) \in X_0$; if $\xi(t) \in X_1$, or $\xi(t) \in X_2$, the operation takes place in the system at the moment t.

The sets X_1 and X_2 consist of the sets of types $(l,j,k,u) \in X_1$ and $(m,v,u) \in X_2$. Here l and m are the indices of the operations that began at the moment $t-u$; u is the time that has passed from the beginning of the operation of the l-type or the m-type; j-some discrete parameters that characterize the random processes $\xi(t)$ at the beginning moment of the l-type operation, hereinafter called the l-type operation market; k and v-some discrete parameters that characterize the random process $\xi(t)$ at the moment t, hereinafter we will call them the markers of the random process $\xi(t)$ – RPM-l and RPM-m, respectively, and at the operation beginning moment we will denote them as $\widetilde{k}$ and $\widetilde{v}$, respectively.

We should note that, depending on the character of random internal or external intervention, the $l \in X_1$ type operation in the process of its fulfillment can be interrupted for some random time with subsequent resumption from the beginning or from the interruption moment, or its fulfillment ceases. Unlike the $l \in X_1$ type operation, the $m \in X_2$ type operation is not interrupted, i.e. the operation of that type ends without interruption.

Such classification of the operations is more natural from the aspect of convenience of applications, as it will be evident from the following discussion. For more complete description of the random process $\xi(t)$, let us introduce other discrete parameter η and let us call it the index of system loading at the beginning moment of the $l \in X_1$ yupe operation, i.e. SLI-l. Suppose that there exist the integer dependence of η on the $j \in X_1$ operation marker and on the RPM-l at the beginning of the operation $\widetilde{k} \in X_1$, i.e. $\eta = \eta(j, \widetilde{k}) \in X_1$; hereinafter we will consider the set of type $(l,j,\eta) \in X_1$ as the initial state of the operation.

Let $H_{lj\eta}(u)=P\{x<u \mid \{ \xi(t-x)=(l,j,\widetilde{k},0)\in X_1\}\}$ represent the function of the probability (FDP) that the $l\in X_1$ type operation with the SLI $- \eta$, which began at the moment t-x, at the process state $\xi(t)=(l,j,\widetilde{k},0)$ will end during the time $x(x<u)$ within the time interval $(t,t+dt)$. $H^{(i)}_{lj\eta}(u)$ is the probability that, at the ending moment of the operation with the initial state $(l,j,\eta)\in X_1$, the operation marker will be equal to $i(i\in X_0$ or $i\in X_1)$; similarly, $G_m(u)$ is the FDP that the $m\in X_2$ type operation, which began at the moment t-x and has been proceeding at the moment t at the process state $\xi(t)=(m,\widetilde{v},0)\in X_2$, will end during the time $x(x<u)$ within the time interval $(t,t+dt)$. It is evident that:

$$H_{lj\eta}(u)= \sum_{i\in x_1} H^{(i)}_{lj\eta}(u), \ H_{lj\eta}(+\infty)=1, \ G_m(+\infty)=1.$$

So, at $\xi(t)=(l,j,k,u)$ or at $\xi(t)=(m,v,u)$, the probability of the operation ending during the Time dt, independent of the process state, at the operation ending moment is equal to:

$$[H_{lj\eta}(u+dt)- H_{lj\eta}(u)]/\overline{H}_{lj\eta}(u) \text{ and } [G_m(u+dt)-Gm(u)]/\overline{G}_m(u),$$

And at the state $i\in X_0$ or $i\in X_1$ at the ending moment of the l-type operation, the above probability is equal to:

$$[H^{(i)}_{lj\eta}(u+dt)- H^{(i)}_{lj\eta}(u)]/\overline{H}_{lj\eta}(u).$$

Here $$\overline{H}_{lj\eta}(u)=1-\overline{H}_{lj\eta}(u); \ \overline{G}_m(u)=1-G_m(u)$$

Suppose $H_{lj\eta}(u)$, $H^{(i)}_{lj\eta}(u)$, and $G_m(u)$ are absolutely continuous functions, then we can Write:

$$r_{lj\eta}(u).\overline{H}_{lj\eta}(u)=h_{lj\eta}(u); \ r^{(i)}_{lj\eta}(u).\overline{H}_{lj\eta}(u)= h^{(i)}_{lj\eta}(u) ; \ \mu_m(u) \overline{G}_m(u)=g_m(u).$$

Here $$h_{lj\eta}(u)=\frac{dH_{lj\eta}(u)}{du} ; \ h^{(i)}_{lj\eta}(u)= \frac{dH^{(i)}_{lj\eta}(u)}{du} ; \ g_m(u)=\frac{dG_m(u)}{du} ;$$

$r_{lj\eta}(u)du$ and $r^{(i)}_{lj\eta}(u)du$ are the probabilities that the $l\in X_1$ type operation with he SLI-η, which began at the process state $\xi(t-u)=(l,j,\widetilde{k},0)$ and has been proceeding during the time u, will end within the time interval $(u,u+du)$, irrespective of the process state $\xi(t-u)$ at the operation ending moment in the first case, and at the state $i\in X_1$ or $i\in X_0$ in the second case. Similarly, $\mu_m(u)du$ is the probability that the $m\in X_2$ operation, which began at u and has been proceeding during the time $\xi(t-u)=(m,\widetilde{v},0)$, will end within the time interval $(u,u+du)$.

Let us note that:

$$h_{lj\eta}(u) = \sum_{i \in X_1} h_{lj\eta}^{(i)}(u); \quad r_{lj\eta}(u) = \sum_{i \in X_1} r_{lj\eta}^{(i)}(u);$$

Definition. The random process $\xi(t)$, the set of the states X of which consists of the three subsets: the discrete states set $(i,k) \in X_0$ and the element sets of type $(l,j,k,u) \in X_1$ and $(m,v,u) \in X_2$, will be the embedded Markovian process with the supplementary variable if the probabilities of transition from any state $\xi(t) \in X_0$, $\xi(t) \in X_1$, or $\xi(t) \in X_2$ to the state that also belongs to these sets within the time interval $(t,t+dt)$ depend on the stage at which the process $\xi(t)$ was at the moment t and on the stage which the process $\xi(t)$ was at the operation beginning moment. The mentioned process is described as follows. If $\xi(t)=(l,j,v,u) \in X_1$ at the moment t, during the Δt time, the process changes to the state $(i,k) \in X_0$ with the probability $r_{lj\eta}^{(i)}(u)p_{lj\eta}^{(k)}\Delta t + o(\Delta t)$ or to the state $(l',i,\tilde{k},0) \in X_1$ eith the probability $r_{lj\eta}^{(i)}(u)p_{lj\eta}^{(l',\tilde{k})}\Delta t + 0(\Delta t)$. Here l' is the same operation class as l. For simplicity, we assume the number of the discrete parameters in the set X_0 to be equal to two. If $\xi(t)=(m,v,u) \in X_2$, during the Δt time, the process changes to the state $(i,k) \in X_0$ with the probability $\mu_m(u)p_{mv}^{(k,i)}\Delta t + o(\Delta t)$ or to the state $(l,j,\tilde{k},0) \in X_1$ with the probability $\mu_m(u)p_{mv}^{(l,j,\tilde{k})}\Delta t + o(\Delta t)$. Here everywhere it is adopted that the operation marker and the process marker independently change to the operation marker and to the process marker, respectively. For example, the probability $p_{lj\eta}^{(k)}$ means: if the l-type process with the OIS-$lj\eta$ ended at the moment t and, immediately before that the marker of the process $\xi(t)$ was v, it changes to the state $k \in X_0$ with the probability $p_{lj\eta}^{(k)}$ or to the state $\tilde{k} \in X_1$ with the probability $p_{lj\eta}^{(l,\tilde{k})}$.

Besides the transitions associated with the operation ending, there are also possible spontaneous transitions from the state $(c,d) \in X_0$ to the state $(c,k) \in X_0$ with the probability $\lambda_d(k)\Delta t+o(\Delta t)$, to the state $(i,d) \in X_0$ with the probability $\alpha_c(i)\Delta t+o(\Delta t)$, to the state $(l,c,\tilde{k},0) \in X_1$ with the probability $\lambda_d(l,\tilde{k})\Delta t+0(\Delta t)$, or to the state $(m,d,0) \in X_0$ with the probability $\alpha_c(m,d)\Delta t+o(\Delta t)$, or to the state $\xi(t+\Delta t)=(c,d)$ with the probability $1-(\alpha_c+\lambda_d)\Delta t+o(\Delta t)$. Here

$$\alpha_c = \sum_{\substack{j \in X_0 \\ j \neq c}} \alpha_c(j) + \sum_{\substack{d \in X_2 \\ d \neq c}} \alpha_c(m,d); \quad \lambda_d = \sum_{\substack{j \in X_0 \\ j \neq d}} \lambda_d(j) + \sum_{(j,\tilde{k}) \in X_1} \lambda_d(l,j,\tilde{k}).$$

During the Δt time there are also possible the transitions from the state $\xi(t)=(l,j,k,u) \in X_1$ to the state $\xi(t+\Delta t)=(l,j.v,u+\Delta t) \in X_1$ with the probability $\lambda_{lk}(v)\Delta t+o(\Delta t)$ and from the state $\xi(t)=(m,v,u)$ to the state $\xi(t+\Delta t)=(m,k,u+\Delta t) \in X_2$ with the probability

$\lambda_{lmv}(k)\Delta t+o(\Delta t)$, or to $\xi(t+\Delta t)=(l,j.k,u+\Delta t)\in X_1$ with the probability $1-(r_{lj\eta}(u)+\lambda_{lk})\Delta t+o(\Delta t)$, or to $\xi(t+\Delta t)=(m,v,u+\Delta t)\in X_2$ with the probability $1-(\mu_m(u)+\lambda_{mv})\Delta t+o(\Delta t)$.

Here
$$\lambda_{lk}=\sum_{\substack{c\in X_1\\c\neq k}}\lambda_{lk}(c);\ \lambda_{mv}=\sum_{\substack{c\in X_2\\c\neq v}}\lambda_{mv}(c).$$

Hereinafter we will call the transitions which are not associated with the operation ending the transitions of the 1st kind, and the transitions associated with the operation ending – the transitions of the 2nd kind.

General equations. Let us now consider some properties of the probability distribution of the process $\xi(t)$.

Theorem. If the set of the discrete components states of the process $\xi(t)$ is finite and $H_{lj\eta}(u)$ and $G_m(u)$ – are absolutely continuous functions, then, at absolutely continuous initial distribution of probabilities of the states of the process $\xi(t)$, there exist the continuous functions:

$$R_c^{(d)}(t)=\mathbf{P}\{\xi(t)=(c,d)\},\ (c,d)\in X_0;$$

$$P_{lj\eta}^{(k)}(t,u)=\frac{d}{du}\mathbf{P}\{\xi(t)\in\{(l,j,k,y),\ y<u\}\},\ (l,j,k,u)\in X_1;$$

$$P_{mj}^{(\eta)}(t,u)=\frac{d}{du}\mathbf{P}\{\xi(t)\in\{(l,j,k,y),\ y<u\}|\xi(t-y)=\{m,j,0\}\}\},\ (m,v,u)\in X_2.$$

that satisfy the system of integral-differential equations:

$$\frac{dR_i^{(k)}(t)}{dt}+(\alpha_i+\lambda_k)R_i^{(k)}(t)=\sum_{\substack{c\in X_0\\c\neq k}}R_i^{(c)}(t)\lambda_c(k)+\sum_{\substack{c\in X_0\\c\neq i}}R_c^{(k)}(t)\alpha_c(i)+$$

$$+\sum_{v,j\in X_1}\int_0^t p_{lj\eta}^{(v)}(t,u)r_{lj\eta}^{(i)}(u)p_{lv\eta}^{(k)}du+\sum_{v,j\in X_2}\int_0^t p_{m,j}^{(v)}(t,u)\mu_m(u)p_{mv}^{(i,k)}du; \qquad (1)$$

$$\frac{\partial p_{lj\eta}^{(k)}(t,u)}{\partial t}+\frac{\partial P_{lj\eta}^{(k)}(t,u)}{\partial u}+[r_{lj\eta}(u)+\lambda_{lk}]p_{lj\eta}^{(k)}(t,u)=\sum_{\substack{c\in X_1\\c\neq k}}p_{lj\eta}^{(c)}(t,u)\lambda_{lc}(k); \qquad (2)$$

$$\frac{\partial p_{mj}^{(v)}(t,u)}{\partial t} + \frac{\partial p_{mj}^{(v)}(t,u)}{\partial u} + [\mu_m(u) + \lambda_{mv}]p_{mj}^{(v)}(t,u) = \sum_{\substack{c \in X_2 \\ c \neq k}} p_{mj}^{(c)}(t,u)\lambda_{mc}(v), \qquad (3)$$

with boundary conditions:

$$p_{lj\eta}^{(\tilde{k})}(t,0) = \sum_{d \in X_0} R_j^{(d)}(t)\lambda_d(l,\tilde{k}) + \sum_{v,i \in X_1} \int_0^t p_{li\eta}^{(v)}(t,u)r_{li\eta}^{(j)}(u)p_{lv\tilde{\eta}}^{(l,\tilde{k})}du + \sum_{v,i \in X_2} \int_0^t p_{mi}^{(v)}(t,u)\mu_m(u)p_{mv}^{(l,j,\tilde{k})}du;$$

$$\qquad (4)$$

$$p_{mj}^{(\tilde{v})}(t,0) = \sum_{k,i \in X_0} R_i^{(j)}(t)\alpha_i(m)p_{mj}^{(\tilde{v})} \quad \tilde{v} = j \qquad (5)$$

The initial conditions have the form:

$$R_i^{(k)}(t)\Big|_{t=0} = R_i^{(k)}(0); \quad p_{ij\eta}^{(k)}(t,u)\Big|_{t=0} = p_{ij\eta}^{(k)}(0,u); \quad p_{mv}^{(k)}(t,u)\Big|_{t=0} = p_{mv}^{(k)}(0,u).$$

Proof. According to $[2,3]$ it can be proved that the number of transitions of the 2nd kind and the time before the first getting to the set X_0 are finite with the probability equal to *1*. From this statement, it also follows that all the trajectories of the process $\xi(t)$ at the moment $t>0$ that pass through the set of the states *(l,j,k,u+du)* or *(m,v,u+du)* get to it after the finite number of subsequent transitions of the 2nd kind passing or some of them not passing through the state set X_0.

Let us now introduce the equations that satisfy $R_i^{(k)}(t)$, $P_{lj\eta}^{(k)}(t,u)$, and $P_{mj}^{(k)}(t,u)$. As the event $\xi(t + \Delta t) = (i,k) \in X_0$ can be realized as the result of one of the following events:

1) $\xi(t) = (i,k) \in X_0$ and, during the time *(t,t+ Δ t)*, there will not be the transitions of the 1nd kind. The probability of this event is $R_i^{(k)}(t)[1 - (\alpha_i + \lambda_k)\Delta t] + o(\Delta t)$;

2) $\xi(t) = (i,c) \in X_0$ and, during the time *(t,t+ Δ t)*, there occurred the transition that resulted in ξ *(t+ Δ t)=(i,k)* $\in X_0$. The probability of this event is

$$\sum_{\substack{c \in X_0 \\ c \neq k}} R_i^{(c)}(t)\lambda_c(k)\Delta t + o(\Delta t);$$

3) $\xi(t) = (c,k) \in X_0$ and, during the time *(t,t+ Δ t)*, there occurred the transition that resulted in ξ *(t+ Δ t)=(i,k)* $\in X_0$. The probability of this event is:

$$\sum_{\substack{c \in X_0 \\ c \neq i}} R_c^{(k)}(t)\alpha_c(i)\Delta t + o(\Delta t);$$

4) $\xi(t)=(l,j,k,u) \in X_1$, $u>0$, but, during the time $(t,t+\Delta t)$, there occurred the 2nd kind transition that resulted in $\xi(t+\Delta t)=(i,k) \in X_0$. The probability of this event is:

$$\sum_{v,j \in X_1} \int_0^{t+\Delta t} p_{lj\eta}^{(v)}(t,u)r_{lj\eta}^{(i)}(u)\Delta t \cdot p_{lvn}^{(k)}du + o(\Delta t);$$

5) $\xi(t)=(m,v,u) \in X_2$, $u>0$, but, during the time $(t,t+\Delta t)$, there occurred the 2nd kind transition that resulted in $\xi(t+\Delta t)=(i,k) \in X_0$. The probability of this event is:

$$\sum_{v,j \in X_2} \int_0^{t+\Delta t} p_{mj}^{(v)}(t,u)\mu_m(u)\Delta t \cdot p_{mv}^{(i,k)}du + o(\Delta t).$$

Thus, we obtain

$$R_i^{(k)}(t+\Delta t) = R_i^{(k)}(t)[1-(\alpha_i + \lambda_k)\Delta t] + \sum_{\substack{c \in X_0 \\ c \neq k}} R_i^{(c)}\lambda_c(k)\Delta t + \sum_{\substack{c \in X_0 \\ c \neq i}} R_c^{(k)}(t)\alpha_c(i)\Delta t +$$

$$+ \sum_{v,j \in X_1} \int_0^{t+\Delta t} p_{lj\eta}^{(v)}(t,u)r_{lj\eta}^{(i)}(u)p_{\ln\eta}^{(k)}\Delta t du + \sum_{v,j \in X_2} \int_0^{t+\Delta t} p_{mj}^{(v)}(t,u)\mu_m(u)p_{mv}^{(i,k)}\Delta t du + o(\Delta t).$$

Proceeding from the similar reasoning

$$p_{lj\eta}^{(k)}(t+\Delta t, u+\Delta u) = p_{lj\eta}^{(k)}(t,u)(1-\lambda_{lk}\Delta t)(1-r_{lj\eta}(u)\Delta t) +$$

$$+ \sum_{\substack{c \in X_1 \\ c \neq k}} [p_{lj\eta}^{(c)}(t,u)\lambda_{lc}(k)\Delta t \cdot (1-r_{lj\eta}(u)\Delta t)] + o(\Delta t); \tag{7}$$

$$p_{mj}^{(v)}(t+\Delta t, u+\Delta u) = p_{mj}^{(v)}(t,u)(1-\lambda_{mv}\Delta t)(1-\mu_m(u)\Delta t) +$$

$$+ \sum_{\substack{c \in X_2 \\ c \neq k}} [p_{mj}^{(c)}(t,u)\lambda_{mv}(c)\Delta t \cdot (1-\mu_m(u)\Delta t)] + o(\Delta t); \tag{8}$$

$$p_{lj\eta}^{(\tilde{k})}(t+\Delta t, (0,\Delta t))\Delta t = \sum_{j,d \in X_0} R_j^{(d)}(t)\lambda_d(l,\tilde{k})\Delta t + \sum_{v,j \in X_1} \int_0^{t+\Delta t} p_{lj\tilde{\eta}}^{(v)}(t,u)r_{lj\tilde{\eta}}^{(j)}(u)p_{lv\tilde{\eta}}^{(l,\tilde{k})}\Delta t du +$$

$$+ \sum_{v,i \in X_2} \int_0^{t+\Delta t} p_{mi}^{(v)}(t,u)\mu_m(u)p_{mv}^{(j,\tilde{k})}\Delta t du + o(\Delta t); \tag{9}$$

$$p_{mv}^{(\tilde{k})}(t+\Delta t,(0,\Delta t))\Delta t = \sum_{k,i\in X_0} R_i^{(j)}(t)\alpha_i(m)p_{mj}^{(\tilde{k})}\Delta t + o(\Delta t). \qquad (10)$$

Here all the terms also have the evident probability meaning. For example, relationship (9) is the record of the fact that the event $\xi(t+\Delta t)\in(l,j,\tilde{k},(0,\Delta t))$ can be realized as the result of the transition from the states of the X_0 class or the transition from one the states of the sets X_1 or X_2 to X_1.

Using the continuity of the functions $R_c^{(d)}(t)$, $P_{lj\eta}^{(k)}(t,u)$, and $P_{mj}^{(v)}(t,u)$ and changing to the limit in equations (6) to (10) at $\Delta t \to 0$, we obtain equations (1) to(5). The theorem has been proved.

Some Applications of the General Theory of Embedded Markovian Processes with the Supplementary Variable in the Problems of the Theory of Queuing and Reliability

The problem of designing and studying the analytical models of modern complex queueing and reliability systems has two main aspects determining versatility, adequacy, and accuracy of these models: 1)the structural – functional aspect which represents the structural and functional organization; 2) the probable – statistic aspect which represents the degree of the generality of statistic regularities which characterize the random environment effect on the queuing mechanism (QM) of the system.

The structural-functional aspect is concerned with the problem of maximum approximation of the models to the original. The probable-statistic aspect represents the natural aspiration of investigators to maximally approximate the probable-time characteristics of the environment effect random processes employed in the system to the characteristics of real systems used in practice. That means the processes of formation, arrival and output of calls (request), their groups or separate parts, and, also, the processes of failures of hardware and software and of their recovery, the processes of controlling and diagnosis, etc.

As it is concerned, one of the most effective methods is the approximation of empiric (and sometimes theoretical) laws of random values distribution (the method of stages).

In practical applications, two basic numeral characteristics of the probability distribution function (PDF): a mathematical expectation and dispersion, are usually considered. It is known that approximation of the PDF by the specified values of the mentioned characteristics can be efficiently performed with the help of the generalized Erlang distribution in case the variation index is <1, or with the help of the hyper exponential distribution if otherwise. We can proceed to further generalization of the method of stages: to find such organization of stages which would lead to obtaining any values of the variation index. That can achieved by arbitrarily uniting the stages and, thus, obtaining complex structures.

For example, parallel- sequential organization of the stages consist of n parallel branches, each of which includes l_i sequential stages $(i=\overline{1,n})$; duration of each stage is distributed by the exponential law with the λ_{ij} parameter (the time of arrival of the i branch to the j stage is determined by the rate of coming calls arrivals- λ_{ij}). The corresponding process state will be state (i,j) by the coming calls stream. The Laplace transform for the density of the FDP of the calls arrival time – $a(S)$ has the form:

$$a(s) = \sum_{i=1}^{n} P_i \prod_{j=i}^{l_i} (\lambda_{ij} / (s + \lambda_{ij})). \tag{11}$$

Here P_i is the probability of the call entry into the i branch after its arrival ($\sum_i P_i = 1$).

The possible further generalization is also lifting the restriction that the time between the QM failures is exponential. Assume the similar time distribution between neighboring failure to be $b(S)$:

$$b(s) = \sum_{i=1}^{m} q_i \prod_{j=1}^{a_i} \alpha_{ij} / (s + \alpha_{ij}). \tag{12}$$

Here m is the number parallel branches, each of which consists of the $a_i (i = \overline{1,m})$ sequential stages (phases): the time of arrival of the occurring failure at each stage is characterized by the parameter α_{ij} (i is the branch number, and j is the number of the stage in the i branch); q_i is the probability of the occurring failure arrival to the i branch after the QM recovery ($\sum_i q_i = 1$). The corresponding state will be called the QM state by the

occurring failure. So, an effective method for investigating any queuing system of the *G/G/1* with a random law of failure occurrence and a random law of system recovery is offered. We mean here such systems for which the Laplace transforms of the time intervals distribution between the calls and the failures are rational functions *of S* or they can be approximated by fractional - rational functions of *S*. We also mean multichannel QS with a general queue, where all the channels can unite for servicing multiple requirements of one call or a group of calls, i.e. the systems with the united queue and resources [6-1,13].

The Queuing system with Various Operation States:

Let the Erlang stream of calls enter the QS. The time intervals between the sequential call arrivals are random values (RV) with the distribution function

$$A(t) = \int_0^t \frac{(n\lambda)^n u^{n-1}}{(n-1)!} e^{-n\lambda u} du.$$

Each stage duration is the exponentially distributed RV with the $n\lambda$ parameter. The call durations measured in some arbitrary units are RV with FDP of the time *F(u)*. The QM is subject to failures; the time intervals between the sequential occurrences of the failures are

RV and are subject to the m-order Erlang law with the each stage rate $\alpha_i (i = 0, m-1)$. After the failure, the QM is transferred to restoring its serviceable; the recovery time is a RV with the FDP $G(u)$ $(g(u)=G'(u)$, $\mu(u) = g(u)/(1-G(u))$. Servicing of the next call begins immediately if the QM is in the serviceable state; the calls leave the system begin Completely serviced. The waiting time and the queue length in the system are not limited. It is necessary to determine probable characteristics of the QS.

Let's describe the system state by the Markovian process $\xi(t)$. If $\xi(t) \in X_0$, the process state at the moment t should be characterized by two parameters:

(a) $k(k = \overline{0, n-1})$ - the phase number by the coming call;

(b) $i(i = \overline{0, m-1})$ - the phase number by the occurring failure.

Then $R_i^{(k)}(t) = P\{\xi(t) = (i,k) \in X_0\}(k = \overline{0, n-1}, \quad i = \overline{0, m-1})$ is the probability that, at the Moment t, there are no calls in the system, the QM by the failure is in the i phase, and the call that should enter the system has passed the k phase of the Erlang pattern of arrivals. If $\xi(t) = (1, j, k, u) \in X_1$, $(j = \overline{0, m-1}, k = \overline{n, \infty}, u > 0)$, then, at the moment t, the l – type operation of calls servicing takes place in the system; u – the time passed from the beginning of servicing; j – the phase number by the occurring failure at the beginning moment of servicing; $k(k = \overline{n, \infty})$ - the number of phase at the moment t by the occurring failure. For briefness, here $\eta = 1$, that means that only one call is simultaneously being serviced in the system (the index η will be omitted hereinafter).

$$P_{li}^{(k)}(t,u)du = d_u[P\{\xi(t) = \{\{(l,i,k,y), y < u\} \mid \xi(t-y) = \{l,i,\tilde{k},0\}\}], i = \overline{0, m-1}; \tilde{k} = \overline{n, \infty},$$

$$u > 0, \quad k \geq \tilde{k}$$

- is the probability that, at the moment t, one call has been serviced during the time $y(u < y < u + du)$, whereas its servicing began at the i state of the QM by the occurring failure, and the total number of phases passed by all calls available in the system and by the call next to them is $k(k = \overline{n, \infty})$. If $\xi(t) = (m, v, u)$, the m – type operation of the QM recovery after the failure takes place in the system. Here u is the time passed from the beginning moment of the recovery; v – the total number of phases passed by all the calls available in the system and the call next to them by the moment t, provided that, at the beginning moment of the m – type operation, the number of the passed phases by the coming call is less than n;

$$P_{m\tilde{v}}^{(v)}(t,u) = \frac{d}{du}[P\{\xi(t) = \{\{(m,v,y), y < u\} \mid \xi(t-y) = \{m,\tilde{v},0\}\}], v = \overline{0, \infty}, \tilde{v} = \overline{0, n-1}$$

$H_{lj}^{(i)}(u)$ - is the arbitrary probability of the fact that the call servicing time is less than u, and that, at the moment of the servicing ending, the QM will be in the state $i(i = 0, \overline{m-1})$ by the occurring failure, provided that, at the beginning moment of servicing, the QM was in the following phase $j(j = 0, \overline{m-1})$ by the occurring failure;

$$r_{lj}^{(i)}(u) = h_{lj}^{(i)}(u)/\overline{H}_{lj}(u); H_{lj}(u) = \sum_i H_{lj}^{(i)}(u), \overline{H}_{lj} = 1 - H_{lj}(u); h_{lj}^{(i)} = (H_{lj}^{(i)}(u))'; r_{lj} = \sum_i r_{lj}^{(i)}(u);$$

Proceeding from the fact that, in the model under consideration during the time dt, there are possible the following transitions: from the state $(i,k) \in X_0$ to the state $(i+1,k) \in X_0$ at $i \le m-2$ and $k \le n-1$ with the probability $\alpha_i dt$, or to the state $(i,k+1)$ at $i \le m-1$ and $k \le n-2$ with the probability $n\lambda dt$; from the state $(m-1,k) \in X_0$ at k<n to the state $(m,k,0) \in X_2$ with the probability $\alpha_{m-1} dt$; or from the state $(i,n-1) \in X_0$ to the state $(l,i,n,0) \in X_1$ with the probability $\lambda_{n\lambda} dt$, or from the state $(l,j,k+n,u) \in X_1$ to the state $(i,k) \in X_0$ at $k \le n-1$ with the probability $r_{lj}^{(i)}(u)du(p_{l,k+n}^{(k)} = 1)$, or to the state $(l,j,k+1,u+dt) \in X_1$ with the probability $n\lambda dt$, from the state $(m,k,u) \in X_2$ at $k \le n-1$ to the state $(0,k) \in X_0$ with the probability $\mu_m(u)du(p_{mk}^{(0,k)} = 1)$, or to the state $(l,0,k,0) \in X_1$ at $k \ge n$ with the probability $\mu_m(u)du\, p_{mk}^{(l,0,k)}$ $(p_{mk}^{(l,0,k)} = 1)$ or to the state $(m,k+1,u+dt)$ with the probability $n\lambda dt$. Equations (1) to (5) become as follows:

$$\frac{dR_i^{(k)}(t)}{dt} + (\alpha_i + n\lambda)R_i^{(k)}(t) = \sum_{v=0}^{m-1} \int_0^t p_{lv}^{(k+n)}(t,u)r_{lv}^{(i)}(u)du +$$

$$+ \delta_{i0}\int_0^t p_m^{(k)}(t,u)\mu_m(u)du + (1-\delta_{k0})n\lambda R_i^{(k-1)}(t) + (1-\delta_{i0})\alpha_{i-1}R_{i-1}^{(k)}(t),$$

(13)

$$k = \overline{0,n-1}, \ i = \overline{0,m-1}, \ P_m^{(k)}(t,u) = \sum_{j=0}^{n-1} p_{mj}^{(k)}(t,u)\delta(j \le k);$$

$$\frac{\partial p_{mj}^{(k)}(t,u)}{\partial t} + \frac{\partial p_{mj}^{(k)}(t,u)}{\partial u} = -(n\lambda + \mu_m(u))p_{mj}^{(k)}(t,u) + (1-\delta_{k0})n\lambda p_{mj}^{(k-1)}(t,u),$$

(14)

$$j = \overline{0,l-1}, \ k = \overline{0,\infty};$$

$$\frac{\partial p_{li}^{(k)}(t,u)}{\partial t} + \frac{\partial p_{li}^{(k)}(t,u)}{\partial u} = -(n\lambda + r_{li}(u))p_{li}^{(k)}(t,u) +$$

$$+ (1-\delta_{kn})n\lambda p_{li}^{(k-1)}(t,u),$$

(15)

$k = \overline{n,\infty}, \ i = \overline{0,m-1};$

$$p_{mj}^{(\widetilde{v})}(t,0) = R_{m-1}^{(j)}(t)\alpha_{m-1}\delta(j < n) \ (j = \overline{0,n-1}), \ \widetilde{v} = j; \qquad (16)$$

$$p_{li}^{(\widetilde{k})}(t,0) = \sum_{v=0}^{m-1}\int_0^t p_{lv}^{(k+n)}(t,u)r_{lv}^{(i)}(u)du + \delta_{i0}\int_0^t p_{mj}^{(k)}(t,u)\mu_m(u)du + \delta_{\widetilde{k}n}n\lambda R_i^{(n-1)}(t), \qquad (17)$$

$k = \widetilde{k}, \ k,\widetilde{k} = \overline{n,\infty}.$

Here

$$\delta(\cdot) = \begin{cases} 1, if \ the(\cdot) \ statement \ is \ true, \\ 0, if \ the(\cdot) \ statement \ is \ wrong, \end{cases} \quad \delta_{kn} = \begin{cases} 1, at \ k = n, \\ 0, at \ k \neq n, \end{cases} \ (\delta_{kn} \text{ - is the Krenecker symbol})$$

Solutions of (14) and (15) can be written in the form:

$$p_{mj}^{(k)}(t,u) = p_{mj}^{(i)}(t-u,0)(n\lambda u)^{k-j}e^{-n\lambda u}\overline{G}(u)\delta(i < n)\delta(j \leq k)/(k-j)!, \qquad (18)$$

$k = \overline{n,\infty}, \ j = \overline{0,n-1}, \ p_{mj}^{(j)}(t-u,0) = P_m^{(j)}(t-u,0);$

$$p_{li}^{(k)}(t,u) = \sum_{v=n}^k p_{li}^{(v)}(t-u,0)\overline{H}_{li}(u)(n\lambda u)^{k-v}e^{-n\lambda u}/(k-n)! \ k = \overline{n,\infty}. \qquad (19)$$

Assuming the initial values of the probabilities to be found and solving (13) together with (17) and with consideration of (18) and (19), we determine the desired probabilities in the form of the Laplace-Stilties transform both in nonstationary and stationary models [6-13] and also all the numeric characteristics.

Generalization of the Problem of the Virtual Waiting Time

In [16], the integral-differential equation, the solution of which gives the distribution of the virtual waiting time in the queue when the calls are serviced by arrival, is considered. This equation was first obtained by Takacs. He also introduced the notion of the virtual waiting time which is the waiting time duration of the real or probably false call. The Takacs equation envisages that the queuing system is a single-channel (faultless-perfect) one with the Poisson input stream and the arbitrary servicing time distribution . Here, using the obtained results, a system of integral-differential and differential equations relative to the virtual waiting time for an imperfect multichannel queuing system with the united queue and resourced when the stream of calls and failures is not the Poisson one has been set up. The results obtained here are the generalization of the results presented in the works [6-11,13].

Assuming that the stream of calls enters the queuing system (the queuing mechanism-QM) by the Erlang pattern with the FDP by law (11); and the FDP of the time between two neighboring failures has form (12). It is necessary to determine the virtual waiting time in the queue and the time of the call staying in the system.

Let us express as $q_{cd}^{(i,j)}(t,u)du(i = \overline{1,n}, j = \overline{1,l_i}; c = \overline{1,m}, d = \overline{1,a_c})$ the probability of the following event: 1) at the moment t, the QM is not capable to immediately begin to service the call if it arrived at the moment (the current call is being serviced or the QM is not in a serviceable state); 2) after the time within the internal *(u,u+du)*, the QM will be capable to begin servicing the call being in the state *(c,d)* by the occurring failure, provided that, at the moment t, the system was in the state *(i,j)* by the coming call; $R_{cd}^{(i,j)}(t)$ - is the probability that, at the moment t, the QM is free and is in the state *(c,d)* $\in X_0$ by the occurring failure, i.e. in the serviceable state, and in the state *(i,j)* $\in X_0$ by the coming call. $H_{ij}^{(c,d)}(u)$ - describes the process of servicing of separate calls or groups of calls and is the probability that the servicing which began in the QM state *(i,j)* by the occurring failure will end during the time less than *u*, and the QM will be in the state *(c,d)* by the occurring failure. *G(u)* is the FDP of the recovery time of the faulty QM *(g(u)= G' (u)).* Similarly,

$$h_{ij}^{(c,d)}(u)du = dH_{ij}^{(c,d)}(u);$$

$$H_i^{(c,d)}(u) = \sum_j H_{ij}^{(c,d)}(u); \quad H_i(u) = \sum_j \sum_c \sum_d H_{ij}^{(c,d)}(u), \quad i = \overline{0,m}.$$

We would note that $H_i(u)$ is the FDP. Let us express the virtual waiting time in the queue at the time t moment as $\gamma(t)$.

Theorem. If a set of state of discrete components of a random process $\gamma(t)$ is finite, and $H_{ij}^{(c,d)}(u)$ and $G(u)$ are absolutely continuous functions, then, at absolutely continuous initial distribution of probabilities, there exist functions $R_{cd}^{(i,j)}(t)$ and $q_{cd}^{(i,j)}(t,u)$ which are continuous by their arguments and which satisfy the following system of differential and integral differential equations:

$$\frac{\partial q_{cd}^{(i,1)}(t,u)}{\partial t} - \frac{\partial q_{cd}^{(i,1)}(t,u)}{\partial u} + \lambda_{i1}q_{cd}^{(i,1)}(t,u) = \sum_{v=1}^{m}\sum_{b=1}^{a_v}\sum_{\tau=1}^{n}\int_0^u q_{vb}^{(\tau,l_\tau)}(t,v)\lambda_{d_\tau}P_ih_{vb}^{(c,d)}(u-v)dv + \quad (20)$$

$$+ \sum_{v=1}^{m}\sum_{b=1}^{a_v}\sum_{\tau=1}^{n}\lambda_{d_\tau}R_{vb}^{(\tau,l_\tau)}(t)P_ih_{vb}^{(c,d)}(u) + \delta_{d1}\sum_{v=1}^{m}a_{va_v}g(u)q_cR_{va_v}^{(i,1)}(t), \quad i=\overline{1,n}, \; c=\overline{1,m}, \; d=\overline{1,a_c}$$

$$\frac{\partial q_{cd}^{(i,j)}(t,u)}{\partial t} - \frac{\partial q_{cd}^{(i,j)}(t,u)}{\partial u} = -\lambda_{ij}q_{cd}^{(i,j)}(t,u) + \delta_{d1}\sum_{v=1}^{m}\alpha_{va_v}g(u)q_cR_{va_v}^{(i,j)}(t) + \quad (21)$$

$$+ (1-\delta_{j1})\lambda_{i,j-1}q_{cd}^{(i,j-1)}(t,u), \quad i=\overline{1,n}, \; j=\overline{2,l_i}, \; c=\overline{1,m}, \; d=\overline{1,a_c};$$

$$\frac{dR_{cd}^{(i,j)}(t)}{dt} + (a_{cd} + \lambda_{ij})R_{cd}^{(i,j)}(t) = (1 - \delta_{0,d-1})a_{c,d-1}R_{c,d-1}^{(i,j)}(t) + (1 - \delta_{0,j-1})\lambda_{i,j-1}R_{cd}^{(i,j-1)}(t) +$$

$$+ q_{cd}^{(i,j)}(t,0), \ i = \overline{1,n}, \ j = \overline{2,l_i}, \ c = \overline{1,m}, \ d = \overline{1,a_c}.$$

$$(22)$$

Here $\qquad q_{vb}^{(\tau,l_\tau)}(t,v) = \delta(l_\tau = 1)q_{vb}^{(\tau,1)}(t,v) + \delta(l_\tau \neq 1)q_{vb}^{(\tau,l_\tau)}(t,v);$

$$\delta_{v\widetilde{v}} = \begin{cases} 1, & \text{if the}(\cdot)\text{ statement is true}, \\ 0, & \text{if the}(\cdot)\text{ statement is wrong}, \end{cases} \qquad q_{cd}^{(i,j)}(t,0) = \lim q_{cd}^{(i,j)}(t,u), \text{ in } u \to 0.$$

Suppose, at the initial moment $q_{cd}^{(i,j)}(0,u) = 0$, $R_{cd}^{(i,j)}(0) = \delta_{c\widetilde{c}}\delta_{d\widetilde{d}}\delta_{i\widetilde{i}}\delta_{j\widetilde{j}} = R(0)$

Proof. According to the definition $q_{cd}^{(i,j)}(t,u)$ we can write

$$q_{cd}^{(i,j)}(t,u)du = P\{ u < \gamma(t) < u + du, \ \varepsilon(t) = (i,j), \ \eta(t + \gamma(t)) = (c,d) \ \}.$$

Here $\varepsilon(t) = (i,j)$ is the system state by the coming call at the moment t; $\eta(t + \gamma(t)) = (c,d)$ -is the process state by the occurring failure at the moment $t + \gamma(t)$. Let us consider two neighboring moments t, $t + \Delta t$. Suppose $j=1$ (equation (20)). The probability that servicing of the call arrived within the time interval $(t, t + \Delta t)$ will begin within the time interval $u - \Delta t < \gamma(t + \Delta t) < u - \Delta t + du$ at the QM state (c,d) by the occurring failure is equal to the probabilities of the following incompatible events:

1) $u < \gamma(t) < u + du$; $\varepsilon(t) = (i,1)$; $\eta(t + \gamma(t)) = (c,d)$; within the interval (t, $t + \Delta t$), no new calls have arrived. The probability of that event is

$$q_{cd}^{(i,1)}(t,u)(1 - \lambda_{i1}\Delta t)du + o(\Delta t);$$

2) $v < \gamma(t) < v + du$;

$$\varepsilon(t) = (\tau,l_\tau)(\tau = \overline{1,n}); \ \eta(t + \gamma(t)) = (v,b)(v = \overline{1,m}, b = \overline{1,a_v});$$

within the interval (t, $t + \Delta t$), by the I branch a new call with the probability $P_i\lambda_{d\tau}\Delta t + 0(\Delta t)$ will arrive. Its servicing will require the time $\xi(u - v < \xi < u - v + dv)$ and $\eta(t + \gamma(t) + \xi) = (c,d)/\eta(t + \gamma(t)) = (v,b)$. the probability of this complex event is

$$\sum_{v=1}^{m} \sum_{b=1}^{a_v} \sum_{\tau=1}^{n} \int_0^u q_{vb}^{(\tau,l_\tau)}(t,v)\lambda_{d_\tau}\Delta t \cdot P_i h_{vb}^{(c,d)}(u-v)dv + o(\Delta t);$$

3) $\gamma(t) = 0$; $\varepsilon(t) = (\tau, l_\tau)(\tau = \overline{1,n})$; $\eta(t) = (v,b)$ $(v = \overline{1,m}, b = \overline{1,a_v})$; within the interval $(t,\ t+\Delta t)$, with the probability $\lambda_{d_\tau} \Delta t\ P_i + 0(\Delta t)$, the call well be arrive by the i branch. Its servicing will require the time $\xi(u \le \xi \le u + du)$ and $\eta(t + \xi) = (c,d)/\eta(t) = (v,b)$. The probability of this event is:

$$\sum_{v=1}^{m}\sum_{b=1}^{a_v}\sum_{\tau=1}^{n} R_{cd}^{(\tau,l_\tau)}(t)\lambda_{d_\tau}\Delta t \cdot p_i h_{vb}^{(c,d)}(u)du + o(\Delta t);$$

4) $\gamma(t) = 0$; $\varepsilon(t) = (i,1)$; $\eta(t) = (v,a_v)$ $(v = \overline{1,m})$; within the interval $(t,\ t+\Delta t)$, with the probability $\alpha_{va_v}\Delta t + 0(\Delta t)$, the QM transferred to the failure state, and with the probability $g(u)du + 0(u)$, the QM recovery has ended during the time $\xi(u \le \xi \le u + du)$, and the QM will be in the state $(c,1)$ by the occurring failure with the probability P_c. The probability of this event is:

$$\delta_{d1}\sum_{v=1}^{m} R_{va_v}^{(i,1)}(t)a_{va_v}\Delta t \cdot g(u)q_c du + o(\Delta t);$$

So, we obtain

$$q_{cd}^{(i,j)}(t+\Delta t, u - \Delta t) = q_{cd}^{(i,1)}(t,u)(1 - \lambda_{i1}\Delta t) + \sum_{v=1}^{m}\sum_{b=1}^{a_v}\sum_{\tau=1}^{n}\int_{0}^{u} q_{vb}^{(\tau,l_\tau)}(t,v)\lambda_{d_\tau}\Delta t \cdot p_i \times$$

$$\times h_{vb}^{(c,d)}(u-v)dv + \sum_{v=1}^{m}\sum_{b=1}^{a_v}\sum_{\tau=1}^{n} R_{vb}^{(\tau,l_\tau)}(t)\lambda_{d_\tau}\Delta t \cdot p_i h_{vb}^{(c,d)}(u) + \delta_{d1}\sum_{v=1}^{m} R_{va_v}^{(i,1)}(t)\Delta t \cdot g(u)q_c + o(\Delta t),$$

$$i = \overline{1,n},\ c = \overline{1,m},\ d = \overline{1,a_c}.$$

Form similar reasoning

$$q_{cd}^{(i,j)}(t+\Delta t, u - \Delta t) = q_{cd}^{(i,j)}(t,u)(1 - \lambda_{ij}\Delta t) + \delta_{d1}\sum_{v=1}^{m} R_{va_v}^{(i,j)}(t)a_{va_v}\Delta t \cdot g(u)q_c +$$

$$(1 - \delta_{j1})q_{cd}^{(i,j-1)}(t,u)\lambda_{i,j-1}\Delta t,$$

$$i = \overline{1,n},\ j = \overline{2,l_i},\ c = \overline{1,m},\ d = \overline{1,a_c}.$$

$$R_{cd}^{(i,j)}(t+\Delta t) = [1 - (a_{cd} + \lambda_{ij})\Delta t]R_{cd}^{(i,j)}(t) + (1 - \delta_{0,d-1})R_{c,d-1}^{(i,j)}(t)a_{c,d-1}\Delta t +$$

$$+ (1 - \delta_{0,j-1})R_{cd}^{(i,j-1)}(t)\lambda_{i,j-1}\Delta t + q_{cd}^{(i,j)}(t,\Delta t) + o(\Delta t).$$

Here all the terms have the exponent probable meaning.

Using the continuity of $R_{cd}^{(i,j)}(t)$ and $q_{cd}^{(i,j)}(t,u)$, and proceeding to the limit in these equations at $\Delta t \to 0$, we obtain equations (20), (21) and (22). The theorem has been proved. The normalizing condition has the form:

$$\sum_{i=1}^{n}\sum_{j=1}^{l_i}\sum_{c=1}^{m}\sum_{d=1}^{a_c}(R_{cd}^{(i,j)}(t) + \int_0^t q_{cd}^{(i,j)}(t,u)du) = 1.$$

Analysis of Feasibility of Calls Servicing Cycles

Now, let's determine $H_{l(i,j)\eta}^{(c,d)}(u)$ for three models of calls servicing at $\eta = 1$ i.e. only one call is being serviced at the same time. Hereinafter indicate l and η will be omitted.

Model 1. Suppose the interval duration between two sequential moments of occurring the technical system faults can be approximated by the distribution consisting of exponentially distributed stages with total density $b(s)$; transitions of the stage only two neighboring stage takes place, and form the branch end only to the beginning of any of the $i, (i = \overline{1,m})$ branches with the probability q_i. To minimize the time losses by reservicing of the call interrupted by the QM failures, the processing correctness is controlled continuously and transition to processing of the next block will be performed only when the preceding n block has been processed correctly. At the end of each block, the processing results are stored in the storages where they are not subjected to distortions. After the failure, the QM is immediately transferred to recovery; after the QM recovery, processing of the block interrupted by the failure is repeated from the beginning, that depreciates the failure; the processing time of each block of the perfect QM is a random value with the FDP-$F(u)$. First, let us define the expression for the FDP of the one block processing real time. Let us denote it by $H_{kb}^{(c,d)}(1,t)$ $(c,k = \overline{1,m},\ b = \overline{1,a_k},\ d = \overline{1,a_c}.$

Describing this model as a semi-Markovian process, one can easily make certain that the functions $H_{kb}^{(c,d)}(1,t)$ satisfy the following system of integral equations:

$$H_{kb}^{(c,d)}(1,t) = \delta_{kc}[\delta_{bd}\int_0^t e^{-\alpha_{bk}u}dF(u) + \int_0^t \tilde{I}_{(d,b)}^{(k)}(u)du \int_0^{t-u} e^{-\alpha_{cd}v}dF_v(u+v) +$$

$$+ \sum_{v=1}^{m}\int_0^t I_{(a_k,b)}^{(k)}(u)\overline{F}(u)du \int_0^{t-u} p_v H_{v1}^{(c,d)}(1,t-u-v)dG(v)],\ b \le d; \tag{23}$$

$$H_{kb}^{(c,d)}(1,t) = \delta_{kc}\sum_{v=1}^{m}[\int_0^t I_{(ab)}^{(k)}(u)\overline{F}(u)du \int_0^{t-u} p_v H_{v1}^{(c,d)}(1,t-u-v)dG(v)],\ b > d; \tag{24}$$

$$H_{kb}^{(c,d)}(1,t) = (1 - \delta_{kc}) \sum_{v=1}^{m} [\int_0^t I_{(a_k,b)}^{(k)}(u) \overline{F}(u) du \int_0^{t-u} P_v H_{v1}^{(c,d)}(1,t-u-v) dG(v)]; \qquad (25)$$

$$\overline{F}(u) = 1 - F(u), \quad k,c = \overline{1,m}, \quad b = \overline{1,a_k}, \quad d = \overline{1,a_c}$$

Here δ_{ij} – the Kronecker symbol; $I_{(db)}^{(k)}(u) du$ – the probability that, within the interval $(u, u+du)$, the occurring failure will be at the stage (phase) d by the branch k, provided that, at the moment $u=0$, the failure was at the phase b by same branch.

As an example, let's explain how (23) was obtained. The under integral expression of the first member is the probability that within the interval $(u, u+du)$, servicing of the block (in the branch k-$dF(u)$) was completed without transition to next phase ($e^{-\alpha_{kb}u}$) that will occur only at $k=c$ and $b=d$; the second number – the probability that, within the interval $(u, u+du)$, the occurring failure was in the phase d by the branch $k(I_{(d,b)}^{(k)}(u) du)$, the block servicing was completed during the time $u+v$, provided that it was not completed during the time $u(dF(u+v)F(u))$, the occurring failure did not changed the phase $e^{-\alpha_{cd}v}$ during the v time, that will occur at $b<d$ and $k=c$; the third member – the probability that, within the interval $(u, u+du)$, the failure will occur ($I_{(a,b)}^{(k)}(u) du$ phases will pass - $a_k - b$), serviced of the block ($\overline{F(u)}$) will not be completed, recovery of the faulty QM will be completed within the interval $(v, v+dv)$-$dG(v)$ after that, with the probability $P_v(v = \overline{1,m})$ by failures, the QM will appear at the first phase of the v – branch, and the block servicing will be completed during the time less than t-u-v, starting from the beginning of the stage $(H_{v1}^{(c,d)}(1,t-u-v))$ at $b \leq d$.

Similarly (24) and (25) are obtained. For the FDP of the servicing time of the N-stage problem, the following relations exists:

$$H_{(kb)}^{(c,d)}(N,t) = \sum_{i=1}^{m} \sum_{j=1}^{a_i} \int_0^t H_{ij}^{(c,d)}(N-1,t-u) dH_{kb}^{(i,j)}(1,u).$$

Model 2. This model differs from model 1 by the fact that:

1) The time between the neighboring failures is distributed by the special Erlang law of the m–th order, i.e. it consists of m phases, and each phase is characterizing by parameter β:

$$A(t) = \int_0^t \frac{\beta(\beta u)^{m-1}}{(m-1)!} e^{-\beta u} du;$$

2) After recovery, the QM resumes the block processing from the point of interruption, i.e. the work has been performed by the moment of the failure will not be depreciated.

3) The length of each blocks is distributed by the low $B(u) = 1(t - \tau_\delta)$.

Let us express the FDP of the real time of one block processing $H_i^{(j)}(1,x,u)$ and define it as the probability that processing of one block of a constant length (volume) will be completed during the time less than u, and that the QM will appear at the phase j by the occurring failure (by the Erlang arrival pattern), provided that, at moment $u=0$, the x part of the block was processed without any errors and the QM was in the I state by the occurring failure. With regard to $H_i^{(j)}(1,x,u)$ there are the following systems of integral equations:

$$\psi_{ij}(t,x) = \delta_{ij} \int_0^t e^{-\beta u} dB(x+u) + \int_0^t [\beta(\beta u)^{m-i-1}/(m-i-1)!] e^{-\beta u} du \times$$

$$\times \int_0^{t-u} \psi_j(t-u-v, x+u) dG(v), \quad i \geq j; \tag{26}$$

$$\psi_{ij}(t,x) = \int_0^t \frac{\beta(\beta u)^{j-i-1}}{(j-i-1)!} e^{-\beta u} du \int_0^{t-u} e^{-\beta v} d_v B(x+u+v) +$$

$$+ \int_0^t \frac{\beta(\beta u)^{m-i-1}}{(m-i-1)!} e^{-\beta u} du \int_0^{t-u} \psi_{0j}(t-u-v, x+u) dG(v), \quad i < j; \ i,j = \overline{0, m-1}. \tag{27}$$

Here $\quad \psi_{ij}(t,x) = \overline{B}(x) H_i^{(j)}(1,x,u); \quad \overline{B}(x) = 1 - B(x);$

$$\psi_{0j}(t-u-v, x+u) = \overline{B}(x+u) H_{0j}(1, t-u-v, x+u); \quad \psi_{ij}(t,x) = \begin{cases} 1, & \text{if the } x = \tau_\delta, \\ 0, & \text{if the } x > \tau_\delta. \end{cases}$$

Let us explain the first equation. The first member is the probability that, within the interval $(x+u, \ x+u+du)$, the block processing will be completed, provided that it did not end burring the x time $- dB(x+u)/\overline{B}(x)(u \in (0,t))$ and, during the u time, the QM did not change the phase $e^{-\beta u}$ by the occurring failure. The second member is the probability that, within the interval $(u, \ u+du)$, the failure $(\beta(\beta u)^{m-i-1}/(m-i-1)!) e^{-\beta u} du \ (u \in (0,t))$ occurred and the QM transferred to the recovery state; During the u time, the calls servicing was not completed - $\overline{B}(x+u)/\overline{B}(x)$; within the interval $(v, v+dv)$, the QM recovery was completed and the QM by the occurring failure transferred to the zero phase - $dG(v) (v \in (0, t-u))$; the block processing has started from the point of interruption and will be completed during the time t-u-v, and the QM by the occurring failure will appear at the phase j, provided that, at the moment t=u+v; the x+u part of the block had already been correctly processed and the QM by the occurring failure was at the zero phase. This equation occurs at $i \geq j$.

Similarly, processing from probable reasoning, (27) was obtained.

Model 3. Assume the message transmitted by data transmission channels is divided into N binary blocks. Each block is following by check positions which, with the single probability, ensures error (failure) detection in the process of information block transmission; the length of each block expressed in time unit or in terms of transmitted bits is distributed by an arbitrary law – $F(u)$; the time between the neighboring failures is a random value and can be approximated by m parallel stages (branches; one phase in each branch), each with the parameters $\alpha_i (i = \overline{1, m})$, i.e. the density of distribution between the neighboring failures has the from (hyper exponential distribution):

$$b(u) = \sum_{k=1}^{m} p_k \alpha_k e^{-\alpha_k u}, \quad \sum_{k=1}^{m} p_k = 1.$$

(p_k - the probability of entering the branch k by the occurring failure after the QM recovery); in case of the failure (error) and its detection by the control system at the end of block transmitting, the QM is transferred to recovery; the QM recovery time is distributed by an arbitrary law – $G(u)$; after the QM recovery, retransmission of the block from the beginning will be performed (the depreciating failure); retransmission with previous realization will be carried out until the block is transmitted correctly.

For this model, $H_i^{(v)}(1, u)$ - the probability that transmission of the one binary information block will be completed during the time less than u, and that the QM by the occurring failure will appear in the v-th parallel branch, provided that, at the moment of the block transmission beginning, the QM by the occurring failure was in the i-th branch is the solution of the following systems of integral equations:

$$H_i^{(v)}(1,u) = \delta_{iv} \int_0^t e^{-\alpha_i u} dF(u) + \sum_{k=1}^{m} \int_0^t (1 - e^{-\alpha_i u}) dF(u) \int_0^{t-u} P_k H_k^{(v)}(1, t-u-v) dG(v),$$

$$i, v = \overline{1, m}. \tag{28}$$

Here δ_{iv} is the Kronecker symbol.

System of equations (28), as well as the previous equation, was obtained on the basis of the formula of the composite probability. This is: the under integral expression of the first member is the probability that transmission of the binary information block will be completed within the interval $(u, u+du)$-$dF(u)$ $(u \in (0,t))$, provided that, during this time, there will not occur the transition to the other branch - $e^{-\alpha_i u}$ (that occurs only at i=v). The second member is the probability of the following complex event: 1) within the interval $(u, u+du)$, transmission of the block $dF(u)$ $(u \in (0,t))$, was completed. 2) during the u time, the QM failure occurred - $(1 - e^{-\alpha_i u})$; the QM recovery was completed within the interval $(v, v+dv)$-$dF(v)$ $(v \in (0,t-u))$; after that, the QM by the occurring failure transferred to the k branch with the P_k probability; 4) the block transmission will be completed during the time less than t-u-v, provided that, at the beginning moment of retransmission, the QM by the occurring failure was in the branch k [6].

We should note that, in applications in practice, it is not difficult to obtain expressions for the functions $H^{(i)}_{l_j \eta}(u)$ in most cases, as you could see above. The value of $H^{(i)}_{l_j \eta}(u)$ and

$H_{l_j \eta}(u)$ both stand duty as initial characteristics when investigating the QS and give the comprehensive probable characteristics of the QM for determining its ability of executing single tasks during the time less than the specified one. They are also of independent significant importance. They are necessary for analytical investigation of the QM. They are also necessary for designing the technical systems with time or other kind redundancies. In particular a) for the justified drawing up of requirements by reliability. b) for the best choice of a time diagram of the device (QM) usage, that finally results in considerable improvement of the reliability; c) for the best choice of the control system characteristic – the control type; d) for determination of the relation between the device capacity and the reliability characteristic. Really, we can introduce such device characteristic as the real capacity index which is determined by the relationship of absolute reliability of the device and the mathematical expectation of the task execution time with consideration of the device reliability, division of the task into blocks, and time necessary for control, maintenance and recovery of the device. The real capacity index should be determine for the given value of the task execution probability.

Conclusions

1. All the offered equations are solved to the end of the Laplace transform. The ways of solving the analogues equation are presented in [6-13], therefore, their solution are not given here.
2. An effective method combining the method of the embedded Markovian chains with the supplementary variable method is offered. The method can be applied for investigation of any complex queuing systems (QS) with any external or internal interfering factors, including multichannel QS with a joint queue and resources, for whish the Laplace-Stilties transform of the distribution of time intervals between the neighboring calls and failures are rational functions of the Laplace operator S or their approximation by fractional-rational functions of S. As the arbitrary irrational function of S can be approximated by fractional-rational functions as accurately as you like [14,15], then in principle, in this work, the method with the help of which rather general problem can be solved is offered. Such combination of the above-mentioned methods significantly expands application of the Markovian processes for

analyzing the complex random processes which describe up-tu-date tasks of the QS with accuracy and reliability sufficient for practice.

3. In accordance with the random process $\xi(t) = (l, j, k, u)$, the given OIS at the moment t-u is (l, j, η); then we determine $H_{lj\eta}(u)$ and $H_{lj\eta}^{(i)}(u)$ irrespective of the initial random process $\xi(t)$. Such approach significantly expands application of the offered methods, which allows the possibility of decomposition of the initial random processes $\xi(t)$ into relatively simpler random processes, each of which is subject to investigation by an analytical method, with the possibility of their further joining.

4. As the introduced random processes $H_{lj\eta}(u)$, $H_{lj\eta}^{(i)}(u)$ allow us to consider the process of servicing of separate calls or groups of calls irrespective of each other and the initial random process $\xi(t)$, there can be embedded a lot of arbitrary factors affecting the servicing process such as: a) QM failure types (partial, hard, self-restoring, with accumulation, etc); b) QM serviceability control types (continuous, periodic, combined, reliable, etc.); c) recovery time distribution; d) a place of each servicing channel (if more than one are available) in the QM structure (distribution of the work among the channels); servicing in the mutual aid mode, servicing with advance, servicing with a duplex solution, servicing with reserving, servicing in the mode of situational integration of the QM resources, etc.; discipline of reservicing of the interrupted calls (reservicing from the beginning, reservicing from the point of interruption or from the preset points of continuation, etc.).

References

[1]   D. C. Kendall, Stochastic processes occurring in the theory of queues and their analysis by the method of the embedded Markovian chain. // *Ann Math. Statistic*, **24**, 1953. –p. 338-354.

[2]   Беляев Ю.К. Линейчатые марковские процессы и их приложение к задачам теории надежности. Тр. VI Всесоюзного совещания по теории вероятностей и математической статистике, (1960). – Вильнюс, Гос. Изд-во политической научной литературы СССР, 1962. – с. 309-323;

[3]   Гнеденко Б.В., Коваленко Введение в теорию массового обслуживания. М.: *Наука,* 1987;

[4]   D.R.Cox. The analysis of non-Markovian stochastic processes by the inclusion of supplementary variables. // *Proc. Of the Cambr. Philos. Soc.*, **51**, 3, 1955. –pp. 433-441.

[5]   Королюк В.С., Турбин А.Ф. Полумарковские процессы и их приложения. – Киев, *Наукова думка*, 1976;

[6]   Микадзе И.С. Система обслуживания со многими состояниями функционирования. // *Автоматика и телемеханика*, **№12**, 1987. –с.104-116;

[7] I.S. Mikadze. *A Queuing system with multiple operating states.* 0005-1179/87/4812-1648, 1988, Plenum Publishing Corporation (USA).

[8] Микадзе И.С. Систена обслуживания с ненадежными обслуживающими приборами. // Кибернетика, №3, 1989. –с.102-109;

[9] I.S. Mikadze. *Queuing system with unreliable processors.* 0011-4235/89/2503-0409, 1990, Plenum Publishing Corporation (USA).

[10] Микадзе И.С., Какубава Р.В. Очереди в системах с накапливающимися отказами и эрланговским входящим потоком. // *Сообщения АН Грузии*, т.**152**, №1, 1995, -с.146-151;

[11] Какубава Р.В., Микадзе И.С. Анализ дублированной системы массового обслуживания. // *Автоматика и телемеханика*, №**1**, 1984. –с.160-166;

[12] R.V. Kakubava and I.S. Mikadze. *Analysis of a redundant queuing system.* 0005-1179/84/4501-0138, 1985, Plenum Publishing Corporation (USA);

[13] Микадзе И.С., Какубава Р.В. Групповое обслуживание в системах со многими работоспособными состояниями // *Сообщения АН Грузии*, т.**152**, №2, 1995, с.362-368;

[14] I.S. Mikadze, R.V. Kakubava. Time-characteristics analysis of the queuing system with cumulative faults and general Erlang arrivals. //*Bulletin of the Georgian Academy of Sciences,* v.154, No.2, 1996. –pp.241-244.

[15] Микадзе И.С. Вероятностная характеристика производительности цифровой вычислительной машины с учетом ее надежности. // *Автоматика и телемеханика*, №**2**, 1979. –с.175-186 (USA);

[16] Микадзе И.С. Периодически контролируемая система обслуживания с ненадежным прибором. // *Кибернетика*, №1, **1988**. –с.56-61 (USA);

[17] Клейнрок Л. Теория Массового обслуживания. –М.: *Машиностроение*, 1979.

[18] Кокс Д., Смит В. *Теория восстановления.* -М.: Советское радио, 1967.

[19] Takacs. On a stochastic process concerning some waiting time problems. // *Теория вероятности и ее применение*, **2**, 1, 1957. –с.92-105.

In: Focus on Science and Technology...
Editor: Sergo Gotsiridze, pp. 37-55

ISBN: 978-1-61209-970-5
© 2011 Nova Science Publishers, Inc.

Chapter 4

POST-COMMUNIST CAPITALISM AND FINANCIAL CRISIS, OR THE MIXING OF THE NECROECONOMICS AND THE ZOMBIE-NOMICS

Vladimer Papava[*,1,2,3]

[1]Georgian Foundation for Strategic and International Studies
[2]Paata Gugushvili Institute of Economics
[3]Joint Transatlantic Research and Policy Center at the Central Asia-Caucasus Institute

Abstract

After the collapse of the Communist regimes and their command economies, the countries of the former Soviet Union found themselves with only a very small amount of goods to supply to the global market. In fact, no markets existed for many types of products. There was no way that they could have existed in that an economy of this type is nothing more than a corpse or a so-called "necroeconomy." The purpose of this paper is to distinguish the various economic foundations of post-Communist capitalism and to examine the key economic problems of this type of society in the context of the modern financial crisis. As international experience shows, dead firms do exist and "successfully" function in the most developed of economies as well with Japan being the most obvious example. These insolvent and, in fact, bankrupt firms which continue to operate despite their "mortality" are commonly referred to as "zombie-firms." Unlike developed economies, which are exposed to the threat of the zombie-ing of the economy under the conditions of a financial crisis, this threat is even greater for the countries of post-Communist capitalism owing also to their exposure to necroeconomy.

Almost twenty years have passed since the beginning of the period of post-Communist transition to a market economy. Naturally, this has resulted in the accumulation of a rather

[*] E-mail address: papavavladimer@gfsis.org. Dr Vladimer Papava is a Senior Fellow at the Georgian Foundation for Strategic and International Studies, a Principal Research Fellow at the Paata Gugushvili Institute of Economics and a Senior Associate Fellow of the Joint Transatlantic Research and Policy Center at the Central Asia-Caucasus Institute (Johns Hopkins University-SAIS). He was the Minister of Economy of the Republic of Georgia (1994-2000) and a Member of Parliament (2004-2008) (Tbilisi, Republic of Georgia).

rich experience overall and one which allows us to make some generalisations. It may be asserted, that market economies, as such, have been established in almost all of the countries of the former Socialist bloc with the period of transition over and the individual newly-independent states having passed through this period with with varying degrees of success.[1]

Some of the countries were so successful within their movement towards a market economy that they achieved EU membership (Papava, 2006) whereas others—in fact, all of the post-Soviet nations with the exception of the Baltic states—became "prisoners" of their own product; that is, post-Communist capitalism (Kennedy and Igen, 2007; Papava, 2005) which is a very special phenomenon of modern times and which includes in itself a wide range of different forms of capitalism (for example, Coates, 2000; Crouch and Streeck, eds., 1997; Gwynne Klak and Shaw, 2003; Hall and Soskice, 2001).

The purpose of this paper is to distinguish the various economic foundations of post-Communist capitalism and to examine the key economic problems of this type of society in the context of the modern financial crisis.

Necroeconomy – Heritage of Command Economy

In order to understand the essence of the economic foundations of post-Communist capitalism, one has to analyse one of the key peculiarities of a command economy; that is, the quality of its material and technical bases.

It is common knowledge that a command economy excludes any possibility for the existence of any forms of competitive relationships either domestically or internationally—inside any distinct economy or between different command economies—owing to its very nature. The majority of command economies used to be integrated into one big common economic space. The former Council of Mutual Economic Assistance (CMEA), which existed for approximately 40 years and was governed by a co-ordinating organ, is perhaps the best example of this. Economic co-operation with market economies was maintained upon a very limited basis and exclusively at an inter-governmental level.

The absence of competition in command economies quashed the only effective stimulus for economic development. As a result, the quality of products, as a rule, was very low—as were their prices—which were maintained artificially by means of national budgetary subsidies. The key sources of the former Soviet Union's national budget revenues included the sales of alcoholic beverages and the export of raw materials (basically those of oil) which represented the only stable channel for the accumulation of foreign currency reserves.

Based upon the studies and generalisations of key aspects of the economic system of the Communist regime in Poland, Adam Lipowski came to a conclusion that when the whole world was divided between the "developed" and the "developing" countries, those with command economies could not qualify for either of the foregoing and so he invented the term of "misdeveloped" countries specifically for command economies (Lipowski, 1998, p. 9). In such cases of "misdevelopment," Lipowski asserted that:

- the share of industries in GDP was too high because of a low percentage of domestic and foreign trade and services,

---

[1] Belarus is perhaps the only exception (Antachak, Guzhinski and Kozarzhevski, eds. 2001).

- a significant portion of industrial production accounted for manufacturing production as opposed to the comparatively small output of consumer goods,
- the volume of high quality competitive products capable of meeting international standards was very limited,
- the major part of industrial output included goods which were generally useless to customers and
- the share of outdated products in industrial output was too high.

After the collapse of the Communist regimes and their command economies, the countries of the former Soviet Union found themselves with only a very small amount of goods to supply to the global market. With few exceptions, such as some hydro energy outputs, oil and gas extraction and the primary processing of raw materials, the goods manufactured they manufactured failed to meet the high international standards as a result of their overall low quality and or high prices. In fact, no markets existed for these particular products. Moreover, in principle, there was no way that they *could* have existed in that an economy of this type is nothing more than a corpse or a so-called "necroeconomy" (Papava, 2001) or, similarly, a "virtual economy" (Gaddy and Ickes, 1998, 2002; Woodruff, 1999). The economic theory which deals with this kind of economy is called "necroeconomic theory" or "necroeconomics" (Papava, 2002, 2005).[2]

Lipowski uses the term "divesting" (as an antonym to "investing") in order to describe the process of a command economy's being "stripped" (for example, Drucker, 1985; Teylor, 1988) which amounts to "liberating" the post-Communist economy from the list of pathologies characterising a "misdeveloped economy" (Lipowski, 1998, pp. 31-32) which, in our opinion, is the way in which a necroeconomy develops.

Naturally, even if one part of an economy is dead, the rest of it may still be alive which can be referred to as its "vital economy" or "vitaeconomy." Further accordingly, the economic theory which deals with this kind of economy is called "vitaeconomic theory" or "vitaeconomics." By its substance, this is nothing more than economic theory itself—or economics in its common meaning—because economic theory, as such, is something which deals with the economy as a vital system.

Collectively, all of the aforementioned leads us to the question: What does necroeconomy have in common with vitaeconomy and how do they differ?

In a necroeconomy, like in a vitaeconomy, some goods may be produced which in fact means that there may exist *supply*. In contrast to those goods produced in a vitaeconomy, however, those emanating from a necroeconomy are in a situation of no *demand* owing to their low quality and or high prices. Consequently, a necroeconomy *excludes any reasonable act of sale and purchase and, as a result—equilibrium prices.*

If any segment of an economy is "dead," then theoretically there should be no problems within. Common sense tells us that a necroeconomy cannot have any influence over its vital parts. Under the conditions of a market economy, economic theory prescribes that uncompetitive productions must disappear and, at the same time, should not create any significant problems for the rest of economy. This explains the limited focus from the side of economic theory upon the problems of such a post-Communist market economy in which necrocompanies are exist.

---

[2] Political science uses the terminology "necropolitics" (Mbembe, 2003).

Specifically, in the countries which are still undergoing the process of post-Communist transformation—as well as those in which post-Communist capitalism has already been established—*necroeconomy has grown on top of the roots of the command economy's material and technical bases*. We can conclude, therefore, that the necroeconomy is *exactly* that which differentiates the economy of post-Communist capitalism from all other models of capitalism.

The economy of post-Communist capitalism consists of the following groups of necroeconomy and vitaeconomy:

Group 1. Necroeconomy in the public sector
Group 2. Vitaeconomy in the public sector
Group 3. Privatised necroeconomy
Group 4. Privatised vitaeconomy
Group 5. Vitaeconomy developed by means of new private investments.

The majority of the first group, as a rule, consists of large- and medium-size processing industry enterprises which, depending upon the types of goods they produce, are labeled "strategic" ones even though they are dead under the conditions of a market economy owing to their low or lack of competitive powers.

On the other hand, enterprises within the energy sector (first of all, electricity generation and transmission and the extraction and distribution of oil and gas), as well as those of transport and communications constitute a basis for the vitaeconomy in the public sector. When privatised, they move to the fourth group comprising the privatized vitaeconomy which may also include some medium- although mostly small-size industrial enterprises (prior to their privatization).

The third group consists of the former first group enterprises following their privatization. The change in ownership by itself does not automatically entail the restarting of formerly idle enterprises in that a "corpse's" status does not depend upon whether it is owned by the government or a private firm. *Disregarding this fact is the key reason that the process of privatization has been relatively discredited*. Privatization, especially during its initial phases—irrespective and taken separately of any investments—has often been believed to be a universal remedy capable of restarting any inoperable enterprise, dead or alive. As we will seek to demonstrate, the institute of private ownership alone is not able to create sufficient conditions for the abolition of necrofirms.

The fifth and last group embraces the "healthiest" segment of the post-Communist economy which is based upon the principles of a market economy maintained by private investments. Some problems, however, may be discovered herein as well which will require adequately address. In particular, this refers to some foreign investments by means of which post-Communist countries receive relatively old technologies which have become obsolete from the standpoint of modern international standards. In our opinion, these could be labelled as "second-hand investments" with goods manufactured by means of this capital only being competitive in "emerging markets" and only for a limited period of time until the arrival of highly competitive goods which meet all of the international standards.

Zombie-Economy—Heritage of Financial Crises

We have already stated that necrocompanies are found within the countries of post-Communist capitalism but the question arises of whether or not this is a problem confined to such countries in transition alone or if these sorts of enterprises also exist in developed and or developing economies.

As international experience shows, dead firms do exist and "successfully" function in the most developed of economies as well with Japan being the most obvious example (Hoshi, 2006). These insolvent and, in fact, bankrupt firms which continue to operate despite their "mortality" are commonly referred to as "zombie-firms."

A system of continued lending is the key source of the sustainability of these zombie-firms (Caballero, Hoshi and Kashyap, 2008; Smith, 2003) with their loans granted by so-called "zombie-banks" which extend beneficial credits to the firms (in particular, interest rates for such loans are lower than average rates at the market level) (Hoshi and Kashyap, 2005; Smith, 2003). In full risk of stating the obvious, these unreasonable loans can only lead these banks to direct and inevitable losses (Ahearne and Shinada; 2005, p. 368).

This, therefore, is the maintenance plan for zombie-firms but how, one may ask, do zombie-banks manage to survive under such circumstances? As a rule, such banks are backed by their country's governments (Kane, 2000b, p. 301) which grant them all kinds of guarantees and assure their deposits, amongst other things, which eventually means that zombie-banks exist at the expense of taxpayers (Kane, 2000a, p. 164). To a certain degree, such a financial system even encourages "healthy" firms to turn into zombies (Hoshi, 2006, p. 40).

As a result of the aforementioned relationships between zombie-firms, zombie-banks and their governments, a "zombie-economy," then, develops which becomes a heavy burden for the "healthy" segments of the economy. In particular, zombie- firms, by mere their existence—and which enjoy guaranteed beneficial loans from zombie-banks—block the emergence of new "healthy" firms in the market (Hoshi, 2006, p. 33) as they have to borrow at rather higher interest rates (Caballero and Kashyap, 2002). In addition, because of their access to guaranteed beneficial loans, zombie-firms, in their fight for market shares, are at liberty to drop prices (Smith, 2003, p. 288) and raise the salaries of their employees (Hoshi, 2006, p. 33). The limited market access for "healthy" and, even more so, productive firms eventually leads to the reduced productivity of the whole economy (Ahearne and Shinada, 2005, p. 364).

A zombie-economy takes its roots in times of a financial crisis (Ahearne and Shinada, 2005; Hoshi and Kashyap, 2004). Under the conditions of stagnation, the economy becomes characterized by a stoppage of production and trade for a relatively long period of time which, in turn, gives rise to unemployment, a reduction in wages and salaries and the overall decline of the standard of living. During these times, governments, as a rule, are called to help the economy to overcome such difficult conditions through the provision of bailouts and other attempts at keeping the banking sector (to avoid a banking crisis) and the entire economy on the surface. After the end of a financial crisis, the economy receives its own lifeless portion as a legacy of the difficulties and continues to try to preserve the old system of the government's financial aid which was readily available to it during the crisis. A zombie-economy, therefore, can be viewed as a legacy of a financial crisis. It must be emphasized that a zombie-economy is a phenomenon peculiar not only to Japan and other developed market economies (for example, Holle, 2005) but also to those countries with developing economies as international experience has shown (for example, Kane, 2000a).

What Are the Similarities and Differences between a Necroeconomy and a Zombie-Economy?

One might have the impression that the terms necroeconomy and zombie-economy refer to one and the same phenomenon; that is, a dead economy which continues to subsist despite its lethal status. In fact, such an impression is both superficial and misleading in that whilst the two "economies" do indeed share one commonality—there is no doubt that they are both dead—there is a wide spectrum of significant differences between them.

First of all, both economies developed in essentially different economic systems. A necroeconomy grew forth from a command economy whilst a zombie-economy is the offspring of a market economy. Further, a necroecoomy, in fact, has nothing to do with a financial crisis whereas a zombie-economy is the immediate end result thereof. It is important to note once again that the existence of zombie-firms depends in the main upon zombie-banks whereas necroeconomic agents subsist by means of immediate and direct subsidies from national budgets or tax exemptions. Moreover, the differences between a necroeconomy and a zombie-economy have also a lot to do with the sectors within which are mostly directly and readily exposed to their influences.

A necroeconomy, therefore, generally expands to large- and medium-size manufacturing industries as opposed to zombie-firms which show no traces of existence therein as evidenced by the situation in Japan's economy (Hoshi, 2006). Moreover, the large manufacturing enterprises in a post-Communist capitalist country, have the greater probability of becoming a part of a necroeconomy. On the other hand, as the same Japanese experience demonstrates, most large-size firms, due to their great financial powers, are not zombie-firms but may also often be encountered in those so-called small businesses which are relatively "larger" than others (Hoshi, 2006).

It is important to note that whilst all international financial institutions, such as the International Monetary Fund (IMF) and the World Bank, insist that the post-Communist capitalist governments eliminate all kinds of national budget subsidies and tax exemptions, all lobbying efforts are mobilized towards pushing those bailout programs into providing some extensive financial support to the national governments in order to enable them to build up some favourable lending systems under the circumstances of the financial crisis (Kane, 2000a, p. 163; 2000b, p. 288).

One may arrive at the conclusion, therefore, that necroeconomies and zombie-economies are related to each other but still differ to a great degree from each other as individual economic phenomena. Unfortunately, however, these differences are not always given due consideration. In some studies, the peculiarities of a necroeconomy are overlooked which means, as a result, that the problems of dead firms in the post-Communist countries (mostly China[3] and Russia) are examined within the context of zombie-economies rather than

---

[3] Despite its Communist ruling party, China's market reforms render it more within the group of post-Communist market economies and, as such, enable us to call it a post-Communist entity.

necroeconomies (for example, Kane, 2000b, pp. 300-301; Lindsey, 2002, pp. 126, 153; Shleifer, and Treisman, 2000, pp. 106-107).

Reproductive Mechanisms of a Necroeconomy and a Zombie-Economy

A key question with respect to a necroeconomy and a zombie-economy is what ensures their stable existence.

The answer may be found in an evolutionary theory of economic changes (Nelson and Winter, 1982) wherein the key tool is the concept of "routine" which implies a certain set of rules and ways of a firm's conduct which regulates the reproduction (of such a conduct) (Murrell, 1992a, 1992b).

It is this very routine, which has developed over a period of several decades upon the roots of a command economy, which pushes dead companies in the countries of post-Communist capitalism *to work in the no-longer-existing regime of a command economy.* Without any special governmental support, therefore, the warehouses of these companies become filled with uncompetitive goods for which there is neither demand nor market. Given the fact that as a matter of principle these goods cannot be sold to anyone, the companies find themselves further and further in arrears as regards wages, salaries and debts to national budgets, social funds, energy sector industries and other businesses which eventually creates a network of mutually indebted businesses (for example, Åslund, 1995, Ch. 6, 2002, pp. 244-248, 328-330, 333-334, 2007a, pp. 132-133).

It was a long-established tradition in command economies that when an enterprise accumulated (often very deliberately) huge debts, its director raised the question before his country's superior governmental institutions (such as Communist Party governing bodes, the Gosplan, the Ministry of Finance, etc.) to write off those debts and, as a rule, such requests were usually granted. Consequently, because of the almost unlimited (or much rather guaranteed) opportunity to have one's own debts removed, enterprise managers did not regard debt accumulation as any kind of danger for their existence.

Such a mechanism of off-writing of debts represented a firmly established routine which, however unfortunate it may have been, reappears over and over in the countries of post-Communist capitalism and within various forms such as "tax amnesties" (for example, Nikolaev, 2002; Shul'ga, 2002).

As for the routine of a zombie-economy, it develops under the conditions of financial crises wherein governments and banks collaborate with each other towards a common purpose of developing and implementing bailout programs for insolvent firms and, thereby, avoiding a greater economic decline and a further growth of unemployment. In the case of a relatively long period of stagnation, such collaboration grows into an established order which establishes the rules of a firm's conduct and, in turn, ensures the reproduction of such a conduct.

In other words, the activities grow into a routine. Herein one must underline the continued character of stagnation which is a condition precedent for the formation of routine; that is, it must have enough time to develop.

When a financial crisis comes to an end, the routine undertakes the mission to ensure the subsistence of a zombie-economy. The reason is that under all other equal conditions, and irrespective of the crisis, no government would tolerate the growth of unemployment which comes as an inevitable result of the closing down of zombie-firms. Undoubtedly, "healthy" firms are never able to instantaneously react to the disappearance of zombie-firms and quickly address the problem of creating new jobs (for example, Caballero and Kashyap, 2002; Lindsey, 2002, p. 235).

Homo Transformaticus and *Zombie Economicus*—Carriers of the Necroeconomic and Zombie-Economic Routines

The human factor, as a rule, is a matter of decisive importance for practically all kinds of economic developments. At the same time, an economic approach, as such, may be applied to any spheres of human behaviour (Becker, 1976).

The routine which reproduces a necroeconomy is conditioned by the behaviour of such a human being who is in the middle of the transition from the state of a *homo etaticus* (Hanhinen, 2000, p. 224) or *homo soveticus* (Buzgalin, 1994, pp. 250-253)—that is, one who is suppressed by the state and who totally depends upon the state—to *homo economicus*[4] or a human being whose driving force is to get the maximum utility at a household and a maximum profit at a firm which is typical of a market economy.

The type of human being who manages the process of post-Communist transformation or, in other words, one who is a "central character" in post-Communist capitalism is someone we call *homo transformaticus*; that is, a human being who failed to completely release himself from both the fear of the state and the habit of living at the government's expense even though he gradually gets used to the situation owing to his private interest in obtaining a maximum utility and profit (Papava, 1996, 1999).

By analogy with Y.A. Levada's concept of *homo adapticus* (Levada, 1999), *homo transformaticus* is a human being who is gradually becomes accustomed to the immanent rules of a market economy and, furthermore, gets involved in the process of setting and developing those rules.

In business, *homo transformaticus* assumes a special form whose roots may be traced back to a command economy.

Even in the times of a command economy's reign, a market economy (or, much rather, certain elements of it) was not eradicated in its entirety. Simply, it was oppressed by the state to the degree that it could only subsist within a "shadow" sector (for example, Shokhin, 1989, pp. 57-83).[5] Under the conditions of a command economy, no director (perhaps with few exceptions) could possibly manage his enterprise without breaking a law adopted by the existing regime. In exceptional cases, therefore, elements of a market economy were actually applied. Such activities were referred to as "shadow activities" and the managers of such

---

[4] The modern understanding of the features of the *homo economicus* has undergone significant changes in the aftermath of Smith's epoch (for example, Avtonomov, 1998, pp. 57-201; Brockway, 1995, Bunkina and Semionov, 2000; Ferber and Nelson, 1993).

[5] It must be noted that unlike a command economy, whose "shadow" sector may make room for some elements of a market economy, the "shadow" sector primarily embraces such activities which enable tax evasion (for example, Bunkina and Semionov, 2000, pp. 50-52; Svensson, 1983).

enterprises were called "shadowists" (Samsin 2003, p. 184). Nevertheless, under the conditions of a command economy and given their suppression by the former Communist regime, such directors never became—and could not have possibly become—truly market-type entrepreneurs. For this reason, those directors who applied some elements of the market economy within their managerial practices were not called "entrepreneurs" (Gins 1992, p. 119) but, rather, "*del'tsy*" (the plural of the Russian word "*delets*" which means "businessman" in its literal translation but is typically rendered as "labourer" given its derogatory connotation).

Even after the collapse of a command economy, most of former *del'tsy* managed to preserve themselves as directors in public sector enterprises within various capacities (Åslund, 2007b, pp. 137-140). Moreover, after the privatisation of those particular enterprises, the same *del'tsy* exploited the rights of "labourers' collectives" and became the owners of the same enterprises (Åslund, 1996). At the same time, whether or not they hired professional managers—especially in the initial phases of the post-privatisation periods—they still tried to manage their enterprises by their own way of thinking following a so-called *delets* way (for example, Nekipelov, 2003, p. 127).

Just as *homo transformaticus* has not yet become *homo economicus*, so is the case with the former *del'tsy* who have not yet developed into entrepreneurs. *Homo transformaticus*, therefore, takes on the title of "post-*delets*"[6] in his entrepreneurship.

It is exactly these post-*del'tsy* who make up the spine of a necroeconomy in both the public and private sectors as they are the carriers of the routine of a command economy. By using their old connections, the post-*del'tsy* manage to break into governmental organs (parliaments, executive offices, etc.) and use their influence to try to politically justify and prolong the existence of the necroeconomy.

It can be stated without any reservation that a necroeconomy best serves the interests of post-*del'tsy* in that this type of economy will always enable the continuation of their existence until such time that they have been entirely replaced by true entrepreneurs through the means of appropriate institutional reforms.

The key to the understanding of a zombie economy's routine may be found in the theories of public choice by James Buchanan in which politics is interpreted as a special variety of a market (Buchanan, 1997), and of "animal spirits" (Akerlof and Shiller, 2009). During a financial crisis, a type of economic policy develops which proposes the government's interference in the economy with its uppermost goal of rescuing it from a critical state, on the one hand, and encourages the addressing of the private interests of those economic agents who have found themselves on the verge of bankruptcy as a result of the said crisis, on the other hand.

As previously mentioned, a zombie-economy's routine is the product of a market economy in the state of crisis. Consequently, it is not a simple *homo economicus* who is the carrier of this routine but, rather said, a mutation which was formed in the process of his adaptation to the deformed conditions of the market economy as a result of the financial crisis. Conventionally, a *homo economicus* mutant may be called *zombie economicus* (Papava, 2009) in that he is the carrier of a zombie-economy's routine.

---

[6] The original Russian-language variant is retained even in English discourse owing to the exclusively Soviet nature of this phenomenon (Papava and Khaduri, 1997).

From *zombie economicus*, as an already accomplished phenomenon, *homo transformaticus* differs by the fact that he is still in the process of adaptation to a market economy and, as such, he has not yet been established as a type of human being. Because *homo transformaticus* is still developing, one may ask the question of whether or not it is possible for a *homo transformaticus* to grow into a *zombie economicus* under modern conditions of the global financial crisis.

How to Get Rid of Necroeconomies and Zombie-Economies?

There is no doubt that a necroeconomy has a very negative influence over the countries of post-Communist capitalism which requires the development and utilisation of a mechanism which would solve the problem of necro-firms. This mechanism must enable the universal dissemination of the market-based principles of economic order. The key to the solution of this problem rests in the abovementioned evolutionary theory of economic changes.

Government must pay particular attention to group No. 5 on the abovementioned list of firms—those within a vitaeconomy developed by means of new private investments—within post-Communist capitalist economies and must provide for the further strengthening and expansion of this group as well as for the development of such a stable political and macroeconomic environment which will encourage private investments and the emergence of new healthy firms. Account must also be taken that routine formed in this group will be one which is market-based by its nature and, therefore, the community will not be exposed to any necroeconomic danger.

The narrowing of the areas of expansion of group No. 1 and group No. 3 on the list—that is, necroeconomy in the public sector and privatised necroeconomy—must become a key priority for the economic policy of post-Communist governments. This process must come from the overall expansion of a vitaeconomy created through the inflow of new private investments. Despite the natural non-attractiveness of such firms which become established through so-called "second-hand investments," they will not be able to create the necessary conditions for the emergence of necroeconomic agents under the conditions of an appropriate legal framework. Specifically, to the extent that such firms were created according to the principles of a vitaeconomy, their respective routine must ensure a painless disappearance of such firms from the market even if they have lost their competitive qualities.

As concerns the list's group No. 2 and group No. 4—a vitaeconomy in the public sector and a privatised vitaeconomy and irrespective of whether or not the owner is the state or a private individual after its sale—these economies will inevitably be in need of more investments which would be attracted by the means of a partial sale of assets or, in the very least, by the long-term concession of a company's management to a strategic investor. Without these inflows, there is a strong probability that the vitaeconomies of group No. 2 and group No. 4 will correspondingly turn into the same necroeconomies of group No. 1 and group No. 3.

As previously noted, privatisation alone does not result in the automatic elimination of a necroeconomy. Consequently, a government may have only one solution to solving the problem of ensuring the operation of the strategically important enterprises of group No. 1; that is, holding an open international tender with the purpose of choosing a strategic investor which may take over the management of a specific dead enterprise (or, much rather, obtain the right to start some strategically and domestically important production within the enterprise) upon a long-term basis. It should not be excluded that this step might not satisfy a strategic investor in which case the government would have to agree to privatise the enterprise even at a symbolic price given that a necroenterprise can naturally not be sold for a high price.

There is literally no future for the enterprises with group No. 3; that is, within a privatised necroeconomy whose only fair estimation of their material and technical base would be nothing more than scrap metal.

Theoretically, it must be made clear that the effective elimination of a necroeconomy is unthinkable without an effective bankruptcy law. As the experience of many post-Communist countries has shown, most of the past attempts at formally adopting bankruptcy laws have unfortunately produced only "stillborn babies" (Sánchez-Andrés and March-Poquet, 2002). In other words, they are "necro-laws" because the occurrence of factually bankrupt enterprises being proclaimed as legally bankrupt is something which happens only very seldom, if at all. In these countries, bankruptcy procedures are blocked by all possible means which, inter alia, may be explained by the assumption that bankruptcy does not "fit" into the institutional environments of those countries.

Post-Soviet Georgia, as one of the countries which carries the heavy burden of a necroeconomy (Papava and Tokmazishvili, 2007), is one of the most infamous examples of the ineffectiveness of a post-Communist bankruptcy law in which new institutions are very often created through the immediate and shallow imitation of their Western originals (Shavans and Manian, 1999). As a result, many institutions of a developed market economy are often extremely ineffective in the countries of post-Communist capitalism. Moreover, they may also lead to some extremely negative consequences such as, for example, the frequent and deserved criticisms lodged against the IMF for its hasty and even simplified approaches towards institutional reforms in post-Communist countries which have ultimately harmed—rather than helped—the process of the establishment of a market economy (Stiglitz, 1999).

The elimination of a necroeconomy may only be accomplished by establishing such institutions (Polanyi, 2001) which will boost the process of democratic reform and raise the efficiency of post-Communist transformation (Hare and Davis, 2006). The dying out of the phenomenon of a necroeconomy may be considered as an economic indicator of a country's successful overcoming of the stage of post-Communist capitalism.

An effective bankruptcy law is another effective tool in the fight against zombie-firms and zombie-banks. Unfortunately, however, one has to admit that the situation in developed countries is not any better than it is in the countries of post-Communist capitalism. Specifically, there is a clearly observed tendency that the legitimate bankruptcy of many firms is not readily documented by means of appropriate legal procedures (for example, Cussen, 2008).

Contemporary Financial Crisis and the Danger of the Transformation of *Homo Transformaticus* into a *Zombie Economicus*

The contemporary global financial crisis (for example, Krugman, 2008) has created complex problems the world over and including the economic development of the countries of post-

Communist capitalism. It is precisely within the context of the current crisis that the subject of the attack of zombie-firms upon the global economy has become so topical (Stepek, 2008; Willie CB, 2009) which resulted in the emergence of multiple research works within the so-called new economic field of "zombie-nomics" (LeLaulu, 2009).

It is an established fact that many developed countries have resorted to certain special governmental bailout programs in support of their financial institutions and real estate businesses (Mau, 2009, pp. 22-23) which creates a threat for the development of a new zombie-economic routine. This threat may become quite real if the financial crisis continues long enough to enable the zombie-economic routine to take solid root.

If only a few short years ago, the existence of a zombie-economy in the United States was categorically negated (Bonner and Wiggin, 2003, p. 120),[7] most recently, however, many have started talking about the threat of the emergence of zombie-firms as a direct result of the financial crisis (Coy, 2009; Krugman, 2009b, Pesek, 2008; Rajiva, 2009; Wong, 2008). As a consequence, the US government's bailout plan has been referred to as a zombie-program whereas the US Treasury Department has been disparagingly called the "mother" of a zombie-economy (Kunstler, 2008).

To do justice, however, one has to underline that the first symptoms of the movement towards the zombie-ing of the US banking sector appeared long before the emergence of the present financial crisis when the US government extended subsidies to American banks in order to stimulate their mortgage lending insurance plans for the benefit of low-profit segments of population which failed to meet general banking standards (Holmes, 1999).

Remarkably, experts assert that an interesting difference exists between the mechanisms of the emergence of zombie-firms in Japan and those in the US. In Japan, for example, the government supports zombie firms through zombie-banks whereas the bailout plans envisaged in the United States are intended to be implemented without the mediation of banks even though the results in both countries are expected to be the same (Hoshi, 2009). Obviously, such an approach does not correspond to the reality which is being shaped in the US in relation to its government's domestic zombie-banks' bailout program (Willie CB, 2008).

Furthermore, the danger of the development of a zombie-economy in Canada (Poschmann, 2009) and some EU countries (Kelly, 2008, Schnittger, 2009) has become a real threat as well.

In the countries of post-Communist capitalism, the present financial crisis also encourages the development of such governmental programs which are aimed at the government's providing its financial support to perishing banks and firms (for example, Illarionov, 2008c). Various proposals have been put forth which would establish some special governmental institutions (banks) and implement long-term beneficial lending and investing (for example, Danilishin, 2009).

Despite the fact that the economic theory has been proven a long time ago and that global experience has shown just how harmful a government's interventions into the economy can be when it introduces favourable lending plans, both developed economies and those of post-Communist capitalism resort to this remedy again and again during the times of a financial crisis (Woods, 2009). Unfortunately, however, and as we have already seen, such a move eventually leads to the emergence of a zombie-economy routine.

---

[7] Higher officials ignoring the problem of the existence of zombie institutions in the US financial system which is typical at the present time (Krugman, 2009a).

The current modern financial crises bring the principle of "privatization of profits and nationalisation of losses" up to date (Mau, 2009, p. 10).

Unlike developed economies, which are exposed to the threat of the zombie-ing of the economy under the conditions of a financial crisis, this threat is even greater for the countries of post-Communist capitalism owing also to their exposure to necroeconomy. There were no doubts, for example, that many Russian companies, amongst those in other economies, were able to get rid of their debts under the financial crisis without the receipt of governmental support (Mau, 2009, p. 5).

If one bears the nature of the material and technical base a necroeconomy in mind alongside the fact that it resides primarily in the industrial sector, then there is no surprise that it is exactly the drastic shrinking of industrial output under the conditions of the economic crisis through which a necroeconomy states its existence (Illarionov, 2008a, 2008b).

In the absence of a financial crisis, post-Communist countries with a large-scale market feel the pain of a necroeconomy far less than those with relatively smaller ones given that there is an opportunity to compete with each other within the market's large dimensions and governmental support (for example, Schaffer, and Kuznetsov, 2008) which creates the illusion that a necroeconomy does not exist. In smaller countries, however, this problem is rather severe due to the fact that many enterprises may have no domestic competitors at all (Papava and Tokmazishvili, 2007).

This difference between the large and small markets has an influence upon the governmental policies supporting a necroeconomy. In particular, the illusion of the absence of a necroeconomy caused by domestic competition blocks, to a great degree, a governments' will to get rid of a necroeconomy. In countries with small-scale markets, however, the absence of analogous enterprises eliminates the conditions for domestic competition which enables the governments of those countries to retain a greater stimulus to fight necroeconomy even though such a stimulus is not always utilized within a proper way (Papava and Tokmazishvili, 2007).

Under the conditions of a financial crisis, the governments of post-Communist capitalist countries may retain far less illusions that a necroeconomy exists. It comes as no surprise, therefore, that they witness a catastrophic decrease in their industrial output.

In view of the fact that it is precisely the financial crisis which creates the favourable conditions for the establishment of zombie-economy routines—that is, the zombie-ing of an economy—the zombie-ing of a necro*economy* is what happens in the countries of post-Communist capitalism which, in fact, is even worse than the simple economic zombie-ing which takes place in developed economies. If in Japan, for example, the zombie-economy never touched the processing industries, as we have seen, then one of the qualities of the necroeconomy is to concentrate *exactly* upon this sector of the economy. Consequently, the zombie-ing of a necroeconomy inevitably amounts to the zombie-ing of this already dead sector as well.

It is important to note that in Russia, for example, the first symptoms of necroeconomic zombie-ing emerged in the immediate aftermath of the August 1998 crisis in the country (Lindsey, 2002, p. 210) which gave rise to the phenomenon of the post-Communist zombie-economy (Lindsey, 2002, p. 211). The present financial crisis which has developed in Russia and which, in addition, has been "strengthened" by that at the global level, may continue to exist and, therefore, is able to have far more serious negative effects for Russia than its predecessor had. There is a very real possibility that *homo transformaticus* will eventually grow into *zombie economicus,* rather than into *homo economicus,* which means that the economic future of the post-Communist capitalist countries could be even more dubious than it is today.

Instead of Conclusions

The dead enterprises which the countries of post-Communist capitalism received as their legacy of the command economy have proven to be quite "tenacious of life." As a consequence, the market economies of many post-Communist countries have been loaded by the burden of a necroeconomy. It is the society's necroeconomic foundations that make up the façade of post-Communist capitalism. *Homo transformaticus*, therefore, is the carrier of a necroeconomy's routine who transforms economy and society in the direction of capitalist values and, at the same time, transforms his own self in the same direction, too.

The phenomenon of a dead economy is also familiar to some developed economies as well. The occurrence of financial crises has encouraged the emergence of a kind of routine which guarantees the stability of a government's bailout programs implemented thorough the banking sector in support of de-facto bankrupt firms. As a result, a network of zombie-banks and zombie-firms develops upon which the entire system of a zombie-economy rests.

Unlike a necroeconomy, whose routine is carried by a human being which is "still-to-be-formed," a zombie-economy's routine is carried by the "gone and departed" man, the so-called *zombie economicus.*

Under the conditions of the present financial crisis, the threat of a zombie-economy is also aggressively knocking on the doors of those developed economies which, until recently, seemed to have escaped the zombie-ing of their economies.

This threat of an economy's zombie-ing is even greater in the countries of post-Communist capitalism given that this zombie-ing also has a great deal to do with a necroeconomy which is a factor that will make it rather difficult to improve an economy's health after the end of the financial crisis.

The only effective mechanism to get rid of both a necroeconomy and a zombie-economy is to adopt a sound bankruptcy law which, in turn, requires the strong political will of the ruling elite.

References

Ahearne, Alan G., and Naoki Shinada. 2005. "Zombie firms and economic stagnation in Japan." *International* **Economics** *and* **Economic** *Policy*, 2(4):363-381.

Akerlof, George A., and Robert J. Shiller. 2009. *Animal Spirits: How Human Psychology Drives the Economy, and why it Matters for Global Capitalism.* Princeton, NJ: Princeton University Press.

Antachak, Rafal, Michail Guzhinski, and Petr Kozarzhevski, eds. 2001. *Belorusskaia ekonomika: ot rynka k plany (Economy of Belarus from Market to Planned Economy)*, Vol II. Warsaw: CASE. (In Russian).

Åslund, Anders. 1995. *How Russia Became a Market Economy.* Washington, D.C.: The Brookings Institution.

Åslund, Anders. 1996. "'Rentoorienrirovannoe povedenie' v rossiiskoi perekhodnoi ekonommike ('Rent-oriented Behavior' in Russian Transitional Economy)". *Voprosy ekonomiki* (*Economic Issues*), **8**:99-108. (In Russian).

Åslund, Anders. 2002. *Building Capitalism. The Transformation of the Former Soviet Bloc.* Cambridge, UK: Cambridge University Press.

Åslund, Anders. 2007a. *How Capitalism Was Built: The Transformation of Central and Eastern Europe, Russia, and Central Asia.* New York: Cambridge University Press.

Åslund, Anders. 2007b. *Russia's Capitalist Revolution: Why Market Reform Succeeded and Democracy Failed.* Washington, D.C.: Peterson Institute for International Economics.

Avtonomov, V. S. 1998. *Model' cheloveka v ekonomicheskoii nauke* (*Human Model in Economic Science*). St. Petersburg, Russia: Ekonomicheskaia shkola. (In Russian).

Becker, Gary S. 1976. *The Economic Approach to Human Behavior.* Chicago: The University of Chicago Press.

Bonner, William, and Addison Wiggin. 2003. *Financial Reckoning Day: Surviving the Soft Depression of the 21st Century.* New Jersey: John Wiley and Sons.

Brockway, George P. 1995. *The End of Economic Man.* New York: W. W. Norton and Company.

Buchanan, James M. 1987. "The Constitution of Economic Policy." *American Economic Review*, **77**(3):243-50.

Bunkina, M. K., and A. M. Semionov. 2000. *Ekonomicheskii chelovek* (*The Economic Man*). Moscow, Russia: Delo. (In Russian).

Buzgalin, A. V. 1994. *Perekhodnaia ekonomika: kurs lektsii po politicheskoi ekonomii* (*Transitional Economy: A Course of Lectures in Political Economy*). Moscow, Russia: Taurus. (In Russian).

Caballero, Ricardo J., Takeo Hoshi, and Anil K. Kashyap, 2008. "Zombie Lending and Depressed Restructuring in Japan." *American Economic Review*, **98**(5):1943–1977, available at http://econ-www.mit.edu/files/3770.

Caballero, Ricardo, and Anil K. Kashyap. 2002. "Japan's Indian Summer." *The Wall Street Journal*, July 18, available at http://faculty.chicagogsb.edu/anil.kashyap /research/awsj.pdf.

Coates, David. 2000. *Models of Capitalism: Growth and Stagnation in the Modern Era.* Cambridge, UK: Polity Press.

Coy, Peter. 2009. "A New Menace to the Economy: ''Zombie' Debtors." *BusinessWeek*, January 15, available at http://www.businessweek.com/magazine/content/ 09_04/b4117024316675.htm.

Crouch, Colin, and Wolfgang Streeck, eds. 1997. *Political Economy of Modern Capitalism: Mapping Convergence and Diversity.* London, UK: SAGE Publications.

Cussen, Mark P. 2008. "Changing The Face Of Bankruptcy." Investopedia, available at http://www.investopedia.com/articles/pf/08/bankruptcy-act.asp?partner=NTU.

Danilishin, Bogdan. 2009. "Ekonomika Ukrainy: zhizn' posle krizisa (Ukraine's Economy: A Life after Crisis?)." *Zerkalo Nedeli*, No. **1** (729), 17-23 January, available at http://www.zn.ua/2000/2020/65131/. (In Russian).

Drucker, Peter F. 1985. *Innovation and Enterpreneurship: Practice and Principles.* New York: Harper and Row.

Ferber, Marianne A., and Julie A. Nelson, eds. 1993. *Beyond Economic Man: Feminist Theory and Economics.* Chicago: The University of Chicago Press.

Gaddy, Clifford G., and Barry W. Ickes. 1998. "Russia's Virtual Economy." *Foreign Affairs*, **77**(5):53-67.

Gaddy, Clifford G., and Barry W. Ickes. 2002. *A Russia's Virtual Economy*. Washington, D.C.: Brookiongs Institutions Press.

Gins, G.K. 1992. *Predprinimatel' (The Entrepreneur)*. Frankfurt a Main: Posev. (In Russian).

Gwynne, Robert N., Thomas Klak, and Denis J. B. Shaw. 2003. *Alternative Capitalisms. Geographies of Emerging Regions*. London, UK: ARNOLD.

Hall, Peter A., and David Soskice, eds. 2001. *Varieties of Capitalism: The Institutional Foundations of Comparative Advantage*. New York: Oxford University Press.

Hanhinen, Sari. 2000. *Social Problems in Transition: Perceptions of Influential Groups in Estonia, Russia, and Finland*. London, UK: Aleksanteri Institute, available at http://books.google.com/books?id=RAzoAAAAIAAJandq=homo+etaticusanddq=homo+etaticusandpgis=1.

Hare, Paul G., and Junior R. Davis. 2006. "Institutions and Development: What We (Think We) Know, What We Would Like to Know." *IPPG Discussion Paper Series*, No 3, April, available at http://www.sml.hw.ac.uk/cert/wpa/2006/dp0603.pdf.

Holle, Peter. 2005. "The Evolution of a Zombie Economy," *The Frontier Centre For Public Policy*, March 15, available at http://www.fcpp.org/main/publication_detail.php?PubID=979.

Holmes, Steven A. 1999. "Fannie Mae Eases Credit To Aid Mortgage Lending." *The New York Times*, September 30, available at http://query.nytimes.com/gst/fullpage.html?res=9c0de7db153ef933a0575ac0a96f958260andsec=andspon=andpagewanted=all.

Hoshi, Takeo. 2006. "Economics of the Living Dead." *The Japanese Economic Review*, **57**(1):30-49.

Hoshi, Takeo. 2009. "Year of the Zombie." *Roubini Global Economics*, January 21, available at http://www.rgemonitor.com/asia-monitor/255220/year_of_the_zombie.

Hoshi, Takeo, and Anil K. Kashyap. 2004. "Japan's Economic and Financial Crisis: An Overview." *The Journal of Economic Perspectives*, Winter, available at http://faculty.chicagogsb.edu/anil.kashyap/research/japancrisis.pdf.

Hoshi, Takeo, and Anil K. Kashyap. 2005. "Solutions to Japan's Banking Problems: What Might Work and What Definitely will Fail." In *Reviving Japan's Economy: Problems and Prescriptions*, eds. Takatoshi It, Hugh Patrick, and David E. Weinstein, 147-195. Cambridge, MA: The MIT Press.

Illarionov, Andrei. 2008a. *Eto—katastrofa. Bespretsdentnyi promyshlennyi spad noiabria (It's a Catastrophe. An Unprecedented Drop of Industrial Output in November)*. Moscow, Russia: Institute of Economic Analysis, available at http://www.iea.ru/econom_rost.php?id=26. (In Russian).

Illarionov, Andrei. 2008b. *Noiabr'skii spad promyshlennogo proizvolstva— katastrofa, kotoraia usugubliaetsia deistviavi vlastei (The November Drops in Industrial Output— Catastrophe which Is Aggravated by Government's Actions)*. Moscow, Russia: Institute of Economic Analysis, available at http://www.iea.ru/macroeconom.php?id=14 (in Russian).

Illarionov, Andrei. 2008c. *Priroda rossiiskogo krizisa (The Nature of Russian Crisis)*. Moscow, Russia: Institute of Economic Analysis, 2 October, available at http://www.iea.ru/macroeconom.php?id=8. (In Russian).

Kane, Edward J. 2000a. "Capital Movements, Banking Insolvency, and Silent Runs in the Asian Financial Crisis." *Pacific-Basin Finance Journal*, **8**(2):153-175.

Kane, Edward J. 2000b. "The Dialectical Role of Information and Disinformation in Regulation-Induced Banking Crises." *Pacific-Basin Finance Journal*, **8**(3-4):285-308.

Kelly, Morgan. 2008. "Bank Guarantee Likely to Deal a Crippling Blow to the Economy." *Machine Nation*, October 28, available at http://machinenation.forumakers.com/economy-business-and-finance-f8/morgan-kelly-our-zombie-nation-and-zombie-banks-t1496.htm#43586.

Kennedy, Michael D., and Elizabeth Igen. 2007. "Post-Communist Capitalism and Transition Culture in Georgia," *Caucasus and Globalization*, **1**(2):53-56.

Krugman, Pual. 2008. *The Return of Depress on Economics and the Crisis of 2008*. New York: W.W. Norton and Company.

Krugman, Pual. 2009a. "The Big Dither." *The New York Times*, March 05, available at http://www.nytimes.com/2009/03/06/opinion/06krugman.html.

Krugman, Pual. 2009b. "Wall Street Voodoo," *The New York Times*, January 18, available at http://www.nytimes.com/2009/01/19/opinion/19krugman.html?_r=2andpartner=rssnytan demc=rss.

Kunstler, James Howard. 2008. "The Inevitable Fate Of Our 'Zombie' Economy." *Contrarian Stock Market Investing News*, November 28, available at http://www.contrarianprofits.com/articles/the-inevitable-fate-of-our-zombie-economy/9233.

Levada, Y.A. 1999. ""Chelovek prisposoblennyi" ("Homo Adapticus")." *Monitoring Public Opinion: Economic and Social Changes*, **5**(43):7-17, September-October, available at http://www.ecsocman.edu.ru/images/pubs/2006/12/02/0000296966/02levada-7-17.pdf. (In Russian).

LeLaulu, Lelei. 2009. "Zombie Economics?" *The Development Executive Group*, 13 January, available at http://www.devex.com/articles/zombie-economics.

Lindsey, Brink. 2002. *Against the Dead Hand: The Uncertain Struggle for Global Capitalism*. New York: John Wiley and Sons.

Lipowski, Adam. 1998. *Towards Normality. Overcoming the Heritage of Central Planning Economy in Poland in 1990-1994*. Warsow: CASE.

Mau, V. 2009. "Drama 2008 goda: ot ekonomicheskogo chyda k ekonomicheskomu krizisu" ("2008 Drama: From the Economic Miracle to the Economic Crisis"). *Voprosy ekonomiki (Economic Issues)*, **2**:4-23.

Murrell, Peter. 1992a. "Evolution in Economics and in the Economic Reform of the Centrally Planned Economies." In *The Emergence of Market Economies in Eastern Europe*, eds. Cristopher Clague, and Gordon C. Rausser, 35-53. Cambridge: Basil Blackwell, 1992a.

Murrell, Peter. 1992b. "Evolutionary and Radical Approaches to Economic Reform," *Economics of Planning*, **25**(1):79-95.

Nekipelov, A. 2003. "Problemy upravlenia ekonomikoi v transformiruemom obshchestve" (Problems of Economic Management in a Transformed Economy"). In *Social-Economic Models in Modern World and Russia's Way*, Vol. 1: *Transformation of Post-Socialist Society*, ed. K. I. Mikulski, 124-136. Moscow, Russia: Ekonomika. (In Russian).

Nelson, Richard R., and Sidney G. Winter. 1982. *An Evolutionary Theory of Economic Change*. Cambridge: The Belknap Press of Harvard University press.

Nikolaev, I. 2002. "Perspektivy nalogovoi amnistii v Rossii" ("Perspectives of Tax Amnesties in Russia"). *Obshchestvo i ekonomika*, (*Society and Economy*), **6**:49-67. (In Russian).

Papava, Vladimer. 1996. "The Georgian Economy from «Shock Therapy» to "Social Promotion." *Communist Economies and Economic Transformation*, **8**(8):251-267.

Papava, Vladimer. 1999. "The Georgian Economy: Main Directions and Initial Results of Reforms." In *Systemic Change in Post-Communist Economies*, ed. Paul G. Hare, 266-292. London, UK: Macmillan Press.

Papava, Vladimer. 2001. "Necroeconomics—A Phenomenon of the Post-Communist Transition Period." *Problems of Economic Transition*, **44**(8):77-89.

Papava, Vladimer. 2002. "Necroeconomics—the Theory of Post-Communist Transformation of an Economy." *International Journal of Social Economics*, **29**(9/10):796-805.

Papava, Vladimer. 2005. *Necroeconomics: The Political Economy of Post-Communist Capitalism*. New York: iUniverse.

Papava, Vladimer. 2006. "Economic Transition to European or Post-Communist Capitalism?" *EACES Working Papers*, No. **1**, March, available at http://www.eaces.net/news/WP-1-06.pdf.

Papava, Vladimer. 2009. "Is Zombie Economicus Coming?" *The Market Oracle*, February 07, available at http://www.marketoracle.co.uk/Article8736.html.

Papava, Vladimer, and Nodari Khaduri. 1997. "On the Shadow Political Economy of the Post-Communist Transformation. An Institutional Analysis." *Problems of Economic Transition*, 40(6):15-34.

Papava, Vladimer, and Micheil Tokmazishvili. 2007. "Necroeconomic Foundations and the Development of Business in Post-Revolution Georgia." *Caucasus and Globalization*, 1(4):84-95.

Pesek, William. 2008. "Jay Leno Tells Asia about Detroit's Zombies." *Livemint.com, The Wall Steer Journal*, November 17, available at http://www.livemint.com/2008/11/17233803/Jay-Leno-tells-Asia-about-Detr.html.

Polanyi, Karl. 2001. *The Great Transformation. The Political and Economic Origins of Our Time*. Boston: Beacon Press.

Poschmann, Finn. 2009. "Beware of Zombies." *Financial Post*, January 26, available at http://network.nationalpost.com/np/blogs/fpcomment/archive/2009/01/26/beware-of-zombies.aspx.

Rajiva, Lila. 2009. "Nightmare on Wall Street." *LewRockwell.com*, April 01, available at http://www.lewrockwell.com/rajiva/rajiva16.html.

Samsin, A.I. 2003. *Osnovy filosofii ekonomiki* (*Basics of Economic Philosophy*). Moscow, Russia: Iuniti. (In Russian).

Sánchez-Andrés, Antonio, and José M. March-Poquet. 2002. "The Construction of Market Institutions in Russia: A View from the Institutionalism of Polanyi." *Journal of Economic Issues*, **XXXVI**(3):1-16.

Shaváns, B., and E. Manian. 1999. "Postsotsialisticheskie traektorii I zapadnyi kapitalizm" ("Post-Socialist Trajectories and Western Capitalism"). *Mirovaia ekonomika i vezhdunarodnye otnoshenia* (*World Economy and International Relations*), **12**:42-46. (In Russian).

Schaffer, Mark, and Boris Kuznetsov. 2008. "Productivity." In *Can Russia Compete?*, eds. Raj M. Desai and Itzhak Goldberg, 12-34. Washington, D.C.: Brookings Institution Press.

Schnittger, Frank. 2009. "Ireland's Zombie Economy." *European Tribune*, January 14, available at http://www.eurotrib.com/?op=displaystory;sid=2009/1/14/113437/798.

Shleifer, Andrei, and Daniel Treisman. 2000. *Without a Map: Political Tactics and Economic Reform in Russia*. Cambridge: The MIT Press.

Shokhin, A.N. 1989. *Sotsial'nyeproblemy perestroiki (Social Problems of Perestroika)*. Moscow: Ekonomika. (In Russian).

Shul'ga, I. 2002. "Opyt nalogovoi amnistii v Kazakhstane" ("Experience of Tax Amnesty in Kazakhstan"). *Obshchestvo i ekonomika* (*Society and Economy*), **6**:69-71. (In Russian).

Smith, David C. 2003. "Loans to Japanese Borrowers." *Japanese International Economies*, 17(3):283-304.

Stepek, John, 2008. "How Zombie Companies Suck the Life from an Economy." *MoneyWeek*, November 18, available at http://www.moneyweek.com/news-and-charts/economics/how-zombie-companies-suck-the-life-from-an-economy-14089.aspx.

Stiglitz, Joseph E. 1999. *Whither Reform? Ten Years of the Transition*. Keynote Address, The Annual Bank Conference on Development Economics, The World Bank, available at http://siteresources.worldbank.org/INTABCDEWASHINGTON1999/Resources/stiglitz.pdf.

Svensson, Bo. 1983. *Economisk kriminalitet*. Göteborg: Tholin/Larrson/Gruppen förlag.

Taylor, M. L. 1988. *Divesting Business Units*. Toronto: Lexington Books.

Willie CB, Jim. 2008. "U.S. Economy Disintegrating as Government Supports Zombie Banks." *The Market Oracle*, December 04, available at http://www.marketoracle.co.uk/Article7619.html.

Willie CB, Jim. 2009. "Gold, Zombie Banking System, Lightning, Earthquakes and Hurricanes." *The Market Oracle*, January 20, available at http://www.marketoracle.co.uk/Article8346.html.

Wong, Tony. 2008. "Can U.S. escape zombie economy's clutch?" TheStar.com, Toronto Star, October 04, available at http://www.thestar.com/comment/columnists/article/511664.

Woodruff, David. 1999. *Money Unmade: Barter and the Fate of Russian Capitalism*. Ithaca, NY: Cornell University Press.

Woods Jr., Thomas E. 2009. *Meltdown: A Free-Market Look at Why the Stock Market Collapsed, the Economy Tanked, and Government Bailouts Will Make Things Worse*. Washington, D.C.: Regnery Publishing.

In: Focus on Science and Technology...
Editor: Sergo Gotsiridze, pp. 57-62

ISBN: 978-1-61209-970-5
© 2011 Nova Science Publishers, Inc.

Chapter 5

SOME CONSIDERATIONS ABOUT DEFINITION OF OPTIMAL HEIGHT OF COMPOSED, WELDED BEAMS ACCORDING TO N. S. STRELETSKY

N.Sh. Berishvili and Kh.Sh. Gorjoladze

The Georgian Technical University, Tbilisi, Georgia

Abstract

The study discusses the task of defining the optimal height of composed welded beams given in the works of different prominent authors. The study points out those inaccuracies and inconsistencies that exist despite preliminary estimates and that make it difficult and often impossible to properly use a mathematical device. The study shows how to define the above-mentioned inconsistencies and namely the coefficient c, how to define the dependence of δ_{well} on h as well as how to avoid them in order to make a correct selection of beams' heights from the perspective of both mathematical and construction mechanics methods.

Keywords: composed, welded beam; moment of inertia, moment of resistance; design resistance; bending moment; minimum condition; optimum height of beam; rational cut.

Introduction

Definition of stiffness and stressing of composed, welded beams is examined by different authors, among which the works of N.S.Streletsky, E.I. Belenia and K.K Mukhanov should be considered as significant. In spite of simplicity of calculation of composed beams on the face of it should be noted that many nuances must be considered therein during designing, as well as during calculation according to the methods of structural analysis and for the making of the mathematical model.

During making and implementation of design model, it is important to observe such preliminary statements and assumptions that make the ground for specific design model and it is not expedient to justify their violation by any practical reasons. This work is about such deviations during the calculation of composed, welded beams.

Basic Part

Lets examine the composed, welded beam with double-T cross-section, the height of which is h, the area of flange cross-section $- A_{flange}$ and the wall thickness $- \delta_{wall}$. The moment of resistance of such cross-section may be represented as the ratio of the moment of inertia and $h/2$ [1]:

$$W = 2\left[A_{1,flange} \left(h/2\right)^2 \right]\frac{2}{h} + \frac{\delta_{wall}h^3}{12} \cdot \frac{2}{h} = A_{1,flange} \cdot h + \frac{\delta_{wall}h^2}{6} = W_{flange} + W_{wall} \qquad (1)$$

From (1) is derived that the area of one flange is - $A_{1,flange} = \dfrac{W_{flange}}{h}$ \qquad (2)

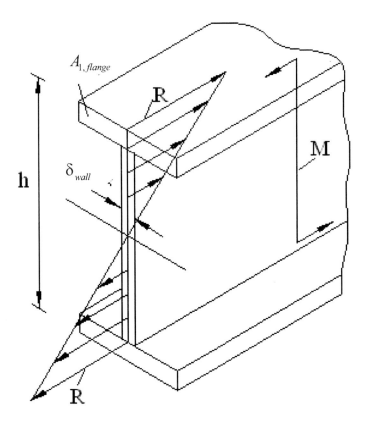

If we represent the W_{flange} through the moment of resistance of the whole beam $-W$, it may be expressed as follows

$$W_{flange} = c \cdot W \tag{3}$$

where, c is such part of the moment of resistance, which is on both flanges. $c < 1$; therefore, the area of both flanges will be:

$A_{flange} = \dfrac{2cW}{h}$, and the area of the wall will be $A_{wall} = \delta_{wall} h$; if multiply A_{flange} and

A_{wall} by volume weight of the beam-γ and take into consideration the structural coefficients

ψ_{flange} and ψ_{wall}, as well as that $W = \dfrac{M}{R}$, through their addition we will derive the unit

length weight of the beam [2]:

$$g_{beam} = g_{flange} + g_{wall} = \frac{2M \cdot c}{hR} \gamma \psi_{flange} + h\delta_{wall}\gamma\psi_{wall} \tag{4}$$

We can derive the form (4) in a different way. The force on $A_{1,flange}$ - will be equal to $A_{1,flange} \cdot R$, the forces on both flanges will generate the moment $2 \cdot A_{1,flange} \cdot R \cdot \dfrac{h}{2} = A_{1,flange} \cdot R \cdot h$; if consider that this moment is the c part of a whole moment- M, then is derived that $M_{flange} = cM = A_{1,flange} \cdot Rh$, where $A_{1,flange} = \dfrac{M \cdot c}{R \cdot h}$;

and $A_{flange} = \dfrac{2M \cdot c}{Rh}$ expression apparently corresponds to first member of (4) expression, and the second member will be the same. By N. S. Streletsky and E.I. Belenia, for determination of optimum, i.e. minimum weight of the beam, the minimum condition shall be applied:

$$\frac{dg_{beam}}{dh} = -\frac{2M \cdot c\gamma\psi_{flange}}{R} \cdot \frac{1}{h^2} + \delta_{wall}\gamma\psi_{wall} = 0 \tag{5}$$

but in derivation it is considered that c, as well as δ_{wall} are constant, i.e. are not depended on the height of the beam $-h$ and therefore takes no part in derivation which is contradicts the fact. The optimum height of the beam is defined from (5):

$$h_{opt.} = \sqrt{\frac{2 \cdot c\psi_{flange}}{\psi_{wall}} \frac{M}{R\delta_{wall}}} = K\sqrt{\frac{W}{\delta_{wall}}} \tag{6}$$

where the coefficient

$$K = \sqrt{\frac{2 \cdot c\psi_{flange}}{\psi_{wall}}} \qquad (7)$$

N.S. Streletsky points to the abovementioned discrepancy and introduces the expression, bonding the wall thickness and the beam height:

$$\delta_{wall} = \nu h^t \qquad (8)$$

where ν is the proportionality coefficient. Then the process of definition of $h_{opt.}$ will be the following [3]:

$$g_{beam} = \frac{2M \cdot c}{hR} \gamma\psi_{flange} + h^{t+1} \cdot \nu\gamma\psi_{wall}$$

$$\frac{dg_{beam}}{dh} = -\frac{2M \cdot c\gamma\psi_{flange}}{Rh^2} + (t+1)\nu\gamma\psi_{wall}h^t = 0 \text{ from this}$$

$$h^{t+2} = \frac{W}{\nu(t+1)} \cdot \frac{2c\psi_{flange}}{\psi_{wall}} \; ;$$

$$h_{opt.} = \sqrt[t+2]{\frac{W}{\nu(t+1)}} \cdot \sqrt[t+2]{\frac{2c\psi_{flange}}{\psi_{wall}}} = K\sqrt[t+2]{\frac{W}{\nu(t+1)}} \qquad (9)$$

According to N. S. Streletsky, based on practical data, t may be taken as being equal to $\frac{1}{2}$ -i.e. $\delta_{wall} = \nu\sqrt{h}$ and $\nu = \left(\frac{1}{10} \div \frac{1}{13}\right)cm^{\frac{1}{2}}$; in this case

$$h_{opt.} = K\sqrt[2,5]{\frac{W}{1,5\nu}} \qquad (10)$$

and as 2.5 is nearer to 2, it is possible to write finally:

$$h_{opt.} = K\sqrt{\frac{W}{1,5\nu}} \qquad (11)$$

As can be seen, during determination of $\delta_{კედ.}$ expression, because of these sufficiently inaccurate conversions, the issue of variation of c -coefficient is still undetermined. Moreover, in the expression of K-coefficient, the c-value is which corresponds to the rational cut, but then determination of rational cut is made by consideration of minimum condition. To avoid abovementioned discrepancies, we consider expedient to determine c as the function, depended on h:

$$c = \frac{W_{flange}}{W} = \frac{W - W_{well}}{W} = \frac{W - \dfrac{\delta_{wall} h^2}{6}}{W} = \frac{\dfrac{6W - v\sqrt{h} \cdot h^2}{6}}{W} = \frac{6W - vh^{2,5}}{6W} = 1 - \frac{vh^{2,5}}{6W} \quad (12)$$

c may be expressed through the areas of the beam cross-section and relate c to $\alpha = \dfrac{A_{wall}}{A}$:

$$c = \frac{W_{flange}}{W} = \frac{A_{flange} \cdot h}{A_{flange} \cdot h + \dfrac{\delta_{wall} \cdot h^2}{6}} = \frac{6A_{flange} \cdot h}{6A_{flange} \cdot h + \delta_{wall} h^2} = \frac{6A_{flange}}{6A_{flange} + \delta_{wall} h} = \frac{3\left(A - h\delta_{wall}\right)}{3\left(A - h\delta_{wall}\right) + h\delta_{wall}} =$$

$$= \frac{3A - 3h\delta_{wall}}{3A - 2h\delta_{wall}} = \frac{3A - 3A \cdot \alpha}{3A - 2A \cdot \alpha} = \frac{3A\left(1 - \alpha\right)}{A\left(3 - 2\alpha\right)} = \frac{3\left(1 - \alpha\right)}{3 - 2\alpha} \quad (13)$$

If enter α - value into (13) expression, while the area of beam flanges equals to the wall area, i.e. $\alpha = 0,5$, we will derive: $c = \dfrac{3\left(1 - 0,5\right)}{3 - 2 \cdot 0,5} = 0,75$

As well as, if determine the value of c from (7), when $K = 1,1$, taken according to N. S. Streletsky, we will derive: $1,1 = \sqrt{\dfrac{2 \cdot c \cdot 1}{1,2}}$; from this $c = 0.75$, i.e. comes out that the c - value is entered into (5) formula previously, which defines the section rationality and then the beam optimum height will be defined by means of the minimum condition.

Enter the c-value, defined from (12) into (4) and then use the section minimum condition:

$$g_{beam} = \frac{2M}{R} \gamma \psi_{flange} \left(\frac{6W - vh^{2,5}}{6Wh}\right) + vh^{1,5} \cdot \gamma \psi_{wall} \quad (14)$$

$$\frac{dg_{beam}}{dh} = \frac{2M\gamma \psi_{flange}}{R} \left[\frac{-2,5vh^{1,5} \cdot 6Wh - 6W\left(6W - vh^{2,5}\right)}{36W^2 h^2}\right] + 1,5vh^{0,5}\gamma \psi_{flange} = 0 \quad (15)$$

As a result of final simplification and conversion we will get the following:

$$h_{opt.} = 2,5\sqrt{\frac{2W\psi_{flange}}{v\left(1,5\psi_{wall} - 0,5\psi_{flange}\right)}} \quad (16)$$

Conclusion

i.e. according to E.I. Belenia, $h_{opt.}$ is calculated by means of the following formula:

$$h_{opt.} = K\sqrt{\frac{W}{t_{wall}}} \tag{17}$$

where, as abovementioned K contains the idea of rationality of section, and t_{wall} is considered constant, despite the fact that it is depended on h and applied in the practical calculations in such a way.

According to K. K. Mukhanov, the optimum height is selected by means of the following formula:

$$h_{opt.} = \sqrt[3]{\frac{3}{2}K_{wall} \cdot W} \tag{18}$$

where, K_{wall} – coefficient is depended on h itself. According to N.S. Streletsky, $h_{opt.}$ is calculated through (11) expression, where K - coefficient is selected as in (17) formula.

Thus, during selection of $h_{opt.}$, all methods have its fault which is theoretically in conflict with the given statements somehow or other.

Moreover, after the selection of section optimal height, it is necessary to redesign the beam cross-section, i.e. to bring into compatibility the wall thickness of the section elements, the thickness and width of flanges etc. and at the same time to consider the required value of moment of resistance, required moment of inertia, achievement of section rationality, that is laborious and often unattainable. Therefore, we consider expedient that the cross-section shall be designed previously according to the criteria given in the theory and shall be formed the correspondent assortment that will significantly simplify the selection of the cross-section for the beam [4].

References

[1]   Belenia E.I. and the others. "Metal Structures" –Moscow, "Stroyizdat" 1986 p. 138-145.
[2]   Mukhanov K.K.. "Metal Structures" –Moscow, "Stroyizdat" 1978 p. 191-203.
[3]   Streletsky N.S. and the others. "Metal Structures" –Moscow, "Gostyizdat" 1962 p. 251-256, p. 271-273.
[4]   Berishvili N.Sh., Gorjoladze Kh.Sh.. "The new method for calculation of steel composed, welded beams with the cross-section ".// "Energy", #3, 2007, p.

In: Focus on Science and Technology...
Editor: Sergo Gotsiridze, pp. 63-71

ISBN: 978-1-61209-970-5
© 2011 Nova Science Publishers, Inc.

Chapter 6

SOLUTION OF SPATIAL TASKS OF RODS WITH CROSS-SECTIONS OF DIFFERENT FORMS USING THE DOUBLE TRIGONOMETRIC SERIES

Irakli Gudushauri, Revaz Tskhvedadze, Aguli Sokhadze and Demuri Tabatadze

Georgian Technical University,
77, M. Kostava Str., 0175, Tbilisi, Georgia

Abstract

The integration of differential equations of classic theory of elasticity with partial derivatives for concrete task considering the given boundary conditions is the intractable problem. The new direction, elaborated by I. Gudushauri /1/ gives possibility to overcome successfully difficulties of this kind.

Keywords: elasticity theory, differential equation, tangent strength, normal strength.

1. Introduction

According to mentioned theory, the mode of deformation of any T body due to effect of different kinds can be represented as the result of the combined work of three "one-dimensional" fictitious systems T_α, T_β T_γ. They are geometrically adequate to T. Physical characteristics ensure bringing of any task to integration of "common" differential equation with the precision admissible for theory of elasticity. Each of infinitesimal elements of this fictitious system, cut out near to point $M(\alpha, \beta, \gamma)$ of curvilinear system of orthogonal coordinates (α, β, γ) are able to work on tension (compression) in one only direction that coincides with tangent of coordinate corresponding to its indexes. On these points such "working" directions of fictitious systems T_α, T_β T_γ are mutual perpendicular, that causes the corresponding redistribution of inner strengths of body T under examination (Figure 1,a) between T_α, T_β T_γ fictitious systems (Figure 1,b, Figure 1, a,b).

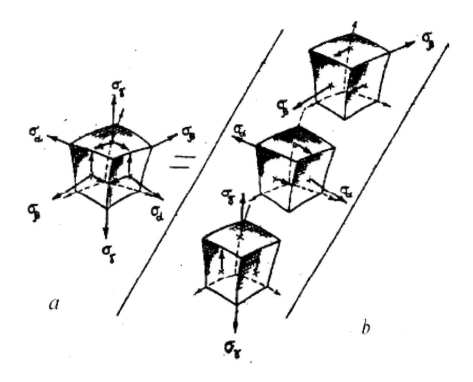

Figure 1.

By such representation (Figure 1,a) of the Figure 1 analytical model of classic theory of elasticity (Figure 1,b), any three-dimensional task can be brought to the "three-dimensional contact" task of entirely new kind about combined work of fictitious systems T_α, T_β T_γ. In this task the required three-dimensional response strength (forces and moments) are determined from necessary and sufficient conditions, considering that the combined work of fictitious systems T_α, T_β T_γ shows the work of given T body with precision admissible for classic theory of elasticity. In examination of three-dimensional task, the mentioned conditions are maximum and equal to nine. Among them three - m^*_α, m^*_β, m^*_γ are moments acting in tangent planes $\alpha = const$, $\beta = const$, $\gamma = const$ correspondingly, and the rest six - $P^*_{\alpha\beta}$, $P^*_{\alpha\gamma}$, $P^*_{\beta\alpha}$, $P^*_{\gamma\alpha}$, $P^*_{\beta\gamma}$, $P^*_{\gamma\beta}$ are forces. Each of them shows the interaction of those two fictitious systems, the working directions of which coincide with their indexes and are directed along the tangent of coordinate corresponding to first index. For instance, the force - $P^*_{\alpha\beta}$ shows the interaction of systems T_α and T_β and is directed along the tangent of α coordinate. In final writing down of equation, the action of mentioned forces are reflected by three response forces P^*_α, P^*_β, P^*_γ, each of which shows action on one fictitious system (the working direction of which coincides with force index) of rest two fictitious systems. For instance, P^*_α response force expresses the action of T_β and T_γ fictitious systems on T_α fictitious system $P^*_\alpha = P^*_{\alpha\beta} + P^*_{\alpha\gamma}$. Analogically can be said about response forces P^*_β and P^*_γ.

In examination of obtained equilibrium equations in total form the exclusion of response moments for infinitesimal elements of fictitious system m^*_α, m^*_β, m^*_γ leads to the law of twoness of inner tangent forces, known from classic theory of elasticity $\tau_{\alpha\beta} = \tau_{\beta\alpha}$, $\tau_{\alpha\lambda} = \tau_{\gamma\alpha}$, $\tau_{\beta\gamma} = \tau_{\gamma\beta}$.

Evidently, inner tangent forces arisen in elastic body T - $\tau_{\alpha\beta}$, $\tau_{\beta\gamma}$, $\tau_{\alpha\gamma}$ really show the interaction of fictitious systems T_α, T_β T_γ. For instance, $T^*_{\alpha\beta}$ expresses the interaction of fictitious systems T_α and T_β.

The values of required response forces and tangent strengths, analogically to contact tasks, are represented in appearance of analytical series:

$$Y^*_1 = \sum_m \sum_n \sum_k A^i_{mnk} \varphi^i_m(x)\psi^i_n f^i_k(z) +$$
$$\sum_{i=1} \left[\xi^i_j(x)\eta^i_j(y,z) + \eta^i_j(y)\mu^i_j(x,z) + \lambda^i_j(z)g^i_j(x,y) \right], \tag{1}$$

where: Y^*_i is the "i" force of interaction of fictitious systems T_α, T_β T_γ,

A $^i_{mnk}$ - required permanent coefficient, $\varphi'_m(\alpha), \psi'_n(\beta), f'_k(\gamma)$ - arbitrary interpolation functions, continuous with their derivatives in limits of body T. Three members of its single series must satisfy following demands: according to character of boundary conditions given in each concrete task, first function of each member must satisfy conditions imposed to interpolation function, and second, i.e. the function, depended on two coordinates, is determined by corresponding boundary conditions. The comply with conditions of equilibrium for T_α, T_β T_γ fictitious systems is verified by way of their examination separately from each other.

So, the practical use of represented new analytical model (Figure 1,b) gives possibility to simplify significantly the solution of any task, as far as the calculation always has been brought to integration of differential equation of simplest structure (that conventionally can be called as "common"). Particularly, inner normal stresses σ_x, σ_y and σ_z are determined immediately from equilibrium equations, and displacements u, v, w - from equations of deformation those are obtained from unified system of Cauchy differential equations and dependences of Hooke's generalized law, by way of exception all components of deformation.

Unknown A^i_{mnk} coefficients all series of form (1), used for response forces P^*_α, P^*_β, P^*_γ and inner tangent strengths $\tau_{\alpha\beta}$, $\tau_{\beta\gamma}$, $\tau_{\alpha\gamma}$, are determined from identical equalities, showing the conditions of above mentioned "spatial contact task". For satisfaction of mentioned identical equalities there is used the collocation method, by means of which the system of algebraical equations with regard to $A^i{}_{mnk}$, solving the task, is obtained.

2. Solution of the Spatial Task of Prismatic Rod with Rectangular Cross-Section by the Mentioned Theory

In examination of spatial task, the classic theory of elasticity, as a rule, uses the hypothetic assumptions of many kinds. For instance, in calculation of the rod on torsion, from six components of strengths four ones are ignored, and only two components of stress τ_{xz} and τ_{yz} are left, by means of which the displacement components are determined later on.

The same has place in examination of rod bending spatial task /2/.

The accurate solution of given spatial task (and not only) is easily achieved using the above mentioned new direction. As a case study, there will be examined the prismatic rod with rectangular cross-section, the ends of which are subjected to action of external mutually antithetical torsion H moments that is realized by means of absolutely rigid disks, fixed on ends of the rod (Figure 2).

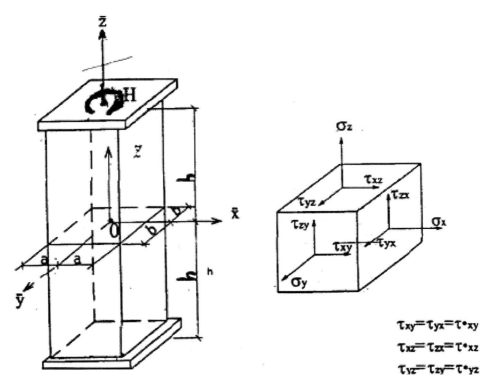

Figure 2.

As far as the spatial forces are not examined ($q_\alpha = q_\beta = q_\gamma = 0$), the equilibrium and physical equations, describing the mode of deformation of the rod, considering the dimensionless quantities $x = \dfrac{\overline{x}}{a}$, $y = \dfrac{\overline{y}}{b}$, $z = \dfrac{\overline{z}}{h}$, $\eta_1 = \dfrac{h}{a}$, $\eta_2 = \dfrac{h}{b}$, have the following appearance:

1. Equilibrium equations:

$$\frac{\partial \sigma_x}{\partial x} + \frac{\eta_2}{\eta_1} \frac{\partial \tau^*_{xy}}{\partial y} + \frac{1}{\eta_1} \frac{\partial \tau^*_{xz}}{\partial z} = 0,$$

$$\frac{\partial \sigma_y}{\partial y} + \frac{\eta_1}{\eta_2} \frac{\partial \tau^*_{xy}}{\partial x} + \frac{1}{\eta_2} \frac{\partial \tau^*_{yz}}{\partial z} = 0, \qquad (2)$$

$$\frac{\partial \sigma_z}{\partial z} + \eta_1 \frac{\partial \tau^*_{xz}}{\partial x} + \eta_2 \frac{\partial \tau^*_{yz}}{\partial y} = 0.$$

2. Physical equations:

$$\frac{\partial u}{\partial x} = \frac{1}{E}\left[\sigma_x - \upsilon(\sigma_y + \sigma_z) + \alpha t^0 E\right], \quad \frac{\partial v}{\partial y} = \frac{1}{E}\left[\sigma_y - \upsilon(\sigma_x + \sigma_z) + \alpha t^0 E\right],$$

$$\frac{\partial w}{\partial z} = \frac{1}{E}\left[\sigma_z - \upsilon(\sigma_x + \sigma_y) + \alpha t^0 E\right], \qquad (3)$$

$$\frac{\partial u}{\partial z} + \frac{\partial w}{\partial x} = \frac{2(1+\upsilon)}{E}\tau_{xz},$$

$$\frac{\partial v}{\partial x} + \frac{\partial u}{\partial y} = \frac{2(1+\upsilon)}{E}\tau_{xy}, \quad \frac{\partial w}{\partial y} + \frac{\partial v}{\partial z} = \frac{2(1+\upsilon)}{E}\tau_{yz}.$$

Proceeding from given task, the inner tangent strengths, representing the forces of interaction of fictitious systems, must satisfy the boundary and initial conditions, given in task (Figure2).

When x $=\pm$ 1, $\tau^*_{xy} = 0$, $\tau^*_{xz} = 0$;

when y $= \pm$ 1, $\tau^*_{yx} = 0$, $\tau^*_{yz} = 0$; when z $= \pm 1$, $\tau^*_{xz} \neq 0$, $\tau^*_{yz} = 0$.

The same time evenly the equality /2/

when z $= \pm 1$, $\qquad\qquad$ H $= \iint (x\tau^*_{yz} - y\tau^*_{xz})dF$ $\qquad\qquad$ (4)

where F is the square of cross-section (Figure 2).

Considering the boundary conditions, the inner tangent strengths, according to (1), can be represented in following appearance:

$$\tau_{xy}^* = \sum_m \sum_n \sum_k A_{mnk}(\sin mx - c\sin m)(\sin ny - y\sin n)f_1(z);$$

$$\tau_{xz}^* = \sum \sum \sum B_{mnk} f_1(x)\sin ny \cdot \sin kz - \frac{D}{\eta_1}y(x^2 - 1);$$

$$\tau_{xz}^* = \sum \sum \sum C_{mnk} f_1(y)\sin mx \sin kz + \frac{D}{\eta_2}x(y^2 - 1);$$

(5)

where $f_1(y)$, and $f_1(z)$ are known functions. For instance:

$$f_1(x) = (\cos mx - \cos m) + \frac{3}{2}\left(\frac{1}{m}\sin m - \cos m\right)(x^2 - 1)$$

D is the constant value and it is determined from conditions (4).

The inner normal stresses σ_x, σ_y and σ_z are determined immediately from equilibrium equations, and u, v, w displacements - from equations of deformation, considering the following real boundary conditions:

When $x = \pm 1$, $\sigma_x = 0$,

when $y = \pm 1$ $\sigma_y = 0$,   (6)

when $z = \pm 1$, $u = v = w = 0$.

Considering the equality (5), from the expressions (3) there can be obtained following:

$$\sigma_x = -\int\left[\frac{\eta_2}{\eta_1}\frac{\partial \tau_{xy}^*}{\partial y} + \frac{1}{\eta_1}\frac{\partial \tau_{xz}^*}{\partial z}\right]dx = \overline{\sigma}_x(x, y, z) + B_1(y, z)$$

$$\sigma_y = \overline{\sigma}_y(x, y, z) + C_1(x, z)$$

$$\sigma_z = \overline{\sigma}_z(x, y, z) + D_1(x, y)$$

Integration functions $B_1(x, y)$ and $C_1(x, z)$ are determined from boundary conditions (6), and for definition of integration function $D_1(x, y)$, the additional boundary conditions, imposed to function w , is used:

From the equality $\dfrac{\partial w}{\partial z} = \dfrac{h}{E}[\sigma_z - v(\sigma_x + \sigma_y)]$ can be obtained:

$$w = \overline{w}(x,y,z) + D_1(x,y)z + D_2(x,y)$$

where , the unknown integration functions $D_1(x,y)$ and $D_2(x,y)$ are determined from conditions:

when z=0, w = 0,

when z = 1, w = $-\dfrac{h}{E}c$, z = -1, w = $\dfrac{h}{E}c$,

where c is the constant value and it is determined from contact conditions: when z = ± 1,

$\iint \sigma_2 df = 0$.

Analogically can be obtained displacements u and v.

So, the normal strengths σ_x, σ_y, σ_z and displacements u, v, w, the expressions of which accurately complies with all equations of equilibrium, and only three - of six equations of deformation, are determined unambiguously. The rest three equations satisfy the conditions of confluence of fictitious systems T_x, T_y, T_z and have appearance of identical equalities:

$$\frac{\partial u}{\partial x} = \frac{h}{E}\frac{1}{h_1}\left[\sigma_x - v(\sigma_y + \sigma_z)\right];$$

$$\frac{\partial v}{\partial y} = \frac{h}{E}\frac{1}{h_2}\left[\sigma_y - v(\sigma_x + \sigma_z)\right];$$

$$\eta_2 \frac{\partial u}{\partial y} + \eta_1 \frac{\partial v}{\partial x} = \frac{h}{E}2(1+v)\tau_{xy}^*.$$

Bringing the expressions of strengths and displacements into these equalities, after corresponding transformations we will have the system of algebraically equations with regard to coefficients A_{mnk}, B_{mnk} and C_{mnk}:

$$\sum_m\sum_n\sum_k\left[A_{mnk}\phi_{mnk}^i(A) + B_{mnk}\phi_{mnk}^i(B) + C_{mnk}\phi_{mnk}^i(C)\right] = \phi_{mnk}^i(q)\cdot D, \quad (i = 1,2,3)$$

After determination of A_{mnk}, B_{mnk} and C_{mnk} coefficients, all desirable values σ_x, σ_y, σ_z, u, v, w are determined.

The mathematical algorithm and computer program are elaborated. For different numerical meanings of values η_1 and η_2 all strengths and deformations are calculated. The results are represented in appearance of corresponding diagrams. On the Figure 3, 4 and Figure 5 the diagrams of variation of values τ_{xy}^* and w are represented, correspondingly, in different cross-sections along the rod height, when $\eta_1 = \eta_2 = 6$.

$$W = \overline{W}\, \frac{H}{Eh^2}$$

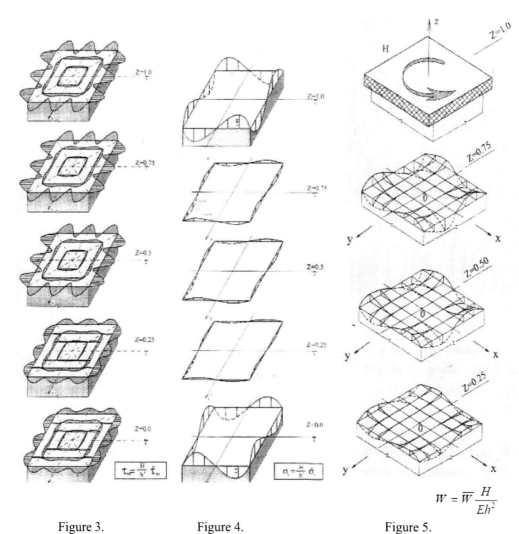

$$W = \overline{W}\, \frac{H}{Eh^2}$$

Figure 3.   Figure 4.   Figure 5.

References

[1]   I.I. Gudushauri. The Theory of Elasticity in Common Differential Equations. Tbilisi,"*Metsniereba*", 1990, 448 pp.
[2]   V.V. Novojhilov. The Theory of Elasticity. M., "*Sidpromgiz*", 1958, 371 pp.

Review -Presentation

of the paper by I. Gudushauri, R. Tskhvedadze, A.Sokhadze, D. Tabatadze

"Solution of Spatial Tasks of Rods with Cross-Sections of different Forms using the Double Trigonometric Series"

In calculation of the rods, using the Saint Venant principle, the mentioned spatial task is brought to "corresponding" two-dimensional task. This approach is recognized as comprehensively established by known scientist of the world those have brought the great part into creation of the classic theory of elasticity in its modern appearance and development of its many directions, by way of own fundamental research works.

By such consideration of the issue there is established that the goal of all research works on torsion of rods, conducted up to present day, was the determination of mode of deformation of the rod, not in full area, occupied by it, but only in limits, "excluding places located near the ends of the rod". However, as it is known, the solution of the rod torsion task in practice is carried out with the aim of its strength calculation, at which there is necessary the knowledge of the law of stress redistribution, not only in separate "homogenous" districts, but everywhere, i.e., where the real spatial mode of deformation, the concentration of stresses and deformations can have place. The ends of rod represent such districts, the influence of which is particularly significant in calculation of rods of small and average length.

Given terse analysis shows, that accurate solution of spatial task of rod torsion is considered as a problem up to present day.

In the paper, without using of Saint Venant principle, i.e. by the method of I. Gudushauri, two cases of given spatial task are examined:

a) the vertical displacement of the rod is restricted entirely;

b) the mentioned displacement is not restricted.

There is examined the prismatic rod with cross-section of rectangular form, the ends of which are subjected to action of external mutually antithetical torsion H moments that is realized by means of absolutely rigid disks, fixed on ends of the rod.

Results, represented in paper contains the novelty, hence, it is worthy to be published in "Georgian International Journal of Science, Technology and Medicine".

In: Focus on Science and Technology...
Editor: Sergo Gotsiridze, pp. 73-87

ISBN 978-1-61209-970-5

Chapter 7

ON INFINITE VERSIONS OF SOME CLASSICAL RESULTS IN LINEAR ALGEBRA AND VECTOR ANALYSIS

Gogi Pantsulaia[*]
Department of Mathematics, Georgian Technical University,
77 Kostava Str., 0175 Tbilisi , Georgia

Abstract

Infinite versions of Cramer's, Ostrogradsky's and Liouville's theorems were considered in some infinite-dimensional topological vector spaces.

2000 Mathematics Subject Classification: Primary 28Axx, 28Cxx, 28Dxx; Secondary 28C20, 28D10, 28D99

Key words and phrases: Translation-invariant Borel measure, Cramer' rule, Ostrogradsy's formula, Liouville's theorem

1. Introduction

It is well known that a behavior of various Dynamical systems with finite-dimensional phase spaces are described by classical results in Linear Algebra and Vector Analysis(see, for example, Cramer's rule, Ostrogradsky's formula, Liouville's theorem and so on). These constructions essentially employ the technique of Lebesgue measures defined on finite-dimensional Euclidean vector spaces.

Let us consider such an example from Vector Analysis.

Example 1.1 (Ostrogradsky's formula) In 1826, M. Ostrogradsky proved that if D is a non-empty region in $\mathbf{R}^3$ with the boundary ∂D and $A = (A_x, A_y, A_z)$ is a continuously

[*]E-mail address: gogi_pantsulaia@hotmail.com. The designated project has been fulfilled by financial support of the Georgia National Science Foundation (Grants: # GNSF /ST 07/3-178, # GNSF /ST 08/3-391).

differentiable vector field, then the volume integral of the divergence div$A$ over D and the surface integral of $\mathbf{A} \cdot \mathbf{n}$ over the boundary ∂D are related by

$$\int_D \text{div}\mathbf{A} dv = \int_{\partial D} \mathbf{A} \cdot \mathbf{n} ds, \tag{1.1}$$

where $\text{div}\mathbf{A} = \frac{\partial A_x}{\partial x} + \frac{\partial A_y}{\partial y} + \frac{\partial A_y}{\partial y}$, $\mathbf{n}(x, y, z) = (\cos(\alpha), \cos(\gamma), \cos(\beta))$ denotes an external normed normal of the boundary ∂D at the point $(x, y, z) \in \partial D$ and $\mathbf{A} \cdot \mathbf{n} = A_x \cos(\alpha) + A_y \cos(\gamma) + A_z \cos(\beta)$, respectively. This mathematical statement describes the physical fact that, in the absence of the creation or destruction of matter, the density within a region of a space can change only by having its flow into or away from the region through its boundary.

In 1834, M. Ostrogradsky generalized his result in the case of n-dimensional Euclidean space $\mathbf{R}^n$. In particular, he proved the validity of the following formula

$$\int_D \sum_{k=1}^n \frac{\partial A_k}{\partial x_k} db_n = \int_{\partial D} \frac{\sum_{k=1}^n A_k \frac{\partial f}{\partial x_k}}{\sqrt{\sum_{k=1}^n (\frac{\partial f}{\partial x_k})^2}} ds_{n-1}, \tag{1.2}$$

where b_n denotes the n-dimensional classical Borel measure on $\mathbf{R}^n$, f is a continuously differentiable function on $\mathbf{R}^n$ defining the boundary ∂D of the region $D \subset \mathbf{R}^n$ by the equation

$$f(x_1, \cdots, x_n) = 0, \tag{1.3}$$

$A = (A_k)_{1 \le k \le n}$ is a continuously differentiable vector field on $\mathbf{R}^n$ and s_{n-1} denotes the $n-1$-dimensional surface measure defined on the boundary ∂D.

Ostrogradsky's formula is valid in all cases when the field $\mathbf{A}$ and its divergence div$\mathbf{A}$(equivalentl $\sum_{k=1}^n \frac{\partial A_k}{\partial x_k}$ do not approach infinity in D. It is also valid when the divergence approaches infinity in such a way that the integral in the left -hand of the formula (1.2) is convergent.

The physical meaning of the divergence of a field depends on the nature of the vector field $\mathbf{A}$. For instance, for the velocity field $\mathbf{v}$ of a gas flow div$\mathbf{v}$ is equal to the rate of relative expansion of an infinitesimal volume of gas and div$\rho\mathbf{v}$ is equal to the density of mass sources. If the mass of the gass remains constant in the process of its flow we must have div$\rho\mathbf{v} = 0$ (in general case the mass can receive an increment, positive or negative, resulting from a chemical or some other reaction in which the mass can change). At the same time we can have div$\mathbf{v} > 0$, div$\mathbf{v} < 0$ or div$\mathbf{v} = 0$ depending on whether the gas expands, contracts or does not change its density in the process of flow.

Example 1.2 (Liouville theorem) From the point of view of measure theory, there exists an interesting way to study the behavior of a dynamical system (see, for example, [1]). This way employs Liouville's theorem giving a possibility to control whether the phase-space distribution function is constant along trajectories of the observing system.

Let $v = A(x)$ be a linear continuously differentiable vector field of velocities defined on $\mathbf{R}^n$ $(n \in \mathbb{N})$.

We recall the reader that a divergence div$v$ of the vector field of velocities v on $\mathbf{R}^n$ is defined by

$$\text{div}v = \sum_{1 \le i \le n} \frac{\partial v_i}{\partial x_i}, \tag{1.4}$$

where $(v_i)_{1 \le i \le n}$ is the family of components of the vector velocity v and $x_i \in \mathbf{R}$ for all $1 \le i \le n$.

Liouville's theorem asserts that if $v = A(x)$ is a linear continuously differentiable vector field of velocities defined on $\mathbf{R}^n$, and $D(0)$ is some Borel subset of $\mathbf{R}^n$, then the formula

$$\frac{db_n(D_t)}{dt} = \int\limits_{D(t)} \operatorname{div}(A) db_n \qquad (1.5)$$

is valid, where $D(t)$ denotes the state of the subset $D(0)$ at the moment t under the action of the phase flow defined by the vector field of velocities $v = A(x)$.

The proof of Liouville's theorem essentially employs the technique of the Lebesgue measure (see, [1]).

Note that proofs of all results presented in Examples 1.1–1.2, essentially employ the technique of Lebesgue measures. A Problem whether these results can be generalized in some infinite-dimensional vector spaces is directly connected with a problem of the existence of partial analogs of the Lebesgue measure on corresponding topological vector spaces. Hence, this problem itself is interesting and it is so important that has been studied for more than a half century ago by many people using various approaches. Among them the result of I.V.Girsanov and B.S. Mityagin [5] should be mentioned especially. Their result asserts that an arbitrary σ-finite quasi-invariant Borel measure defined on infinite-dimensional locally convex topological vector space is identically zero. This result asserts that the properties of the σ-finiteness and the translation-invariance are not consistent for non-zero Borel measures in infinite-dimensional topological vector spaces. A.B. Kharazishvili [7] ignoring the property of translation-invariance, constructed an example of a such non-zero σ-finite Borel measure in the Hilbert space ℓ_2 which is invariant with respect to everywhere dense (in ℓ_2) linear manifold. Ignoring the property of the σ-finiteness, various examples of non-zero non-atomic translation-invariant Borel measure were constructed on the vector space of all real valued sequences $\mathbf{R}^{\mathbb{N}}$ equipped with Tychonoff Topology (see, for example, [2], [3], [12],[14], [15]).

The paper is organized as follows.

In Section 2 we consider some auxiliary notions and propositions established in [13].

In Section 3 we give an infinite version of Cramer's rule in $\mathbf{R}^{\mathbb{N}}$.

An infinite version of Ostrogradsky's formula in ℓ_2 is considered in Section 4.

In Section 5 we give an infinite version of Lioville's theorem in $\mathbf{R}^{\mathbb{N}}$.

2. Preliminaries

In the sequel we need some notions and facts.

Recall that Solovay model is the following system of axioms:

$$(ZF)\&(DC)\&(\text{every subset of } \mathbf{R} \text{ is measurable in the Lebesgue sense}),$$

where (ZF) denotes the Zermelo-Fraenkel set theory and (DC) denotes the axiom of Dependent Choices (cf. [4]).

Let (E_1, S_1, μ_1) and (E_2, S_2, μ_2) be two measure spaces. The measures μ_1 and μ_2 are called isomorphic if there exists a measurable isomorphism from E_1 onto E_2 such that

$$(\forall X)(X \in S_1 \rightarrow \mu_1(X) = \mu_2(f(X))).$$

Lemma 2.1 *Let E_1 and E_2 be any two Polish topological spaces. Let μ_1 be a probability diffused Borel measure on E_1 and let μ_2 be a probability diffused Borel measure on E_2. Then there exists a Borel isomorphism $\varphi : (E_1, B(E_1)) \rightarrow (E_2, B(E_2))$ such that*

$$\mu_1(X) = \mu_2(\varphi(X))$$

for every $X \in B(E_1)$.

The proof of Lemma 2.1 can be found in [4].

Lemma 2.2([14], Lemma 7.2, p. 118) *Let E be a Polish space and let μ be a probability diffused Borel measure on E. Then, in Solovay's model the completion $\bar{\mu}$ of the μ is defined on the powerset of E.*

Corollary 2.1 *Let $\mathbb{J}$ be any non-empty subset of the set all natural numbers $\mathbb{N}$. Let, for $k \in \mathbb{J}$, S_k be the unit circle in the Euclidean plane $\mathbf{R}^2$. We may identify the circle S_k with a compact group of all rotations of $\mathbf{R}^2$ around its origin. Let $\lambda_{\mathbb{J}}$ be the probability Haar measure defined on the compact group $\prod_{k \in \mathbb{J}} S_k$. Then in Solovay's model the completion $\overline{\lambda_{\mathbb{J}}}$ of the $\lambda_{\mathbb{J}}$ is defined on the powerset of the $\prod_{k \in \mathbb{J}} S_k$.*

For $k \in \mathbb{J}$, define the function f_k by $f_k(x) = \exp\{2\pi x i\}$ for every $x \in \mathbf{R}$. Then the equality

$$\left(\prod_{k \in \mathbb{J}} f_k\right)(z + w) = \left(\prod_{k \in \mathbb{J}} f_k\right)(z) \circ \left(\prod_{k \in \mathbb{J}} f_k\right)(w)$$

holds for every $z, w \in \mathbf{R}^{\mathbb{J}}$, where $\mathbf{R}^{\mathbb{J}}$ denotes the vector space of all real-valued sequences defined on $\mathbb{J}$, $\prod_{k \in \mathbb{J}} f_k$ denotes the direct product of the family of functions $(f_k)_{k \in \mathbb{J}}$, " $\circ$ " denotes the group operation in $\prod_{k \in \mathbb{J}} S_k$(see, [14], Remark 7.1, p. 118).

Theorem 2.1 ([14], Theorem 7.1, p. 118) *Let $\mathbb{J}$ be any non-empty subset of the set all natural numbers $\mathbb{N}$. For $E \subset \mathbf{R}^{\mathbb{J}}$ and $g \in \prod_{k \in \mathbb{J}} S_k$, put*

$$f_E(g) = \begin{cases} card\left(\left(\prod_{k \in \mathbb{J}} f_k\right)^{-1}(g) \cap E\right), & \text{if this is finite;} \\ +\infty, & \text{in all other cases.} \end{cases}$$

Then a functional $\mu_{\mathbb{J}}$, defined by

$$(\forall E)(E \subset \mathbf{R}^{\mathbb{J}} \rightarrow \mu_{\mathbb{J}}(E) = \int_{\prod_{k \in \mathbb{J}} S_k} f_E(g) d\overline{\lambda_{\mathbb{J}}}(g)),$$

is a translation-invariant measure in Solovay model.

3. On Generalized Fourier Series in Solovay Model

We begin this section by a construction of partial analogs of the Lebesgue measure in infinite-dimensional separable Banach spaces with basis.

Lemma 3.1 ((see, [14], Lemma 14. 1, p. 195) *Let* $C_\beta = h_\beta + [0,1[^{\mathbb{N}\setminus\{k\}}$ *for* $\beta \in [0,x_k[$ $(x_k > 0)$ *and* $h_\beta \in \mathbb{R}^{\mathbb{N}\setminus\{k\}}$. *Then*

$$f_{\sum_{\beta \in [0,x_k[}\{\beta\}\times C_\beta}(g_k,(g_i)_{i\in\mathbb{N}\setminus\{k\}}) = \#(f_k^{-1}(g_k)\cap[0,x_k[) \tag{3.1}$$

for $(g_k,(g_i)_{i\in\mathbb{N}\setminus\{k\}}) \in S_k \times \prod_{i\in\mathbb{N}\setminus\{k\}} S_i$.

Lemma 3.2 ([14], Lemma 14.2, p. 196) *Let B be an infinite-dimensional separable Banach space with absolutely convergent basis $\Gamma = (e_k)_{k\in\mathbb{N}}$. We set*

$$A(h) = (x_k)_{k\in\mathbb{N}} \tag{3.6}$$

for $h = \sum_{k\in\mathbb{N}} x_k e_k \in B$. Then in Solovay's model, a functional μ_Γ defined by

$$\mu_\Gamma(X) = \mu_\mathbb{N}(A(X)) \tag{3.7}$$

for $X \subseteq B$, is a translation-invariant diffused measure which gets the value one on the set $\Delta = \{\sum_{k\in\mathbb{N}} \alpha_k e_k : (\alpha_k)_{k\in\mathbb{N}} \in [0,1[^{\mathbb{N}}\}$.

Let $\mathbb{J} \subseteq \mathbb{N}$. We set $B_\mathbb{J} = \{b = \sum_{k\in\mathbb{N}} \alpha_k e_k : b \in B \ \& \ \alpha_k \geq 0 \text{ if } k \in \mathbb{J} \text{ and } \alpha_k < 0 \text{ if } k \notin \mathbb{J}\}$. It is clear that $B = \sum_{\mathbb{J}\subseteq\mathbb{N}} B_\mathbb{J}$.

Let $(b_k)_{k\in\mathbb{N}}$ be a family of elements in B such that

$$\sum_{k\in\mathbb{N}} ||b_k|| < \infty. \tag{3.8}$$

We set

$$P(b_1,\cdots) = \{\sum_{k\in\mathbb{N}} \alpha_k b_k : (\alpha_k)_{k\in\mathbb{N}} \in [0,1[^{\mathbb{N}}\}. \tag{3.9}$$

Clearly, $P(b_1,\cdots)$ can be considered as an infinite-dimensional parallelepiped in B generated by the family of vectors $(b_k)_{k\in\mathbb{N}}$.

Let $b \in B$. A formal series

$$\sum_{k\in N} (-1)^{\text{Ind}_\mathbb{J}(k)+1} \mu_\Gamma(P(e_1,\cdots,e_{k-1},b,e_{k+1},\cdots)) e_k \tag{3.10}$$

and a sequence of real numbers

$$((-1)^{\text{Ind}_\mathbb{J}(k)+1} \mu_\Gamma(P(e_1,\cdots,e_{k-1},b,e_{k+1},\cdots)))_{k\in\mathbb{N}}, \tag{3.11}$$

where $\mathbb{J}$ is a unique subset of $\mathbb{N}$ for which $b \in B_\mathbb{J}$ and $\text{Ind}_\mathbb{J}(\cdot)$ denotes an indicator of the set $\mathbb{J}$ defined on $\mathbb{N}$, are called a generalized Fourier series and a family of generalized Fourier coefficients of the element b in the basis Γ, respectively.

The following assertion is an infinite version of Cramer's rule in separable Banach space with absolutely convergent basis.

Theorem 3.1 ([14], Theorem 14.1, p. 197) *Let B be an infinite-dimensional Banach space with absolutely convergent basis $\Gamma = (e_k)_{k \in \mathbb{N}}$ and let $b \in B$. Then, in Solovay's model, the generalized Fourier series of the element $b \in B$ in the basis Γ coincides with its representation in the same basis.*

Remark 3.5 A problem asking whether one can generalize Cramer's rule in some infinite-dimensional Polish topological vector spaces was discussed in [10], [11], [9], [17], [16]. The same problem has been investigated for Mankiewicz's and Baker's generators of shy sets on $\mathbf{R}^{\mathbb{N}}$(see, [19]).

4. On Ostrogradsky's Formula in ℓ_2

For $\mathbb{J} \subset \mathbb{N}$, we put

$$\ell_2(\mathbb{J}) = \{ (x_k)_{k \in \mathbb{J}} \in \mathbb{R}^{\mathbb{J}} \ \& \ \sum_{k \in \mathbb{J}} x_k^2 < \infty \}, \tag{4.12}$$

$$T_{\mathbb{J}}((x_k)_{k \in \mathbb{J}}) = (\frac{x_k}{k+1})_{k \in \mathbb{J}}. \tag{4.13}$$

It is clear that, if $\text{card}(\mathbb{J}) < \omega$, then

$$\ell_2(\mathbb{J}) = \mathbb{R}^{\mathbb{J}}. \tag{4.14}$$

One can easily demonstrate, that

$$B(\ell_2(\mathbb{J}_1 \cup \mathbb{J}_2)) = B(\ell_2(\mathbb{J}_1)) \times B(\ell_2(\mathbb{J}_2)) \tag{4.15}$$

for $\mathbb{J}_1, \mathbb{J}_2 \subset \mathbb{N}$ with $\mathbb{J}_1 \cap \mathbb{J}_2 = \emptyset$, where $B(\cdot)$ denotes a Borel σ-algebra of the corresponding space.

Lemma 4.1([14], Lemma 13.4, p. 188) *A functional $\mu_{\mathbb{J}}$, defined by*

$$(\forall X)(X \in B(\ell_2(\mathbb{J})) \to \mu_{\mathbb{J}}(X) = \nu_{\mathbb{J}}(T_{\mathbb{J}}^{-1}(X))) \tag{4.16}$$

for $\mathbb{J} \subseteq \mathbb{N}$, is a translation-invariant Borel measure on $\ell_2(\mathbb{J})$ which gets the value one on the cube

$$\prod_{i \in \mathbb{J}}[0, \frac{1}{i+1}]. \tag{4.17}$$

Let us consider some notions of the theory of vector fields in ℓ_2.

Let D_i be some Borel subset in $\ell_2(\mathbb{N} \setminus \{ i \})$ with $0 < \mu_{\mathbb{N} \setminus \{ i \}}(D_i) < \infty$ and let

$$f_i : D_i \to \ell_2(\{ i \})$$

be any real-valued function on D_i.

The set $\Gamma_{f_i} \subset \ell_2$, defined by

$$\Gamma_{f_i} = \{ (x_0, \cdots, x_{i-1}, f_i((x_0, \cdots, x_{i-1}, x_{i+1}, \cdots)), x_{i+1}, \cdots) :$$

$$(x_0, \cdots, x_{i-1}, x_{i+1}, \cdots) \in D_i\} \tag{4.18}$$

is called a surface generated by the function f_i.

We say that f_i is differentiable at the point $(x_k)_{k \in \mathbb{N} \setminus \{i\}}$ if there exists $(C_k)_{k \in \mathbb{N} \setminus \{i\}} \in \ell_2(\mathbb{N} \setminus \{i\})$ and $\sigma : \ell_2(\mathbb{N} \setminus \{i\}) \to \mathbb{R}$ with the property

$$\lim_{\Sigma_{k \in \mathbb{N}, k \neq i} h_k^2 \to 0} \sigma((h_0, \cdots, h_{i-1}, h_{i+1}, \cdots)) = 0, \tag{4.19}$$

such that an equality

$$f_i((x_0 + \Delta x_0, \cdots, x_{i-1} + \Delta x_{i-1}, x_{i+1} + \Delta x_{i+1}, \cdots)) - f_i((x_0, \cdots, x_{i-1}, x_{i+1}, \cdots))$$

$$= \sum_{k \in \mathbb{N}, k \neq i} C_k \Delta x_k + \sigma(\Delta x_0, \cdots, \Delta x_{i-1}, \Delta x_{i+1}, \cdots) \sqrt{\sum_{k \in \mathbb{N}, k \neq i} (\Delta x_k)^2} \tag{4.20}$$

holds for arbitrary

$$(\Delta x_k)_{k \in \mathbb{N}, k \neq i} \in \ell_2(\mathbb{N} \setminus \{i\}). \tag{4.21}$$

If the function $g_k(x) = f_i(a_0, \cdots, a_{k-1}, a_k + x, a_{k+1}, \cdots))$ is differentiable at $0 \in \mathbb{R}$, then the value $\frac{dg_k}{dx}(0)$ is called a partial derivative of the f_i with respect to the value x_k ($k \in \mathbb{N} \setminus \{i\}$) at point $a = (a_j)_{j \in \mathbb{N} \setminus \{i\}}$ and is denoted by $\frac{\partial f_i}{\partial x_k}(a)$.

Note that if $x_i = f_i((x_0, \cdots, x_{i-1}, x_{i+1}, \cdots))$ is differentiable in the above mentioned sense at the point $(a_j)_{j \in \mathbb{N} \setminus \{i\}}$ then $C_k = \frac{\partial f_i}{\partial x_k}((a_j)_{j \in \mathbb{N} \setminus \{i\}})$ for $k \in \mathbb{N} \setminus \{i\}$.

Note also that if $x_i = f_i((x_0, \cdots, x_{i-1}, x_{i+1}, \cdots))$ is differentiable at the point $(a_j)_{j \in \mathbb{N} \setminus \{i\}}$, then a surface Γ_{f_i} has a tangential plane π_M at the point $M = (a_0, \cdots, a_{i-1}, f_i(a_0, \cdots, a_{i-1}, a_{i+1}, \cdots), a_i$ which is defined by:

$$\pi_M = \{ (x_k)_{k \in \mathbb{N}} | (x_k)_{k \in \mathbb{N}} \in \ell_2(\mathbb{N}) \,\&\, \sum_{k \in \mathbb{N} \setminus \{i\}} \frac{\partial f_i}{\partial x_k}((a_k)_{k \in \mathbb{N} \setminus \{i\}})(x_k - a_k) +$$

$$(x_i - f_i((a_0, \cdots, a_{i-1}, a_{i+1}, \cdots))) = 0\}. \tag{4.22}$$

We say that a Borel subset $D \subset \ell_2(\mathbb{N})$ is simple in the i-th direction if there exists a Borel subset $D_i \subset \ell_2(\mathbb{N} \setminus \{i\})$ with $0 < \mu_{\mathbb{N} \setminus \{i\}}(D_i) < \infty$ and differentiable functions $\Psi_1^{(i)}$ and $\Psi_2^{(i)}$ defined on D_i such that

1) $(\forall x)(x \in D_i \to \Psi_1^{(i)}(x) < \Psi_2^{(i)}(x))$;

2) $D = \{ (x_k)_{k \in \mathbb{N}} | (x_0, \cdots, x_{i-1}, x_{i+1}, \cdots) \in D_i \,\&\,$

$$\Psi_1^{(i)}((x_0, \cdots, x_{i-1}, x_{i+1}, \cdots)) < x_i < \Psi_2^{(i)}((x_0, \cdots, x_{i-1}, x_{i+1}, \cdots))\}. \tag{4.23}$$

A subset $D \subset \ell_2(\mathbb{N})$ is called simple if it is simple in the i-th direction for every $i \in \mathbb{N}$ and $S_i \cap S_j = \emptyset$ for $0 \leq i < j < \infty$, where $S_k = \Gamma_{\Psi_1^{(k)}} \cup \Gamma_{\Psi_2^{(k)}}$ for $k \in \mathbb{N}$.

Let $M \in S_i$ and π_M be a tangential plane of the surface S_i at the point M. A normed vector n_M^+ is called an external normed normal of the surface S_i at the point M, if $\cos(\angle(n_M^+, e_i)) > 0$, $n_M^+ \perp \pi_M$ when $M \in \Gamma_{\Psi_2^{(i)}}$, or $\cos(\angle(n_M^+, e_i)) < 0$, $n_M^+ \perp \pi_M$ when $M \in \Gamma_{\Psi_1^{(i)}}$. The vector $-n_M^+$ is called an inner normed normal of the surface S_i at the point M and is denoted by n_M^-.

A simple Borel set D is called a cube-set if $n^+_{M_i} \perp e_j$ for every different $i, j \in \mathbb{N}$ and $M_i \in S_i$.

Remark 4.1 Note that a set $\prod_{i \in \mathbb{N}}]x_i, x_i + \frac{1}{i+1}[$ is the cube-set in ℓ_2 for $(x_i)_{i \in \mathbb{N}} \in \ell_2$.

Let $D \subset \ell_2(\mathbb{N})$ be a cube-set in the above mentioned sense.

A surface integral of the first order from the function $g : \ell_2(\mathbb{N}) \to \mathbb{R}$ along i-th surface S_i of D is denoted by $\int_{S_i} g ds_i$ and is defined by

$$\int_{S_i} g ds_i = \int_{D_i} \frac{g(M_2)}{\cos(\angle(n_{M_2}^+, e_i))} d\mu_{\mathbb{N}\setminus\{i\}} - \int_{D_i} \frac{g(M_1)}{\cos(\angle(n_{M_1}^+, e_i))} d\mu_{\mathbb{N}\setminus\{i\}}, \quad (4.24)$$

where

$$M_k = (x_0, \cdots, x_{i-1}, \Psi_k^{(i)}(x_0, \cdots, x_{i-1}, x_{i+1}, \cdots), x_{i+1}, \cdots) \text{ for } k = 1, 2,$$

$$n^+_{M_1} = (\frac{\partial \Psi_1^{(i)}}{\partial x_1}, \cdots, \frac{\partial \Psi_1^{(i)}}{\partial x_{i-1}}, -1, \frac{\partial \Psi_1^{(i)}}{\partial x_{i+1}}, \cdots)(M_1),$$

$$n^+_{M_2} = (\frac{\partial \Psi_2^{(i)}}{\partial x_1}, \cdots, \frac{\partial \Psi_2^{(i)}}{\partial x_{i-1}}, 1, \frac{\partial \Psi_2^{(i)}}{\partial x_{i+1}}, \cdots)(M_2).$$

If the series $\sum_{i=1}^{\infty} \int_{S_i} g ds_i$ is convergent then its sum is called a surface integral of the first order from the function $g : \ell_2(\mathbb{N}) \to \mathbb{R}$ along the surface ∂D and is denoted by $\int_{\star \partial D} g ds$, i.e.,

$$\int_{\star \partial D} g ds = \sum_{i=1}^{\infty} \int_{S_i} g ds_i. \quad (4.25)$$

A surface integral of the second order from the function $g : \ell_2(\mathbb{N}) \to R$ along the surface ∂D (in the i-th direction) is denoted by $\int_{\partial D} g d\mu_{\mathbb{N}\setminus\{i\}}$ and is defined by

$$\int_{\partial D} g d\mu_{\mathbb{N}\setminus\{i\}} = \int_{D_i} g((x_0, \cdots, x_{i-1}, \Psi_2^{(i)}(x_0, \cdots, x_{i-1}, x_{i+1}, \cdots), x_{i+1}, \cdots)) d\mu_{\mathbb{N}\setminus\{i\}} -$$

$$\int_{D_i} g((x_0, \cdots, x_{i-1}, \Psi_1^{(i)}(x_0, \cdots, x_{i-1}, x_{i+1}, \cdots), x_{i+1}, \cdots)) d\mu_{\mathbb{N}\setminus\{i\}}. \quad (4.26)$$

The following assertion is an infinite version of Ostrogradski's theorem in $\ell_2(\mathbb{N})$.

Theorem 4.1 ([14], Theorem 13.1, p. 192) *Let D be a cube-set in $\ell_2(\mathbb{N})$ and let $A = (A_i)_{i \in \mathbb{N}}$ be a continuously differentiable vector field in ℓ_2 such that*

$$\int_D \sum_{i=1}^{\infty} \frac{\partial A_i}{\partial x_i} d\mu_{\mathbb{N}} = \sum_{i=1}^{\infty} \int_D \frac{\partial A_i}{\partial x_i} d\mu_{\mathbb{N}}. \quad (4.27)$$

Then the following formula is valid

$$\int_D divA d\mu_{\mathbb{N}} = \int_{\star \partial D} < T_{\mathbb{N}}^{-1}(A), \mathbf{n}_{\mathbf{M}}^+ >_{\ell_2} ds, \quad (4.28)$$

where $< \cdot, \cdot >_{\ell_2}$ denotes the usual scalar product in ℓ_2.

Remark 4.1 Under our notations, the formula (1.2) can be rewritten as follows

$$\int_D divA d\mu_{\{1, \cdots, n\}} = \int_{\star \partial D} < T_{\{1, \cdots, n\}}^{-1}(A), \mathbf{n}_{\mathbf{M}}^+ >_{\ell_2(\{1, \cdots, n\})} ds, \quad (4.29)$$

where $< \cdot, \cdot >_{\ell_2(\{1,\cdots,n\})}$ denotes the usual scalar product in $\ell_2(\{1,\cdots,n\}) = \mathbf{R}^n$. By this reasoning the formula (4.28) can be considered as an infinite version of Ostrogradsky's formula (1.2) for cube-sets in ℓ_2.

In context of Theorem 4.1, the following simple example is of interest.

Example 4.1 Let $\mathbf{A}$ be a field in ℓ_2 defined by

$$\mathbf{A}((x_i)_{i \in N}) = (\frac{\lambda^i x_i}{i!})_{i \in \mathbb{N}}, \tag{4.30}$$

and let $D = \Delta_b$, where

$$\Delta_b = \prod_{i \in N}[b_i - \frac{1}{2(i+1)}, b_i + \frac{1}{2(i+1)}] \tag{4.31}$$

for $b = (b_i)_{i \in N} \in \ell_2$.

Then, following Theorem 4.1, we have

$$\int_{\star \partial \, \Delta_b} < T_{\mathbb{N}}^{-1}(A), \mathbf{n_M}^+ >_{\ell_2} ds = \int_{\Delta_b} \sum_{i=1}^{\infty} \frac{\partial \mathbf{A}_i}{\partial x_i} d\mu_N =$$

$$\int_{\Delta_b} \sum_{i=1}^{\infty} \frac{\lambda^i}{i!} d\mu_N = (e^\lambda - 1) \int_{\Delta_b} d\mu_N = e^\lambda - 1. \tag{4.32}$$

5. On an Infinite Version of Liouville's Theorem in $\mathbf{R}^{\mathbb{N}}$

Let $\mathbf{R}^n (n > 1)$ be an n-dimensional Euclidean space and let μ_n an n-dimensional standard Lebesgue measure in $\mathbf{R}^n$. Further, let T be a linear μ_n-measurable transformation of $\mathbf{R}^n$.

It is obvious that $\mu_n T^{-1}$ is absolutely continuous with respect to μ_n, and there exists a non-negative μ_n-measurable function Φ on $\mathbf{R}^n$ such that

$$\mu_n(T^{-1}(X)) = \int_X \Phi(y) d\mu_n(y) \tag{5.33}$$

for every μ_n-measurable subset X of $\mathbf{R}^n$.

The function Φ plays the role of the Jacobian $J(T^{-1})$ of the transformation T^{-1}(or, rather the absolute value of the Jacobian)(see, e.g., [6]) in the theory of transformations of multiple integrals. It is clear that $J(T^{-1})$ coincides with a Radon-Nikodym derivative $\frac{d\mu_n T^{-1}}{d\mu_n}$, which is unique a.e. with respect to μ_n.

It is clear that

$$\frac{d\mu_n T^{-1}}{d\mu_n}(x) = \lim_{k \to \infty} \frac{\mu_n(T^{-1}(U_k(x)))}{\mu_n(U_k(x))} (\mu_n - \text{a.e.}), \tag{5.34}$$

where $U_k(x)$ is a spherical neighborhood with the center in $x \in \mathbf{R}^n$ and radius $\mathbf{r}_k > 0$ so that $\lim_{k \to \infty} r_k = 0$. The class of such spherical neighborhoods generate the so-called Vitali differentiability class of subsets which allows us to calculate the Jacobian $J(T^{-1})$ of the transformation T^{-1}.

If we consider a vector space of all real-valued sequences $\mathbf{R}^{\mathbb{N}}$(equipped with Tychonoff topology), then we observe that for the infinite-dimensional Lebesgue measure [2] (or [3]) defined on $\mathbf{R}^{\mathbb{N}}$ there does not exist any Vitali system of differentiability, but in spite of non-existence of such a system the inner structure of this measure allows us to define a form of the Radon-Nikodym derivative defined by any linear transformation of $\mathbf{R}^{\mathbb{N}}$. In order to show it, let us consider the following

Example 5.1 Let $\mathcal{R}_1$ be the class of all infinite dimensional rectangles $R \in B(\mathbf{R}^{\mathbb{N}})$ of the form

$$R = \prod_{i=1}^{\infty} R_i, \ \ R_i = (a_i, b_i), -\infty < a_i \leq b_i < +\infty, \tag{5.35}$$

such that

$$0 \leq \prod_{i=1}^{\infty}(b_i - a_i) < \infty. \tag{5.36}$$

Let τ_1 be the set function on $\mathcal{R}_1$ defined by

$$\tau_1(R) = \prod_{i=1}^{\infty}(b_i - a_i). \tag{5.37}$$

R.Baker [2] proved that the functional λ_1 defined by

$$(\forall X)(X \in B(\mathbf{R}^{\mathbb{N}}) \to \lambda_1(X) = \inf\{\sum_{j=1}^{\infty} \tau_1(R_j) : R_j \in \mathcal{R}_1 \ \& \ X \subseteq \cup_{j=1}^{\infty} R_j\}) \tag{5.38}$$

is a quasi-finite [1] translation-invariant Borel measure on $\mathbf{R}^{\mathbb{N}}$.

The following change of variable formula has been established in [2] (cf. p. 1029): *Let $T^n : \mathbf{R}^n \to \mathbf{R}^n, n > 1$, be a linear transformation with Jacobian $\Delta \neq 0$, and let $T^{\mathbb{N}} : \mathbf{R}^{\mathbb{N}} \to \mathbf{R}^{\mathbb{N}}$ be the map defined by*

$$T^{\mathbb{N}}(x) = (T^n(x_1, \cdots, x_n), x_{n+1}, x_{n+2}, \cdots), \ x = (x_i)_{i \in N} \in \mathbf{R}^{\mathbb{N}}. \tag{5.39}$$

Then for each $E \in \mathcal{B}(\mathbf{R}^{\mathbb{N}})$, we have

$$\lambda_1(T^{\mathbb{N}}(E)) = |\Delta|\lambda_1(E). \tag{5.40}$$

We are to show whether one can extend change of variable formula presented in Example 5.1 for wide class of linear transformations of $\mathbf{R}^{\mathbb{N}}$.

Now, let (X_i, M_i, μ_i) be a sequence of measure spaces and for each i let ρ_i be a metric on the set X_i. Assume that the following conditions are satisfied.

(i) Each (X_i, ρ_i) is a locally compact metric space.

(ii) Each M_i contains the family, $B(X_i)$, of Borel subsets of X_i and μ_i is a regular Borel measure on X_i (cf. [20], Def. 2.15).

(iii) For all i, and for every $\delta > 0$, there exists a sequence (A_j) of Borel subsets of X_i such that $d_i(A_j) < \delta$ and $X_i = \cup_{j=1}^{\infty} A_j$, where $d_i(A_j)$ is the diameter of A_j in X_i.

---

[1] A Borel measure μ defined in $\mathbf{R}^{\mathbb{N}}$ is called quasi-finite if $0 < \mu(U) < +\infty$ for any compact set $U \subset \mathbf{R}^{\mathbb{N}}$

Define $X = \prod_{i=1}^{\infty} X_i$ and let equip X with the product topology. Let denote by $\mathcal{R}$ the family of all rectangles $R \subseteq X$ of the form

$$R = \prod_{i=1}^{\infty} R_i, R_i \in \mathcal{B}(X_i), \tag{5.41}$$

and

$$0 \le \prod_{i=1}^{\infty} \mu_i(R_i) := lim_{n \to \infty} \prod_{i=1}^{n} \mu_i(R_i) < +\infty. \tag{5.42}$$

For $R \in \mathcal{R}$, we define

$$\tau(R) = \prod_{i=1}^{\infty} \mu_i(R_i). \tag{5.43}$$

Let τ^* be the set function on the powerset $\mathcal{P}(X)$ defined by

$$\tau^*(E) = \inf\{\sum \tau(R_j) : R_j \in \mathbf{R} \,\&\, E \subseteq \cup R_j\}$$

for $E \subseteq X$.

Also, let use the convention that $0 \cdot +\infty = +\infty \cdot 0 = 0$, $+\infty \cdot +\infty = +\infty$, and that the infimum taken over the empty set has the value $+\infty$.

Theorem 5.1([3], Theorem I, p. 2579)*The set function* τ^* *is an outer measure on* X. *Let M be the* σ*-algebra of subsets of X which are measurable with respect to* τ^*, *and let* μ *be the measure on M obtained by restricting* τ^* *to M. Then* $B(X) \subseteq M$, *and for all* $R = \prod_{i=1}^{\infty} R_i \in \mathcal{R}$, *we have* $\mu(R) = \prod_{i=1}^{\infty} \mu_i(R_i)$. *If each space* X_i *contains disjoint subsets* A_i, B_i *such that* $\mu_i(A_i) = \mu_i(B_i) = 1$, *then the measure* μ *is not* σ*-finite. Finally, assume that each* (X_i, ρ_i) *is an* M_i*-measurable group. If each* μ_i *is left-invariant measure on* M_i, *then* μ *is a left-invariant measure on M. Similarly, if each* μ_i *is right-invariant measure on* M_i, *then* μ *is right-invariant measure on M.*

Let $\mathcal{R}_2$ be the class of all infinite dimensional rectangles $R \in \mathbb{B}(\mathbf{R}^{\mathbb{N}})$ of the form

$$R = \prod_{i=1}^{\infty} R_i, \ \ R_i \in B(\mathbf{R}) \tag{5.44}$$

such that

$$0 \le \prod_{i=1}^{\infty} m(R_i) := \lim_{n \to \infty} \prod_{i=1}^{n} m(R_i) < \infty, \tag{5.45}$$

where m denotes a one-dimensional classical Borel measure on $\mathbf{R}$.

Let τ_2 be a set function on $\mathcal{R}_2$, defined by

$$\tau_2(R) = \prod_{i=1}^{\infty} m(R_i). \tag{5.46}$$

As a simple consequence of Theorem 5.1, R.Baker [3] had obtained that the functional λ_2 defined by

$$(\forall X)(X \in \mathcal{B}(\mathbf{R}^{\mathbb{N}}) \to \lambda_2(X) = \inf\{\sum_{j=1}^{\infty} \tau_2(R_j) : R_j \in \mathcal{R}_2 \,\&\, X \subseteq \cup_{j=1}^{\infty} R_j\}) \tag{5.47}$$

is a quasi-finite translation-invariant Borel measure on $\mathbf{R}^{\mathbb{N}}$.

Theorem 5.2([18], Theorem 2, p. 6) *Let* $\alpha = (n_i)_{i \in N}$ *be the sequence of natural numbers and let* $\mathcal{R}_\alpha$ *be the class of all infinite dimensional rectangles* $R \in B(\mathbf{R}^{\mathbb{N}})$ *of the form*

$$R = \prod_{i=1}^{\infty} R_i, \quad R_i \in B(\mathbf{R}^{n_i}) \tag{5.48}$$

such that

$$0 \le \prod_{i=1}^{\infty} m_{n_i}(R_i) := \lim_{n \to \infty} \prod_{i=1}^{n} m_{n_i}(R_i) < \infty, \tag{5.49}$$

where m_{n_i} *denotes an* n_i-*dimensional classical Borel measure on* $\mathbf{R}^{n_i}$.

Let τ_α *be a set function on* $\mathcal{R}_\alpha$, *defined by*

$$\tau_\alpha(R) = \prod_{i=1}^{\infty} m_{n_i}(R_i). \tag{5.50}$$

Then the functional λ_α *defined by*

$$(\forall X)(X \in B(\mathbf{R}^{\mathbb{N}}) \to \lambda_\alpha(X) = \inf\{\sum_{j=1}^{\infty} \tau_\alpha(R_j) : R_j \in \mathcal{R}_\alpha \,\&\, X \subseteq \cup_{j=1}^{\infty} R_j\}) \tag{5.51}$$

is a quasi-finite translation-invariant Borel measure in $\mathbf{R}^{\mathbb{N}}$ *such that* λ_α *is absolutely continuous with respect to the measure* λ_2. *Moreover, if there exists* i_0 *such that* $n_i = 1$ *for* $i > i_0$, *then* $\lambda_\alpha = \lambda_2$.

Theorem 5.3 ([18], Theorem 3, p. 3) *Let* $\alpha = (n_i)_{i \in N} \in \mathbb{N}^{\mathbb{N}}$, *and let* $T^{(i)} : \mathbf{R}^{n_i} \to \mathbf{R}^{n_i}, i > 1$, *be a family of linear transformations with Jacobians* $\Delta_i \ne 0$ *and* $0 < \prod_{i=1}^{\infty} |\Delta_i| < \infty$. *Let* $T^N : \mathbb{R}^{\mathbb{N}} \to \mathbb{R}^{\mathbb{N}}$ *be a map defined by*

$$T^{\mathbb{N}}(x) = (T^{(1)}(x_1, \cdots, x_{n_1}), T^{(2)}(x_{n_1+1}, \cdots, x_{n_1+n_2}), \cdots), \tag{5.52}$$

where $x = (x_i)_{i \in N} \in \mathbb{R}^{\mathbb{N}}$. *Then for each* $E \in \mathcal{B}(\mathbb{R}^{\mathbb{N}})$, *we have*

$$\lambda_\alpha(T^N(E)) = (\prod_{i=1}^{\infty} |\Delta_i|) \times \lambda_\alpha(E). \tag{5.53}$$

Let consider some corollaries of Theorem 5.3.

Corollary 5.1 *Let* A *be an infinite dimensional cellular matrix such that its* k-*th cell* A_k *is an* n_k-*dimensional quadratic matrix for* $k \in \mathbb{N}$. *Assume that the series* $\sum_{k \in \mathbb{N}} Tr(A_k)$ *is convergent. Let us consider a linear continuously differentiable vector field of velocities in* $\mathbf{R}^{\mathbb{N}}$ *defined by*

$$(\frac{dx_k}{dt})_{k \in \mathbb{N}} = A \times (x_k)_{k \in \mathbb{N}}. \tag{5.54}$$

Then for a motion Φ_t *defined by* (5.55) *and for every* $D_0 \in \mathcal{B}(\mathbf{R}^{\mathbb{N}})$, *we have*

$$\lambda_\alpha(D_t) = e^{t \sum_{k \in \mathbb{N}} Tr(A_k)} \lambda_\alpha(D_0), \tag{5.55}$$

where $D_t = \Phi_t(D_0)$ *for* $t \in \mathbf{R}$.

Proof. Note that a motion Φ_t defined by (5.55) has the following form

$$\Phi_t((x_i)_{i\in\mathbb{N}}) = (T^{(1)}(x_1,\cdots,x_{n_1}), \dot{T}^{(2)}(x_{n_1+1},\cdots,x_{n_1+n_2}),\cdots), \tag{5.56}$$

where

$$T^{(k)}((x_1,\cdots,x_{n_k})) = e^{tA_k} \times (x_1,\cdots,x_{n_k}) \tag{5.57}$$

for every $k \in \mathbb{N}$ and $(x_1,\cdots,x_{n_k}) \in \mathbf{R}^{n_k}$.

Note that for $k \in \mathbb{N}$, the Jacobian $|\Delta_k|$ of the mapping $T^{(k)}$ is equal to $e^{tTr(A_k)}$. Let us show that $0 < \prod_{i=1}^{\infty}|\Delta_i| < \infty$. Indeed, we have

$$\prod_{k=1}^{\infty}|\Delta_k| = e^{t\sum_{k=1}^{\infty} Tr(A_k)}. \tag{5.58}$$

Since the series $\sum_{k=1}^{\infty} Tr(A_k)$ converges, we deduce that $0 < \prod_{i=1}^{\infty}|\Delta_i| < \infty$.

An application of Theorem 5.3 implies that for every $D_0 \in \mathcal{B}(\mathbf{R}^{\mathbb{N}})$ the following formula

$$\lambda_\alpha(D_t) = e^{t\sum_{k\in\mathbb{N}} Tr(A_k)}\lambda_\alpha(D_0) \tag{5.59}$$

is valid. $\square$

Corollary 5.2 *Let* $\alpha = (n_i)_{i\in\mathbb{N}} \in \mathbb{N}^{\mathbb{N}}$, *and let* $T^{(i)} : \mathbf{R}^{n_i} \to \mathbf{R}^{n_i}, i > 1$, *be a family of linear continuously differentiable transformations. Let* $T^N : \mathbf{R}^{\mathbb{N}} \to \mathbf{R}^{\mathbb{N}}$ *be a mapping defined by*

$$T^{\mathbb{N}}(x) = (T^{(1)}((x_1,\cdots,x_{n_1})), T^{(2)}((x_{n_1+1},\cdots,x_{n_1+n_2})),\cdots), \tag{5.60}$$

where $x = (x_i)_{i\in\mathbb{N}} \in \mathbf{R}^{\mathbb{N}}$.

Assume that the series

$$div(T^{\mathbb{N}}) = \sum_{i\in\mathbb{N}} \frac{\partial T_i^{\mathbb{N}}}{\partial x_i}. \tag{5.61}$$

converges to any continuous function on $\mathbf{R}$, *where* $(T_i^{\mathbb{N}})_{i\in\mathbb{N}}$ *denotes components of the mapping* $T^{\mathbb{N}}$.

Let consider a linear continuously differentiable vector field of velocities in $\mathbf{R}^{\mathbb{N}}$ *defined by*

$$\frac{d\Psi}{dt} = T^{\mathbb{N}}(\Psi) \tag{5.62}$$

where $\Psi = (x_k)_{k\in\mathbb{N}} \in \mathbf{R}^{\mathbb{N}}$.

If Φ_t *is a motion defined by* (5.63), *then for every* $D_0 \in \mathcal{B}(\mathbf{R}^{\mathbb{N}})$ *the following equality*

$$\lambda_\alpha(D_t) = e^{\int_0^t div(T^{\mathbb{N}})(\tau)d\tau}\lambda_\alpha(D_0) \tag{5.63}$$

holds, where $D_t = \Phi_t(D_0)(t \in \mathbf{R})$.

Remark 5.1 The formula (5.64) can be rewrite in the equivalent form as follows

$$\frac{d\lambda_\alpha(D_t)}{dt} = \int_{D_t} div(T^{\mathbb{N}})d\lambda_\alpha. \tag{5.64}$$

References

[1] Arnold W.I., *Ordinary differential equations*, Nauka, Moscow (1975) (in Russian).

[2] Baker, R., "Lebesgue measure" on R^∞, *Proc. Amer. Math. Soc.*, **113(4)** (1991), 1023–1029.

[3] Baker, R., "Lebesgue measure" on $\mathbb{R}^\infty$. II. *Proc. Amer. Math. Soc.*, **132(9)** (2004), 2577–2591

[4] Cichon J., Kharazishvili A., Weglorz B., *Subsets of the real line*, Wydawnictwo Uniwersytetu Lodzkiego, Lodz (1995).

[5] Girsanov I. V., Mityasin B. S. , Quasi-invariant measures and linear topological spaces, *Nauchn. Dokl. Vys. Skol.*, **2**(1959), 5–10 (in Russian).

[6] Halmos P.R., *Measure theory*, Princeton, Van Nostrand (1950).

[7] Kharazishvili A.B., On invariant measures in the Hilbert space, *Bull. Acad. Sci. Georgian SSR*, **114(1)** (1984), 41–48 (in Russian).

[8] Nemicki N., Stepanov N., *The qualitative theory of dynamical systems*, Moscow–Leningrad (1949) (in Russian).

[9] Pantsulaia G. R., *An Analogue of Lioville's Theorem in Infinite-Dimensional Separable Hilbert space ℓ_2, Collection of Abstracts, Section 1, Conference "Kolmogorov-100"*, MSU, Moskow, 15-21 June (2003), 79–80.

[10] Pantsulaia G. R., Vector fields of velocities in the infinite-dimensional topological vector space $\mathbf{R}^N$ which preserve a measure μ, *Trans. of GTU*, **1(434)** (2001), 9–16.

[11] Pantsulaia G. R., *Calculation of the Jacobian for measurable linear operators in the Vector Space of all Real-Valued Sequences*. Thesises of Reports of a Conference Dedicated to the 80 Anniversary of GTU, Tbilisi (2002), 211.

[12] Pantsulaia G. R., Relations between shy sets and sets of ν_p-measure zero in Solovay's Model, *Bull. Polish Acad. Sci.*, **52(2)** (2004), 63–69.

[13] Pantsulaia G. R., On an invariant Borel measure in Hilbert space, *Bull. Polish Acad. Sci.* , **52(2)** (2004), 47–51.

[14] Pantsulaia G. R., *Invariant and Quasiinvariant Measures in Infinite-Dimensional Topological Vector Spaces*, Nova Science Publishers, Inc. , New York (2007) , xii+231 pp.

[15] Pantsulaia G. R., On generators of shy sets on Polish topological vector spaces, *New York Journal of Mathematics*, **14** (2008) , 235-261 .

[16] Pantsulaia G. R., Giorgadze G.P., On analogy of Liouville theorem in infinite-dimensional separable Hilbert space, *Georgian International Journal of Science and Technology*, **1 (2)**, (2008), 167–179.

[17] Pantsulaia G.R., On a generalized Fourier μ-series in some infinite-dimensional Polish topological vector spaces, *Georgian International Journal of Science and Technology*, **1(4)**, (2008).

[18] Pantsulaia G.R., Chainge of variable formula for "Lebesgue measures" on $\mathbf{R}^\infty$, *Journal of Mathematical Sciences: Advances and Applications, Scientific Advances Publishers*, **2 (1)** (2009), 1-12.

[19] Pantsulaia G.R., Giorgadze G.P., On Lioville type theorems for Mankiewicz and Preiss-Tier generators in $\mathbf{R}^{\mathbb{N}}$, *Georgian International Journal of Science and Technology, Nova Science Publishers* (submitted).

[20] Rogers, C.A *Hausdorff Measures*, Cambridge Univ. Press (1970).

[21] Solovay R.M., A model of set theory in which every set of reals is Lebesgue measurable,*Ann. Math.*,**92**, (1970), 1–56.

[22] Sudakov V.N., *Dokl. Akad. Nauk SSSR*; **127**(1959), 524-525 (in Russian).

[23] Veršik A. M., Duality in the theory of measure in linear spaces (in Russian), *Dokl. Akad. Nauk SSSR* **170** (1966), 497–500.

[24] Veršik A. M., Sudakov V. N., Probability measures in infinite-dimensional spaces, *Zap. Naučn. Sem. Leningrad. Otdel. Mat. Inst. Steklov. (LOMI)*, **12** (1969), 7–67 (in Russian).

In: Focus on Science and Technology...
Editor: Sergo Gotsiridze, pp. 89-102

ISBN 978-1-61209-970-5
© 2011 Nova Science Publishers, Inc.

Chapter 8

LIOUVILLE-TYPE THEOREMS FOR MANKIEWICZ AND PREISS-TIŠER GENERATORS IN $\mathbb{R}^N$

Gogi Pantsulaia[*] *and Givi Giorgadze* [†]
Department of Mathematics, Georgian Technical University,
77 Kostava Str., 0175 Tbilisi 75, Georgia

Abstract

We describe the class of all infinite-dimensional diagonal matrices $\mathbb{A}$ and the class of all infinite collections of continuous impulses $\mathbb{F}$ such that Mankiewicz generator $\mathbb{G}_M$ preserves a phase flow (in $\mathbb{R}^N$) defined by a non-homogeneous differential equation $\frac{d\Psi}{dt} = A \times \Psi + f$ for $A \in \mathbb{A}$, $f \in \mathbb{F}$. An analogous question is studied for Preiss-Tišer generators in $\mathbb{R}^N$.

2000 Mathematics Subject Classification: Primary 28Axx, 28Cxx, 28Dxx; Secondary 28C20, 28D10, 28D99

Key words and phrases: Mankiewicz generator, Preiss-Tišer generators, infinite-dimensional non-homogeneous differential equation, Liouville theorem, phase flow

The first author's research for this paper was supported by the Georgia National Science Foundation (Grants: # GNSF /ST 07/3-178, # GNSF /ST 08/3-391).

1. Introduction

The present manuscript extends an observation [12] devoted to study Liouville-type theorems for generators of shy sets introduced in [14].

Let $\mathbb{R}^N$ be an infinite-dimensional topological vector space of real-valued sequences equipped with Tychonoff topology. We denote by $\mathcal{B}(\mathbb{R}^N)$ the σ-algebra of Borel subsets in $\mathbb{R}^N$. Let G be a Borel measure defined on $\mathbb{R}^N$. We say that the measure G is *generator* of shy sets (or shortly, generator) in $\mathbb{R}^N$ if a condition $G(X) = 0$ follows that X is shy set(cf. [4, Definition 2, p. 222]). One can easily demonstrate that every translation-invariant and

---

[*]E-mail address: gogi_pantsulaia@hotmail.com
[†]E-mail address: g_givi@hotmail.com

quasi-finite[1] Borel measure is generator. Also, such measures are not σ-finite, but they possess many interesting properties (see, for example, [2], [11], [12], [13], [14]).

In the present paper we focus on a non-homogeneous differential equation defined by

$$\frac{d\Psi}{dt} = A \times \Psi + f, \qquad (1.1)$$

where $\Psi \in \mathbb{R}^{\mathbb{N}}$, A is an infinite-dimensional diagonal matrix with convergent trace and f is a continuous vector-function in $\mathbb{R}^{\mathbb{N}}$ of a parameter $t \in \mathbb{R}$. It is obvious to see that a phase flow defined by (1.1) has the following form $(\mathbb{R}^{\mathbb{N}}, (\Phi^t)_{t \in \mathbb{R}})$ where $(\Phi^t)_{t \in \mathbb{R}}$ is defined by

$$\Phi^t(\Psi) = e^{tA} \times \Psi + \int_0^t e^{(t-\tau)A} f(\tau) d\tau \qquad (1.2)$$

for $t \in \mathbb{R}$ and $\Psi \in \mathbb{R}^{\mathbb{N}}$.

Let G be a generator of shy sets in $\mathbb{R}^{\mathbb{N}}$. The generator G is said to be preserving a phase flow $(\mathbb{R}^{\mathbb{N}}, (\Phi^t)_{t \in \mathbb{R}})$ if $G(\Phi^t(X)) = G(X)$ for all $X \in \mathcal{B}(\mathbb{R}^{\mathbb{N}})$ and $t \in \mathbb{R}$.

Like [6, 13.6 - 2, p. 407], a Green's infinite-dimensional matrix $h_+ = e^{(t-\tau)A}$ is called a reaction on the collection of infinite asymmetrical impulses $f(\tau) = \delta_+(\tau)$ ($\tau \in \mathbb{R}$).

Specialized in the theory of generators of shy sets in $\mathbb{R}^{\mathbb{N}}$, here naturally arises the following

Problem 1. 1. *Let G be any generator in $\mathbb{R}^{\mathbb{N}}$. Describe a class of all infinite-dimensional diagonal matrices $\mathbb{A}$ and a class $\mathbb{F}$ of all collections of infinite continuous impulses under which the generator G preserves the phase flow defined by (1.2).*

It is reasonable to note that Problem 1. 1 has been investigated for R. Baker's generator G_B in [12]. More precisely, it was established that this generator preserves the phase flow defined by (1.2) iff $\mathbb{A}$ is a class of all infinite-dimensional matrices with zero traces. This result asserts that, for an arbitrary infinite-dimensional matrix A with $Tr(A) = 0$, under an arbitrary collection of infinite continuous asymmetrical impulses the Green's reaction always balances the flow in this way that R.Baker's distribution function remains constant along the trajectory of the observing system. By this reasoning this result can be considered as a partial infinite-dimensional generalization of the well-known Joseph Liouville theorem being a key theorem in classical statistical and Hamiltonian mechanics (cf. [1]).

The main goal of the present paper is to study Problem 1.1 for Mankiewicz and Preiss-Tišer generators. The main results have been announced in [13]

The paper is organized as follows.

In Section 2 we give some auxiliary propositions. By using main properties of A.B.Kharazishvil measure [5], we give a solution of Problem 1.1 for Mankiewicz generator in Section 3. In Section 4 we study analogous question for Preiss-Tišer generators in $\mathbb{R}^{\mathbb{N}}$.

2.   Auxiliary Propositions

Let $(a_i)_{i \in \mathbb{N}}$ and $(b_i)_{i \in \mathbb{N}}$ be sequences of real numbers such that $a_i < b_i$ for $i \in \mathbb{N}$.

---

[1]A Borel measure G defined on $\mathbb{R}^{\mathbb{N}}$ is called quasi-finite if $0 < G(Y) < \infty$ for any set $Y \in \mathcal{B}(\mathbb{R}^{\mathbb{N}})$.

We put

$$A_n = \mathbb{R}_0 \times \cdots \times \mathbb{R}_n \times \prod_{i>n} \Delta_i \tag{2.3}$$

for $n \in \mathbb{N}$, where

$$(\forall i)(i \in \mathbb{N} \to \mathbb{R}_i = \mathbb{R} \;\&\; \Delta_i = [a_i; b_i[). \tag{2.4}$$

We put also

$$\Delta = \prod_{i \in \mathbb{N}} \Delta_i. \tag{2.5}$$

For $i \in \mathbb{N}$, let μ_i be the Lebesgue measure defined on $\mathbb{R}_i$ and satisfying the condition $\mu_i(\Delta_i) = 1$. Let us denote by l_i a probability Lebesgue measure on Δ_i.

For $n \in \mathbb{N}$, we put

$$\nu_n = \prod_{i=1}^{n} \mu_i \times \prod_{i>n} l_i \tag{2.6}$$

and

$$(\forall X)(X \in \mathcal{B}(\mathbb{R}^{\mathbb{N}}) \to \bar{\nu}_n(X) = \nu(X \cap A_n)). \tag{2.7}$$

The following assertion is valid.

Lemma 2.1 [11, Lemma 5. 1, p. 93]. *For an arbitrary Borel set $X \subseteq \mathbb{R}^{\mathbb{N}}$ there exists a limit*

$$\nu_\Delta(X) = \lim_{n \to \infty} \bar{\nu}_n(X). \tag{2.8}$$

Moreover, the functional ν_Δ is a non-trivial σ-finite measure defined on the Borel σ-algebra $\mathcal{B}(\mathbb{R}^{\mathbb{N}})$.

We recall the reader that an element $h \in \mathbb{R}^{\mathbb{N}}$ is called an admissible translation in the sense of invariance for the measure ν_Δ if

$$(\forall X)(X \in \mathcal{B}(\mathbb{R}^{\mathbb{N}}) \to \nu_\Delta(X + h) = \nu_\Delta(X)). \tag{2.9}$$

We define

$$I_\Delta = \{h : h \in \mathbb{R}^{\mathbb{N}} \;\&\; h \text{ is an admissible translation for } \nu_\Delta(X)\}. \tag{2.10}$$

Our next proposition gives an information regarding the algebraic structure of the vector subspace I_Δ.

Lemma 2.2 [11, Theorem 5. 1, p. 96]. *The following conditions are equivalent*

$$1)\; g = (g_1, g_2, \cdots) \in I_\Delta; \tag{2.11}$$

$$2)\; \text{the series } \sum_{i \in N} \frac{|g_i|}{b_i - a_i} \text{ is convergent.} \tag{2.12}$$

For $\mu = \nu_{[0,1]^{\mathbb{N}}}$, we have the following assertion.

Lemma 2.3 [11, Theorem 3. 4, p. 56]. *Let consider a differential equation*

$$\frac{d\Psi}{dt} = A \times \Psi, \tag{2.13}$$

where $\Psi \in \mathbb{R}^{\mathbb{N}}$, $t \in \mathbb{R}$ and A is an infinite-dimensional diagonal matrix.

Then the phase flow defined by the vector field of velocities (2.13)(*being one-parameter group of transformations of $\mathbb{R}^{\mathbb{N}}$) preserves the measure μ iff the series $Tr(A)$ is absolutely convergent and $Tr(A) = 0$.*

In the sequel we will apply the following notation.

Let A be an infinite-dimensional diagonal matrix. We put

$$(\forall X)(X \subseteq \mathbb{R}^{\mathbb{N}} \rightarrow A \otimes X \equiv \{A \times x : x \in X\}). \tag{2.14}$$

Remark 2.1. It reasonable to note that if $Tr(A)$ is not absolutely convergent then $\mu(e^{tA} \otimes [0,1[^N) = 0$ for $t \neq 0$ (see [11, p.57]). Using same argument, one can easily show that $\mu(e^{tA} \otimes [0,1[^N+g)$ is equal to zero for all $g \in \mathbb{R}^{\mathbb{N}}$.

Remark 2.2. If A is an infinite-dimensional diagonal matrix and $Tr(A)$ is absolutely convergent, then

$$(\forall X)(X \in \mathcal{B}(\mathbb{R}^{\mathbb{N}}) \rightarrow \mu(e^{tA} \otimes X) = e^{tTr(A)} \times \mu(X)). \tag{2.15}$$

The formula (2.15) admits the following equivalent formulation.

Lemma 2.4 ([11, Theorem 3. 28, p. 59]). *Let $v = A \times \psi$ be a vector field of velocities defined in $\mathbb{R}^{\mathbb{N}}$, where A is an infinite-dimensional diagonal matrix such that $Tr(A)$ is absolutely convergent. Let $D(0)$ be a Borel subset in $\mathbb{R}^{\mathbb{N}}$. Then the formula*

$$\frac{\mu(D(t))}{dt} = \int_{D(t)} div(A) d\mu \tag{2.16}$$

is valid, where $D(t)$ denotes the state of the set $D(0)$ at the moment t under the action of the phase flow $(\mathbb{R}^{\mathbb{N}}, (e^{sA})_{s \in \mathbb{R}})$.

Lemma 2.5 ([11, Theorem 3. 25, p. 58]). *Let us consider a differential equation*

$$\frac{d\Psi}{dt} = A \times \Psi + f, \tag{2.17}$$

where $\Psi \in \mathbb{R}^{\mathbb{N}}$, A is an infinite-dimensional diagonal matrix with absolutely convergent trace and f is a continuous vector-function in $\mathbb{R}^{\mathbb{N}}$ of a parameter $t \in \mathbb{R}$. Then a phase flow defined by (2.17) *preserves the measure μ iff the following condition*

$$(\forall t)(t > 0 \rightarrow \int_0^t e^{(t-\tau)A} f(\tau) d\tau \in \ell_1) \tag{2.18}$$

holds, where ℓ_1 denotes a vector space of all absolutely summable real-valued sequences.

Lemma 2.6. *Let A be an infinite-dimensional diagonal matrix with absolutely convergent trace $Tr(A)$. Then a mapping of ℓ_1 to itself defined by $e^A((x_k)_{k \in \mathbb{N}}) = e^A \times (x_k)_{k \in \mathbb{N}}$ for $(x_k)_{k \in \mathbb{N}} \in \ell_1$, is a linear one-to-one operator.*

Proof. Let us show that

$$(\forall (x_k)_{k\in\mathbb{N}})((x_k)_{k\in\mathbb{N}} \in \ell_1 \to e^A((x_k)_{k\in\mathbb{N}}) \in \ell_1). \tag{2.19}$$

Let $(\lambda_k)_{k\in\mathbb{N}}$ be a sequence of all diagonal elements of the matrix A. Since $Tr(A)$ is absolutely convergent, we deduce that there exists a positive number C such that $|\lambda_k| < C$ for $k \in \mathbb{N}$. Hence, we get

$$(\forall k)(k \in \mathbb{N} \to e^{-C} \le e^{\lambda_k} \le e^C). \tag{2.20}$$

Obviously,

$$(\forall (x_k)_{k\in\mathbb{N}})((x_k)_{k\in\mathbb{N}} \in \ell_1 \to e^{-C}\sum_{k\in\mathbb{N}}|x_k| \le \sum_{k\in\mathbb{N}}|e^{\lambda_k}x_k| \le e^C\sum_{k\in\mathbb{N}}|x_k|). \tag{2.21}$$

The latter relation implies that $e^A \times (x_k)_{k\in\mathbb{N}} \in \ell_1$.

We have show that the mapping $e^A : \ell_1 \to \ell_1$ is injective. Let $(x_k)_{k\in\mathbb{N}}$ and $(y_k)_{k\in\mathbb{N}}$ be different elements of ℓ_1. Then there exists a natural number $k_0 \in \mathbb{N}$ such that $x_{k_0} \neq y_{k_0}$. Correspondingly, $e^{\lambda_{k_0}}x_{k_0} \neq e^{\lambda_{k_0}}y_{k_0}$ which follows that $e^A \times (x_k)_{k\in\mathbb{N}} \neq e^A \times (y_k)_{k\in\mathbb{N}}$.

Finally, let us show that $e^A : \ell_1 \to \ell_1$ is surjective. Indeed, let $(x_k)_{k\in\mathbb{N}} \in \ell_1$. On the one hand, we have $(e^{-\lambda_k}x_k)_{k\in\mathbb{N}} \in \ell_1$, because $e^{-C} \le e^{-\lambda_k} \le e^C$ for $k \in \mathbb{N}$. On the other hand, we have

$$e^A \times (e^{-\lambda_k}x_k)_{k\in\mathbb{N}} = (x_k)_{k\in\mathbb{N}}. \tag{2.22}$$

It is easy to check that the following equality

$$e^A \times ((x_k)_{k\in\mathbb{N}} + (y_k)_{k\in\mathbb{N}}) = e^A \times (x_k)_{k\in\mathbb{N}} + e^A \times (y_k)_{k\in\mathbb{N}} \tag{2.23}$$

holds for all $(x_k)_{k\in\mathbb{N}}, (y_k)_{k\in\mathbb{N}} \in \ell_1$.

□

Let V be a Polish vector space. Let L be a proper vector subspace of V. By Axiom of Choice, we can construct a proper vector subspace $F_0 \subset V$ such that

$$L \cap F_0 = \{\mathbf{0}\} \ \& \ L + F_0 = V,$$

where $\{\mathbf{0}\}$ denotes the zero of V.

Let $\mathcal{F}$ be a class of vector subspaces F in V such that the following relation

$$(\forall F)(F \in \mathcal{F} \to L \cap F = \{\mathbf{0}\} \ \& \ L + F = V)$$

holds.

We set $L^\perp = \tau(\mathcal{F})$, where τ denotes a global operator of choice. A vector subspace $L^\perp$ is said to be a linear complement of the vector space L in V.

The following auxiliary proposition is very important for our further investigations.

Lemma 2.7. *Let A be an infinite-dimensional diagonal matrix with absolutely convergent trace. Let $\ell_1^\perp$ be a linear complement of ℓ_1 in $\mathbb{R}^{\mathbb{N}}$. We set*

$$(\forall h)(\forall g)(h \in \ell_1 \ \& \ g \in \ell_1^\perp \to Pr_{\ell_1^\perp}^{\ell_1}(h+g) = g). \tag{2.24}$$

Then $Pr_{\ell_1^\perp}^{\ell_1} : e^A \otimes \ell_1^\perp \to \ell_1^\perp$ is a linear one-to-one operator.

Proof. A validity of the fact that $e^A \otimes \ell_1^{\perp}$ is a linear vector subspace in $\mathbb{R}^{\mathbb{N}}$ can be proved easily as well the proof of the linearity of the operator $Pr_{\ell_1^{\perp}}^{\ell_1}$.

Let us show that the operator $Pr_{\ell_1^{\perp}}^{\ell_1}$ is injective. Indeed, let g_1, g_2 be two different elements in $e^A \otimes \ell_1^{\perp}$. It means that there exist $f_1, f_2 \in \ell_1^{\perp}$ such that $e^A \times f_1 = g_1$ and $e^A \times f_2 = g_2$. It is clear that $f_1 - f_2 \notin \ell_1$. Indeed, assume the contrary and let $f_1 - f_2 \in \ell_1$. Since $\ell_1^{\perp} \cap \ell_1 = \{0\}^{\mathbb{N}}$, $e^A \times (f_1 - f_2) \in \ell_1$ (cf. Lemma 2.6) and $e^A \times (f_1 - f_2) = g_1 - g_2 \in \ell_1^{\perp}$, we claim that $e^A \times (f_1 - f_2) = \{0\}^{\mathbb{N}}$. It follows that $g_1 = g_2$ and we get the contradiction. Further, we have unique representations $g_1 = x_1 + y_1$ and $g_2 = x_2 + y_2$ such that $x_1, x_2 \in \ell_1$ and $y_1, y_2 \in \ell_1^{\perp}$. Assume the contrary and let $y_1 = y_2$. Then we get $g_1 - x_1 = g_2 - x_2$, equivalently, $g_1 - g_2 = x_1 - x_2$. This means that $g_1 - g_2 \in \ell_1$ and, by using the result of Lemma 2.6, we get that $e^{-A} \times (g_1 - g_2) \in \ell_1$, equivalently, $f_1 - f_2 \in \ell_1$. The latter relation is the contradiction and the injectivity of the operator $Pr_{\ell_1^{\perp}}^{\ell_1} : e^A \otimes \ell_1^{\perp} \to \ell_1^{\perp}$ is proved.

Now let us show that the operator $Pr_{\ell_1^{\perp}}^{\ell_1} : e^A \otimes \ell_1^{\perp} \to \ell_1^{\perp}$ is surjective. In this direction we must note that this operator is a linear one-to-one operator on $\mathbb{R}^{\mathbb{N}}$ such that the linear vector subspace $e^A \otimes \ell_1^{\perp}$ is again a linear complement of ℓ_1. Let us assume that $f \in \ell_1^{\perp}$. Then it admits a unique representation in the form $h_1 + h_2$, where $h_1 \in e^A \otimes \ell_1^{\perp}$ and $h_2 \in \ell_1$. It is obvious that $Pr_{\ell_1^{\perp}}^{\ell_1}(h_1) = Pr_{\ell_1^{\perp}}^{\ell_1}(-h_2 + f) = f$. $\square$

Lemma 2.8. *Let A be any infinite-dimensional diagonal matrix with absolutely convergent trace $Tr(A)$ and $Tr(A) = 0$. Then*

$$\mu(e^{tA} \otimes B_{[0,1[^{\mathbb{N}}} \triangle B_{[0,1[^{\mathbb{N}}}) = 0, \tag{2.25}$$

where $\triangle$, as usually, denotes the symmetrical difference of sets.

Proof. Assume the contrary and let

$$\mu(e^{tA} \otimes B_{[0,1[^{\mathbb{N}}} \triangle B_{[0,1[^{\mathbb{N}}}) > 0. \tag{2.26}$$

Without loss of generality, we can assume that

$$\mu(B_{[0,1[^{\mathbb{N}}} \setminus e^{tA} \otimes B_{[0,1[^{\mathbb{N}}}) > 0. \tag{2.27}$$

Taking into account the result of Lemma 2.3, we get

$$0 < \mu(B_{[0,1[^{\mathbb{N}}} \setminus e^{tA} \otimes B_{[0,1[^{\mathbb{N}}}) = \mu(e^{-tA} \otimes (B_{[0,1[^{\mathbb{N}}} \setminus e^{tA} \otimes B_{[0,1[^{\mathbb{N}}})) = \tag{2.28}$$

$$\mu(e^{-tA} \otimes B_{[0,1[^{\mathbb{N}}} \setminus B_{[0,1[^{\mathbb{N}}}), \tag{2.29}$$

which is the required contradiction, because the measure μ is concentrated on the set $B_{[0,1[^{\mathbb{N}}}$. $\square$

3. Solution of Problem 1.1 for Mankiewicz Generator in $\mathbb{R}^{\mathbb{N}}$

Let λ be the product of the family of probability Lebesgue measures $(l_k)_{k\in\mathbb{N}}$. We say that a Borel subset $X \subset \mathbb{R}^{\mathbb{N}}$ is a standard cube null set in $\mathbb{R}^{\mathbb{N}}$ if

$$(\forall a)(a \in \mathbb{R}^{\mathbb{N}} \to \lambda(X + a \cap [0,1]^{\mathbb{N}}) = 0). \tag{3.30}$$

We denote a class of all standard cube null sets by $S.C.\mathcal{N}.S.(\mathbb{R}^{\mathbb{N}})$.

Let ν be any Borel measure in $\mathbb{R}^{\mathbb{N}}$. We denote the class of all ν-null sets by $\mathcal{N}.S.(\nu)$.

Definition 3.1. We say that ν is a Mankiewicz generator on $\mathbb{R}^{\mathbb{N}}$ if it is a quasi-finite translation-invariant measure such that

$$S.C.\mathcal{N}.S.(\mathbb{R}^{\mathbb{N}}) = \mathcal{N}.S.(\nu). \tag{3.31}$$

We put

$$(\forall X)(X \in \mathcal{B}(\mathcal{R}^{\mathcal{N}}) \to G_M(X) = \sum_{g \in \ell_1^\perp} \mu((X - g) \cap B_{[0,1[^{\mathbb{N}}}). \tag{3.32}$$

Remark 3.1. Following [11, Theorem 15.3.3, p.221], the measure G_M is Mankiewicz generator. Starting from this point, we can say that the class of all standard cube null sets is not implicitly introduced by Mankiewicz(cf. [3]), because it coincides with the class of sets of G_M-measure zero in $\mathbb{R}^{\mathbb{N}}$.

Lemma 3.1. *Let A be any infinite-dimensional diagonal matrix with absolutely convergent trace $Tr(A)$. Then the following formula is valid*

$$(\forall X)(\forall t)(X \in \mathcal{B}(\mathbb{R}^{\mathbb{N}}) \,\&\, t \in \mathbb{R} \to G_M(e^{tA} \otimes X) = e^{tTr(A)} \times G_M(X). \tag{3.33}$$

Proof. Taking into account Remark 2.3 and results of Lemmas 2.2 and 2.7, we get

$$(\forall X)(\forall t)(X \in \mathcal{B}(\mathcal{R}^{\mathcal{N}}) \,\&\, t \in \mathbb{R} \to G_M(e^{tA} \otimes X) =$$

$$\sum_{g \in \ell_1^\perp} \mu((e^{tA} \otimes X - g) \cap B_{[0,1[^{\mathbb{N}}}) =$$

$$\sum_{g \in \ell_1^\perp} e^{tTr(A)} \mu((X - e^{-tA} \times g) \cap e^{-tA} \otimes B_{[0,1[^{\mathbb{N}}}) =$$

$$e^{tTr(A)} \sum_{g \in \ell_1^\perp} \mu((X - Pr_{\ell_1^\perp}^{\ell_1^\perp}(e^{-tA} \times g) - Pr_{\ell_1^\perp}^{\ell_1}(e^{-tA} \times g)) \cap B_{[0,1[^{\mathbb{N}}}) =$$

$$e^{tTr(A)} \sum_{g \in \ell_1^\perp} \mu((X - Pr_{\ell_1^\perp}^{\ell_1}(e^{-tA} \times g)) \cap (B_{[0,1[^{\mathbb{N}}} + Pr_{\ell_1^\perp}^{\ell_1^\perp}(e^{-tA} \times g))) =$$

$$e^{tTr(A)} \sum_{g \in \ell_1^\perp} \mu((X - Pr_{\ell_1^\perp}^{\ell_1}(e^{-tA} \times g)) \cap B_{[0,1[^{\mathbb{N}}}) =$$

$$e^{tTr(A)} \sum_{h \in \ell_1^\perp} \mu((X-h) \cap B_{[0,1[^\mathbb{N}}) = e^{tTr(A)} \times G_M(X)). \qquad (3.34)$$

$\square$

The main result of the present section is formulated as follows.

Theorem 3.1. *Let A be any infinite-dimensional diagonal matrix with convergent trace. Then the phase flow defined by the vector field of velocities* (2.13) *preserves Mankiewicz generator G_M iff $Tr(A)$ is absolutely convergent and $Tr(A) = 0$.*

Proof. A sufficiency follows from Lemma 3.1. Now let us prove a necessity. If $Tr(A)$ is not absolutely convergent, then, taking into account Remark 2.2, we get

$$(\forall t)(t \in \mathbb{R} \to \mu(e^{\pm tA} \otimes [0,1[^\mathbb{N}) = 0). \qquad (3.35)$$

The latter relation implies that

$$(\forall t)(t \in \mathbb{R} \to \mu(e^{\pm tA} \otimes B_{[0,1[^\mathbb{N}}) = 0). \qquad (3.36)$$

Hence, for $t \in \mathbb{R}$ we have

$$1 = G_M(e^{tA} \otimes [0,1[^\mathbb{N}) = \sum_{g \in \ell_1^\perp} \mu((e^{tA} \otimes [0,1[^\mathbb{N}-g) \cap B_{[0,1[^\mathbb{N}}) = 0. \qquad (3.37)$$

This is the required contradiction and thus, $Tr(A)$ must be absolutely convergent. Further, using again the result of Lemma 3.1, we conclude that $Tr(A) = 0$. $\square$

We have the following consequence of Theorem 3.1.

Corollary 3.1. *Let us consider the following non-homogeneous differential equation*

$$\frac{d\Psi}{dt} = A\Psi + f, \qquad (3.38)$$

where $\Psi \in \mathbb{R}^\mathbb{N}$, f is a continuous vector-valued function of a parameter $t \in \mathbb{R}$, A is an infinite-dimensional diagonal matrix with convergent trace. Then Mankiewicz generator G_M preserves the motion defined by (3.38) *iff $Tr(A)$ is absolutely convergent and $Tr(A) = 0$.*

Proof. **Necessity.** Assume that the phase flow defined by (3.38) preserves Mankiewicz generator G_M. Then, for $X \in \mathcal{B}(\mathbb{R}^\mathbb{N})$, we have

$$G_M(e^{tA} \otimes X + \int_0^t e^{(t-\tau)A} f(\tau)d\tau) = G_M(X). \qquad (3.39)$$

Applying the property of translation-invariance of the generator G_M(cf. Definition 3.1), we get

$$G_M(e^{tA} \otimes X) = G_M(X) \qquad (3.40)$$

for $X \in \mathcal{B}(\mathbb{R}^\mathbb{N})$, which implies that $Tr(A)$ is absolutely convergent and $Tr(A) = 0$.

Sufficiency. The proof of sufficiency is obtained by using Theorem 3.1 and Remark 3.1.   □

Let us introduce some notions characterizing the behavior of such dynamical systems which are described by the vector field of velocities $v = A \times \Psi$ in $\mathbb{R}^{\mathbb{N}}$, where A denotes again an infinite-dimensional real-valued diagonal matrix.

Definition 3.1. Let v be a Borel measure on $\mathbb{R}^{\mathbb{N}}$. We say that (in view of the measure v) the phase flow $(\mathbb{R}^{\mathbb{N}}, (e^{tA})_{t \in \mathbb{R}})$ is:

(i) stable if it preserves the measure v;

(ii) pressing if

$$(\forall t_1)(\forall t_2)(\forall D)(0 < t_1 < t_2 < \infty \ \& \ 0 < v(D) < \infty \rightarrow$$
$$v(e^{t_1 A}(D)) < v(e^{t_2 A}(D))); \tag{3.41}$$

(iii) totally pressing if

$$(\forall t)(\forall D)(0 < t < \infty \ \& \ 0 < v(D) < \infty \rightarrow v(e^{tA}(D)) = 0); \tag{3.42}$$

(iv) expansible if

$$(\forall t_1)(\forall t_2)(\forall D)(0 < t_1 < t_2 < \infty \ \& \ 0 < v(D) < \infty \rightarrow$$
$$v(e^{t_1 A}(D)) < v(e^{t_2 A}(D))); \tag{3.43}$$

(v) totally expansible if

$$(\forall t)(\forall D)(0 < t < \infty \ \& \ 0 < v(D) < \infty \rightarrow v(e^{tA}(D)) = \infty). \tag{3.44}$$

The following example demonstrates whether Mankiewicz generator can be applied for a description of the behavior of the infinite continuous Malthusian growths model in $\mathbb{R}^{\mathbb{N}}$.

Example 3.1 (Infinite continuous Malthusian growths model). The differential equation describing the continuous Malthusian growth model says that the derivative of an unknown Population function $P(t)$ is proportional to the unknown Population function. The only function that is equal to the derivative of itself is the exponential function. We try to find a solution of the form $P(t) = ce^{rt}$, where c is an arbitrary constant. But $P'(t)$ is $c \times r \times e^{r \times t}$, which is $r \times P(t)$, so it satisfies the differential equation. If we add an initial condition $P(0) = P_0$, then the unique solution becomes $P(t) = P_0 \times e^{r \times t}$. This is another reason why Malthusian growth is often called exponential growth.

Now, let us consider infinite non-antagonistic family of populations and let $\Psi_k(t)$ be Population function of the k-th Population. Then continuous Malthusian growth model for infinite family of non-antagonistic populations is described by the following linear differential equation

$$\frac{d\Psi(t)}{dt} = A \times \Psi(t) \ (t \in \mathbb{R}) \tag{3.45}$$

with an initial condition $\Psi(0) = \Psi_0 \in (\mathbb{R}^+)^{\mathbb{N}}$, where A is an infinite-dimensional real-valued diagonal matrix, $\Psi(t) = (\Psi_k(t))_{k \in \mathbb{N}} \in (\mathbb{R}^+)^{\mathbb{N}}$ for $t \in \mathbb{R}$.

Let $\mathrm{div}(A)$ is absolutely convergent. Then, applying Lemma 3.1, we conclude that the phase flow $(\mathbb{R}^{\mathbb{N}}, (e^{tA})_{t \in \mathbb{R}})$ (in view of Mankiewicz generator G_M) is:

(i*) stable if $\mathrm{Tr}(A) = 0$;

(ii*) pressing if $\mathrm{Tr}(A) < 0$;

(iii*) expansible if $\mathrm{Tr}(A) > 0$;

Remark 3.3. Let $Tr(A)$ be absolutely convergent and $Tr(A) = 0$. Let A_g be an infinite-dimensional matrix with diagonal elements $(\lambda_{g(k)})_{k \in N}$, where g is any permutation of the set of all natural numbers N. Then, following Dirichlet well known result about invariance of the sum of an absolutely convergent series of real numbers under permutations, we deduce that the phase flow $(\mathbb{R}^{\mathbb{N}}, (e^{tA_g})_{t \in \mathbb{R}})$ is stable again in the sense of Mankiewicz generator G_M. If $Tr(A)$ is not an absolutely convergent series with $Tr(A) = 0$, then, taking into account Riemann's well known result about non-invariance of the sum of a conditionally convergent series of real numbers under all permutations, we can choose such a permutation g of N that the phase flow $(\mathbb{R}^{\mathbb{N}}, (e^{tA_g})_{t \in \mathbb{R}})$ will be stable, pressing, totally pressing, expansible or totally expansible in the sense of R.Baker's generator G_B, respectively. Note here that in such a case the same flow always will be totally pressing in the sense of Mankiewicz generator G_M.

4. On Problem 1.1 for Preiss-Tišer Generators in $\mathbb{R}^{\mathbb{N}}$

Like [13, Definition 2.3.5, p.11], a Borel set $S \subset \mathbb{R}^{\mathbb{N}}$ is said to be an n-dimensional Preiss-Tišer null set if every Lebesgue measure μ_i concentrated on any n-dimensional vector subspace L_i in $\mathbb{R}^{\mathbb{N}}$ is transverse to S.

We denote by $\mathcal{P}.\mathcal{T}.\mathcal{N}.\mathcal{S}.(\mathbb{R}^{\mathbb{N}}, n)$ the class of all n-dimensional Preiss-Tišer null sets in $\mathbb{R}^{\mathbb{N}}$.

Definition 4.1. A quasi-finite translation-invariant Borel measure μ in $\mathbb{R}^{\mathbb{N}}$ is called an n-dimensional Preiss-Tišer generator of shy sets (or shortly, an n-dimensional Preiss-Tišer generator), if the following equality

$$\mathcal{P}.\mathcal{T}.\mathcal{N}.\mathcal{S}.(\mathbb{R}^{\mathbb{N}}, n) = \mathcal{N}.\mathcal{S}.(\mu) \tag{4.46}$$

holds.

Now we are going to show that there exists an n-dimensional Preiss-Tišer generator in $\mathbb{R}^{\mathbb{N}}$ for all natural number $n \geq 1$.

Let $(\Gamma_i)_{i \in \mathbb{N}}$ be a family of all an n-dimensional vector subspaces in $\mathbb{R}^{\mathbb{N}}$ and μ_i be an n-dimensional Lebesgue measure concentrated on Γ_i for $i \in I$. Let $\Gamma_i^{\perp}$ be any linear compliment of Γ_i for $i \in I$.

We put

$$(\forall X)(X \in \mathcal{B}(\mathbb{R}^{\mathbb{N}}) \rightarrow G_{P\&T}^{(n)}(X) = \sum_{i \in I} \sum_{g \in \Gamma_i^{\perp}} \mu_i((X - g) \cap \Gamma_i)). \tag{4.47}$$

The following assertion is valid.

Lemma 4.1. *Let $n \in \mathbb{N} \setminus \{0\}$. Then the functional $G_{P\&T}^{(n)}$ is an n-dimensional Preiss-Tišer generator in $\mathbb{R}^{\mathbb{N}}$.*

Proof. Let us show that the functional $G_{P\&T}^{(n)}$ is a quasi-finite translation-invariant Borel measure in $\mathbb{R}^{\mathbb{N}}$.

A quasi-finiteness is obvious because

$$0 < G_{P\&T}^{(n)}([0,1]^n \times \{0\}^{\mathbb{N}\setminus\{1,\cdots,n\}}) = \mu_{i_0}([0,1]^n \times \{0\}^{\mathbb{N}\setminus\{1,\cdots,n\}}) < \infty, \qquad (4.48)$$

where i_0 is such an index that $[0,1]^n \times \{0\}^{\mathbb{N}\setminus\{1,\cdots,n\}} \subseteq \Gamma_{i_0}$.

Let us show that the functional $G_{P\&T}^{(n)}$ is translation-invariant. Indeed, let $X \in \mathcal{B}(\mathbb{R}^{\mathbb{N}})$ and $h \in \mathbb{R}^{\mathbb{N}}$. For $i \in I$, we have $h = h_i^{(0)} + h_i^{(1)}$, where $h_i^{(0)} \in \Gamma_i$ and $h_i^{(1)} \in \Gamma_i^{\perp}$.

We have

$$G_{P\&T}^{(n)}(X+h) = \sum_{i \in I} \sum_{g \in \Gamma_i^{\perp}} \mu_i((X+h-g) \cap \Gamma_i) =$$

$$\sum_{i \in I} \sum_{g \in \Gamma_i^{\perp}} \mu_i((X+h_i^{(0)}+h_i^{(1)}-g) \cap \Gamma_i) =$$

$$\sum_{i \in I} \sum_{g \in \Gamma_i^{\perp}} \mu_i((X+h_i^{(1)}-g) \cap \Gamma_i + h_i^{(0)}) =$$

$$\sum_{i \in I} \sum_{g \in \Gamma_i^{\perp}} \mu_i((X+h_i^{(1)}-g) \cap \Gamma_i) =$$

$$\sum_{i \in I} \sum_{s+h_i^{(1)} \in \Gamma_i^{\perp}} \mu_i((X-s) \cap \Gamma_i) =$$

$$\sum_{i \in I} \sum_{s \in \Gamma_i^{\perp}-h_i^{(1)}} \mu_i((X-s) \cap \Gamma_i) =$$

$$\sum_{i \in I} \sum_{s \in \Gamma_i^{\perp}} \mu_i((X-s) \cap \Gamma_i) = G_{P\&T}^{(n)}(X). \qquad (4.49)$$

Now let us show that the functional $G_{P\&T}^{(n)}$ is σ-additive. Indeed, let $(X_k)_{k \in \mathbb{N}}$ be a family of pairwise disjoint Borel subsets in $\mathbb{R}^{\mathbb{N}}$. Then we have

$$G_{P\&T}^{(n)}(\sum_{k \in \mathbb{N}} X_k) = \sum_{i \in I} \sum_{g \in \Gamma_i^{\perp}} \mu_i((\sum_{k \in \mathbb{N}} X_k - g) \cap \Gamma_i) =$$

$$\sum_{i \in I} \sum_{g \in \Gamma_i^{\perp}} \mu_i((\sum_{k \in \mathbb{N}}(X_k - g)) \cap \Gamma_i) =$$

$$\sum_{i \in I} \sum_{g \in \Gamma_i^{\perp}} \sum_{k \in \mathbb{N}} \mu_i((X_k - g) \cap \Gamma_i) =$$

$$\sum_{k \in \mathbb{N}} \sum_{i \in I} \sum_{g \in \Gamma_i^{\perp}} \mu_i((X_k - g) \cap \Gamma_i) = \sum_{k \in \mathbb{N}} G_{P\&T}^{(n)}(X_k). \qquad (4.50)$$

Let us show that the following equality

$$\mathcal{P}.\mathcal{T}.\mathcal{N}.\mathcal{S}.(\mathbb{R}^{\mathbb{N}}, n) = \mathcal{N}.\mathcal{S}.(G_{P\&T}^{(n)}) \tag{4.51}$$

holds for $n > 1$.

Suppose that $X \in \mathcal{P}.\mathcal{T}.\mathcal{N}.\mathcal{S}.(\mathbb{R}^{\mathbb{N}}, n)$. It means that every n-dimensional Lebesgue measure concentrated on any n-dimensional vector subspace is transverse to X. In particular, every Lebesgue measure μ_i concentrated on the n-dimensional vector subspace L_i is transverse to X, too. Obviously, $\mu_i((X-g) \cap L_i) = 0$ for $i \in I$ and $g \in \mathbb{R}^{\mathbb{N}}$, which follows that $\mu_i((X-g) \cap L_i) = 0$ for $i \in I$ and $g \in \Gamma_i^{\perp}$. The latter relation implies that the following equality

$$G_{P\&T}^{(n)}(X) = \sum_{i \in I} \sum_{g \in \Gamma_i^{\perp}} \mu_i((X-g) \cap \Gamma_i) = 0 \tag{4.52}$$

holds. Hence, $X \in \mathcal{N}.\mathcal{S}.(G_{P\&T}^{(n)})$.

The validity of the converse inclusion

$$\mathcal{N}.\mathcal{S}.(G_{P\&T}^{(n)}) \subseteq \mathcal{P}.\mathcal{T}.\mathcal{N}.\mathcal{S}.(\mathbb{R}^{\mathbb{N}}, n) \tag{4.53}$$

can be proved analogously.   □

Remark 4.1. One can easily demonstrate that Preiss-Tišer generators are inner regular. Now let $(\alpha_i)_{i \in I}$ be a sequence of positive real numbers. Then a functional $\mu^{(n)}$, defined by

$$(\forall X)(X \in \mathcal{B}(\mathbb{R}^{\mathbb{N}}) \to \mu^{(n)}(X) = \sum_{i \in I} \sum_{g \in \Gamma_i^{\perp}} \alpha_i \mu_i((X-g) \cap \Gamma_i)), \tag{4.54}$$

is again an n-dimensional Preiss-Tišer generator of shy sets in $\mathbb{R}^{\mathbb{N}}$. Thus, in the class of all n-dimensional Preiss-Tišer generators in $\mathbb{R}^{\mathbb{N}}$, there does not exist a generator with the property of uniqueness.

The main result of the present section is formulated as follows.

Theorem 4.1. *For $n \in \mathbb{N} \setminus \{0\}$, there does not exist a non-zero infinite-dimensional matrix A such that the phase flow defined by (2.11) preserves the n-dimensional Preiss-Tišer generator $G_{P\&T}^{(n)}$.*

Proof. Let $(\lambda_k)_{k \in \mathbb{N}}$ be a sequence of all diagonal elements of the matrix A. Note here that for an arbitrary natural number $n \in N$ there exists a finite sequence of diagonal elements $(\lambda_{n_k})_{1 \le k \le n}$ which $\sum_{k=1}^{n} \lambda_{n_k} \neq 0$. We put

$$X = [0,1]^{\{n_1, \cdots, n_n\}} \times \{0\}^{N \setminus \{n_1, \cdots, n_n\}}.$$

We have

$$G_{P\&T}^{(n)}(e^{tA} \otimes X) = \mu_{i_1}(\prod_{k=1}^{n} [0, e^{t\lambda_{n_k}}] \times \{0\}^{N \setminus \{n_1, \cdots, n_n\}}) =$$

$$e^{t \sum_{k=1}^{n} \lambda_{n_k}} \mu_{i_1}([0,1]^{\{n_1, \cdots, n_n\}} \times \{0\}^{N \setminus \{n_1, \cdots, n_n\}}) = e^{t \sum_{k=1}^{n} \lambda_{n_k}} G_{P\&T}^{(n)}(X), \tag{4.55}$$

where i_1 is such an index that

$$[0,1]^{\{n_1, \cdots, n_n\}} \times \{0\}^{N \setminus \{n_1, \cdots, n_n\}} \subset \Gamma_{i_1}.$$

□

References

[1] Arnold W. I., Rownania różniczkowe zwyczajne. (Polish) *Ordinary differential equations*, Rownania różniczkowe zwyczajne. (Polish),Translated from the first Russian edition by Alicja Derkowska and Gabriel Derkowski. Państwowe Wydawnictwo Naukowe, Warsaw, 1975. 264 pp.

[2] Baker R., "Lebesgue measure" on $\mathbb{R}^\infty$. II. *Proc. Amer. Math. Soc.* **132(9)** (2004), 2577–2591.

[3] Csörnyei Marianna, Aronszajn null and Gaussian null sets coincide, *Israel J.Math.*, **111** (1999), 191–201

[4] Hurt B.R., Sauer T., Yorke J.A., Prevalence:A Translation-Invariant "Almost Every" On Infinite-Dimensional Spaces, *Bulletin (New Series) of the American Mathematical Society*, Volume 27, **10(2)** (1992), 217–238.

[5] Kharazishvili A.B., On invariant measures in the Hilbert space. *Bull. Acad. Sc. Georgian SSR*, **114(1)** (1984), 41–48 (in Russian).

[6] Korn, Granino A.; Korn, Theresa M., *Mathematical handbook for scientists and engineers*. Second, enlarged and revised edition McGraw-Hill Book Co., New York-Toronto, Ont.-London (1968), xvii+1130 pp.

[7] Mankiewicz P., On the differentiability of Lipschitz mappings in Fréchet spaces. *Studia Math.*, **45** (1973),15–29.

[8] Pantsulaia G. R., An Analogue of Lioville's Theorem in Infinite-Dimensional Separable Hilbert space ℓ_2, *Collection of Abstracts, Section 1, Conference "Kolmogorov-100"*, MSU, Moskow, 15-21 June (2003), 79–80.

[9] Pantsulaia G. R., Relations between Shy sets and Sets of v_p-Measure Zero in Solovay's Model, *Bull. Polish Acad.Sci.*, **52(1)** (2004), 63–69.

[10] Pantsulaia G. R., On an Invariant Borel Measure in Hilbert Space , *Bull. Polish Acad. Sci.*, **52(2)** (2004), 47–51.

[11] Pantsulaia G. R., *Invariant and Quasiinvariant Measures in Infinite-Dimensional Topological Vector Spaces,* Nova Science Publishers, Inc. (2007), xii+231 pp.

[12] Pantsulaia, G. R., Giorgadze G. P., On analogy of Liouville theorem in infinite-dimensional separable Hilbert space, *Georgian International Journal of Science and Technology*, Nova Science Publishers, Inc., Volume 1, Number 2 (2008), 53-65

[13] Pantsulaia G. R., Giorgadze G. P., On some applications of generators of shy sets in infinite dimensional analysis, *International workshop on variable exponent analysis and related topics*, September 2–5 (2008), 1–46 .

[14] Pantsulaia G. R., On generators of shy sets on Polish topological vector spaces, *New York Journal of Mathematics,* Volume 14, (2008), 235 – 261 .

[15] Hongjia Shi. *Measure-Theoretic Notions of Prevalence,* Ph.D.Dissertation (under Brian S.Thomson), Simon Fraser University, October (1997), ix+165 pp.

In: Focus on Science and Technology...
Editor: Sergo Gotsiridze, pp. 103-108

Chapter 9

ON PLANAR SETS WITHOUT OF VERTICES OF A TRIANGLE OF AN AREA ONE

Gogi Pantsulaia*
Department of Mathematics, Georgian Technical University
0175 Kostava Str. 77, Tbilisi, Georgia
I.Vekua Institute of Applied Mathematics
Tbilisi State University
0143 University Str. 2, Tbilisi, Georgia

Abstract

We show that if S is a planar set of positive two-dimensional Lebesgue measure ℓ_2, then for arbitrary c, the S contains the vertices of a triangle of area c if and only if it is not bounded by a natural norm. Also, we consider a certain generalized problem of P. Erdösh asking whether there exists a positive constant c such that every planar set with outer measure larger than c contains the vertices of a triangle of area one. We prove that in the system of axioms $ZFC + MA$, the answer to this question is no.

2010 Mathematics Subject Classification: Primary 03xx, 03Exx, 05xx, 28xx; Secondary 03E35, 03E50, 11T99

Keywords and phrases: Planar set, Vertices of a triangle, Martin Axiom.

1. Introduction

In [2], P.Erdösh noted that if S is a set in the plane of infinite two-dimensional Lebesgue measure, then it contains for every $c > 0$ the vertices of a triangle of area c.

---

*E-mail address: gogi_pantsulaia@hotmail.com
The designated project has been fulfilled by financial support of the Georgia National Science Foundation (Grants: # GNSF /STO 7/3-178, # GNSF /STO 8/3-105).

The present paper is organized as follows.

In Section 1, we discuss a question asking whether a demand of infinity of two-dimensional Lebesgue measure for planar measurable sets is necessary. We will see that this demand is sufficient but no necessary. More precisely, we give a necessary and sufficient condition under which a measurable set in the plane of positive two-dimensional Lebesgue measure contains for every $c \geq 0$ the vertices of a triangle of area c.

In Section 2, we study a certain generalized problem of P.Erdösh asking whether there exists a positive constant c such that every planar set with outer two-dimensional Lebesgue measure larger than c contains the vertices of a triangle of area one. We prove that in the system of axioms $ZFC + MA$, the answer to this question is no.

2. An Unbounded Planar Set of Positive Two-dimensional Lebesgue Measure ℓ_2 Contains Vertices of a Triangle of an Arbitrary Area

In the sequel we need some notations and definitions.

We denote

a) by $|| \cdot ||$ a norm in the plane $\mathbb{R}^2$ defined by $||(x, y)|| = \sqrt{x^2 + y^2}$ for $(x, y) \in \mathbb{R}^2$;

b) by $|| \cdot ||_{OX}$ a restriction of the norm $|| \cdot ||$ in the real axis $OX = \mathbb{R} \times \{0\}$;

c) by $|| \cdot ||_{OY}$ a restriction of the norm $|| \cdot ||$ in the real axis $OY = \{0\} \times \mathbb{R}$.

Definition 2.1 A set $D \subset \mathbb{R}^2$ is called bounded by norm $|| \cdot ||$ if

$$(\exists R)(R > 0 \rightarrow (\forall(x, y))((x, y) \in D \rightarrow ||(x, y)|| < R)).$$

Otherwise, we say that the D is unbounded (by the norm $|| \cdot ||$).

Definition 2.2 A set $D \subset OX$ is called bounded by the norm $|| \cdot ||_{OX}$ if

$$(\exists R)(R > 0 \rightarrow (\forall(x, 0))((x, 0) \in D \rightarrow ||(x, 0)||_{OX} < R)).$$

Otherwise, we say that the D is unbounded (by the norm $|| \cdot ||_{OX}$).

Definition 2.3 A set $D \subset OY$ is called bounded by norm $|| \cdot ||_{OY}$ if

$$(\exists R)(R > 0 \rightarrow (\forall(0, y))((0, y) \in D \rightarrow ||(0, y)||_{OY} < R)).$$

Otherwise, we say that the D is unbounded (by the norm $|| \cdot ||_{OY}$).

By Fubini theorem it is not hard to prove

Lemma 2.1 Let $D \subseteq \mathbb{R}^2$ be a Borel set of positive two-dimensional Lebesgue measure l_2. Then there exist $x, y \in \mathbb{R}$ such that $l_x(D_x) > 0$ and $l^y(D^y) > 0$, where

(a) $D_x = \{(x, z) : (x, z) \in (\{x\} \times \mathbb{R}) \cap D\}$;

(b) $D^y = \{(x, z) : (x, z) \in (\mathbb{R} \times \{y\}) \cap D\}$;

(c) l_x is a linear Lebesgue measure defined on $\{x\} \times \mathbb{R}$;

(d) l^y is a linear Lebesgue measure defined on $\mathbb{R} \times \{y\}$.

Lemma 2.2([3], Exercise 124, p. 189). *Let F be a Borel subset of $\mathbb{R}$. The a set $F - F$ defined by*

$$F - F = \{x - y : x, y \in F\}$$

contains any interval $(-a, a)$ for some $a > 0$.

The main result is formulated as follows.

Theorem 2.1. *Let $D \subseteq \mathbb{R}^2$ be a Borel set of positive two-dimensional Lebesgue measure l_2. Then the D contains for every $c \geq 0$ the vertices of a triangle of area c if and only if it is not bounded by the norm $\|\cdot\|$.*

Proof. Necessity. We have that the D contains for every $c \geq 0$ the vertices of a triangle of area c. Assume the contrary and let D is bounded by a norm $\|\cdot\|$. Then there will be such a positive real number R that

$$(\forall (x, y))((x, y) \in D \rightarrow \|(x, y)\| < R).$$

Then for every three points w_1, w_2 and w_3 from the set D we have that the area of the triangle Δ with these vertices is not more then $2R^2$. Indeed, let w_1, w_2 and w_3 be three points from the D. Since the area of the D is invariant under rotations of the $\mathbb{R}^2$ about its origin, we can assume that points w_1 and w_2 belong to any horizontal line $y = y_*$, i.e., $w_1 = (z_1, y_*)$, $w_2 = (z_2, y_*)$ and $w_3 = (z_3, y_3)$.

Then we get

$$l_2(\Delta) = \frac{1}{2}\|w_1 - w_2\| \times \|y_3 - y_*\|_{OY} < \frac{1}{2}2R \times 2R = 2R^2.$$

The latter relation means that for $c > R^2$, the D does not contain the vertices of a triangle of area c.

Sufficiency. Suppose that D is not bounded by a norm $\|\cdot\|$. It means that $Pro_X(D)$ is not bounded by the norm $\|\cdot\|_X$ or $Pro_Y(D)$ is not bounded by the norm $\|\cdot\|_Y$, where

$Pro_X(D) = \{(x, 0) : (\exists y)(y \in \mathbb{R} \,\&\, (x, y) \in D\}$

and

$Pro_Y(D) = \{(0, y) : (\exists x)(x \in \mathbb{R} \,\&\, (x, y) \in D\}.$

Without loss of generality we can assume that $Pro_Y(D)$ is not bounded by the norm $\|\cdot\|_Y$. Let (x_k, y_k) be such a sequence of points that $(y_k)_{k \in \mathbb{N}}$ is an increasing sequence of real numbers tending to $+\infty$.

By Lemma 2.1, there exists $y_* \in \mathbb{R}$ such that $l^{y_*}(D^{y_*}) > 0$. By Lemma 2.2 there exists $a > 0$ such that $(-a, a) \times \{0\} \subseteq D^{y_*} - D^{y_*}$. Since a sequence of real numbers $(y_k - y_*)_{k \in \mathbb{N}}$ tends to $+\infty$ for $c > 0$ there exists an index $k_0 \in \mathbb{N}$ such that $0 < f^{-1}(y_{k_0} - y_*) < a$, where $f(x) = \frac{2c}{x}$.

Since

$$(f^{-1}(y_{k_0} - y_*), 0) \subseteq (-a, a) \times \{0\} \subseteq D^{y_*} - D^{y_*},$$

we claim that there exist $(z_1, y_*), (z_2, y_*) \in D^{y_*} \subset D$ such that

$$(z_1, y_*) - (z_2, y_*) = (f^{-1}(y_{k_0} - y_*), 0).$$

Let us show that $w_1 = (z_1, y_*)$, $w_2 = (z_2, y_*)$ and $w_3 = (x_{k_0}, y_{k_0})$ are the vertices of such a triangle Δ whose area is c. Indeed, we have

$$l_2(\Delta) = \frac{1}{2}||w_1 - w_2|| \times ||y_{k_0} - y_*||_{OY} = \frac{1}{2}(z_1 - z_2) \times (y_{k_0} - y_*) = c.$$

If $c = 0$ then three different points w_1, w_2 and w_3 of the linear set D^{y_*} will be the vertices of such a triangle whose area is 0. $\quad\square$

Let us consider some corollaries of Theorem 2.1.

Corollary 2.1. (Erdös [2])*If S is a set in the plane of infinite two-dimensional measure, then it contains for every $c \geq 0$ the vertices of a triangle of area c.*

Proof. Note that every set of infinite two-dimensional measure in the plane is not bounded by the norm $|| \cdot ||$. Hence, by sufficient condition of Theorem 2.1 we deduce that the S contains for every $c \geq 0$ the vertices of a triangle of area c. $\quad\square$

Corollary 2.2. *Let $(x_k)_{k \in \mathbb{N}}$ be a sequence of points in the plane $\mathbb{R}^2$ which is not bounded by the norm $|| \cdot ||$. Further, let Y be a planar set of positive two-dimensional Lebesgue measure. Then the union $\cup_{k \in \mathbb{N}}\{x_k\} \cup Y$ contains for every $c \geq 0$ the vertices of a triangle of area c.*

3.   On a Certain Generalized Problem of P. Erdösh

In 1985 P.Erdösh posed the following problem, see e.g., [2].

Problem 3.1 Is there a positive constant c such that every measurable set in the plane with area larger than c contains the vertices of a triangle of area one?

We delete a requirement of the measurability in Problem 3.1 and consider the following generalized P. Erdösh problem.

Problem 3.2 Is there a positive constant c such that every set in the plane with outer two-dimensional Lebesgue measure larger than c contains the vertices of a triangle of area one?

The purpose of the present paper is to show that the answer to Problem 3.2 is no in some consistent system of axioms.

By **MA** we denote Martin's Axiom (see, e.g., [4])

In the sequel we need some auxiliary propositions.

Lemma 3.1 (Martin & Solovay [4]) *Let ℓ_n be an n-dimensional Lebesgue measure defined on R^n. If $(E_i)_{i \in I}$ is a family of ℓ_n-measure zero subsets of R^n, such that $card(I) < c$, then in the system of axioms*

$$\textbf{ZFC \& MA}$$

the set $E = \bigcup_{i \in I} E_i$ is Lebesgue null in $\mathbb{R}^n$.

Theorem 3.1 In the system axioms $ZFC + MA$ the exists a planar set D such that the following two conditions are fulfilled:

    a) the D intersects every measurable set of positive measure;

b) the D does not contain the vertices of a triangle of area one.

Proof. Let $(F_\xi)_{\xi<\phi}$ be a sequence of planar Borel measurable sets of positive two-dimensional Lebesgue measure ℓ_2, where ϕ is a first ordinal number of a cardinality of the continuum.

Assume that for $\theta < \xi$ we have constructed a sequence of points $(x_\theta)_{\theta<\xi}$ such that the following two conditions are fulfilled:

(i) For $\theta < \xi$ the point $x_\theta \in F_\theta$;

(ii) A set $\{x_\theta : \theta < \xi\}$ does not contain the vertices of a triangle of area one.

For every pair $x, y \in \{x_\theta : \theta < \xi\}$ we denote by $L_{(x,y)}(1)$ and $L_{(x,y)}(2)$ lines on the plane which are parallel to the line drawing on points x, y and are separating from this on a distant $\frac{2}{\|x-y\|}$, respectively.

Let consider a set

$$F_\xi \setminus \left(\cup_{(x,y)\in\{x_\theta:\theta<\xi\}^2}\left(L_{(x,y)}(1) \cup L_{(x,y)}(2)\right) \cup \cup_{\theta<\xi}\{x_\theta\}\right).$$

By Lemma 3.1, this difference is not empty set. We choice a point x_ξ from this set. Thus, a transfinite sequence of points $(x_\xi)_{\xi<\phi}$ is constructed such that conditions a) and b) are fulfilled.

A validity of the condition a) is obvious.

Let us check the validity of the condition b). Assume the contrary and let $x_{\xi_1}, x_{\xi_2}, x_{\xi_3}$ be three points of the set $\{x_\xi : \xi < \phi\}$ that $\xi_1 < \xi_2 < \xi_3$ and an area of the triangle with vertices at points $x_{\xi_1}, x_{\xi_2}, x_{\xi_3}$ is equal to one.

Following our construction, x_{ξ_3} does not belong to lines $L_{(x_{\xi_1},x_{\xi_2})}(1)$ and $L_{(x_{\xi_1},x_{\xi_2})}(2)$, and hence, we get a contradiction. Finally, we set $D = \{x_\xi : \xi < \phi\}$.   □

We have the following corollary of Theorem 3.1.

Corollary 3.1. *In the system axioms $ZFC + MA$, the answer to Problem 3.2 is no. More precisely, for an arbitrary positive constant c there exists a planar set E with outer two-dimensional Lebesgue measure larger than c which does not contain the vertices of a triangle of area one.*

Proof. Let A be an arbitrary planar set of measure larger than c. We set $E = A \cap D$, where D is a set constructed in Theorem 3.1. Then outer measure of the set E is larger than c and following Theorem 3.1, it, as a subset of the set D, does not contain the vertices of a triangle of area one. □

References

[1] Halmos P. R., *Measure theory, Princeton, Van Nostrand (1950)*.

[2] Erdos P., Problems and results in combinatorial geometry. Discrete geometry and convexity (New York, 1982), 1–11, *Ann. New York Acad. Sci.*, **440**, New York Acad. Sci., New York (1985).

[3] Kirillov A.A., Gvishiani A.D., *Theorems and exercises in functional analysis*, Nauka, Moscow (1979) (in Russian).

[4] Martin D.A., Solovay R.M., Internal Cohen extensions, *Ann. Math. Logic*, **2** (1970).

[5] Pantsulaya G.R., On Komjath conjecture in Martin-Solovay model, *Georgian International Journal of Science and Technology*, **1(1)** (2008).

Received: October 10, 2009

In: Focus on Science and Technology...
Editor: Sergo Gotsiridze, pp. 109-117

ISBN 978-1-61209-970-5
© 2011 Nova Science Publishers, Inc.

Chapter 10

ON A CERTAIN PARTITION OF THE LEBESGUE NULL SET IN $\mathbb{R}^n$

Gogi Pantsulaia[*]

Department of Mathematics, Georgian Technical University
0175 Kostava Str. 77, Tbilisi, Georgia
I.Vekua Institute of Applied Mathematics
Tbilisi State University
0143 University Str. 2, Tbilisi, Georgia

Abstract

Specialized in Solovay model [9], we prove that any Haar null set in $R^n (n > 2)$ can be decomposed into two at most 1-dimensionally Haar null sets. Moreover, we describe an algorithm which gives a partition $\{E_k : 1 \leq k \leq n\}$ of the Lebesgue null set $E \subset \mathbb{R}^n$ such that e_k is transverse to E_k for $1 \leq k \leq n$. This partition answers negatively on H. Shi's question [6]. In addition, we consider J.Mycielski's extension of this result in the theory (ZF) & (DC).

2010 Mathematics Subject Classification: Primary 03xx, 28xx; Secondary 03E35, 28Axx.

Keywords and phrases: Christensen's null sets, s-null sets, one dimensional probe, Solovay model

In 1970, Solovay [9] demonstrated that the existence of a non-measurable set for Lebesgue measure is not provable within the framework of Zermelo-Frankel set theory in the absence of the Axiom of Choice. He also showed that (assuming the consistency of an inaccessible cardinal) there is a model of ZF(cf.[4]) in which countable choice holds, every set is Lebesgue measurable and in which the full axiom of choice fails. However, the axioms of Determinacy AD and dependent choice DC(cf.[4]), together, are sufficient

---

[*]E-mail address: gogi_pantsulaia@hotmail.com
The designated project has been fulfilled by financial support of the Georgia National Science Foundation (Grants: # GNSF /STO 7/3-178, # GNSF /STO 8/3-105).

for most geometric measure theory, potential theory, Fourier series and Fourier transforms, while making all subsets of the real line Lebesgue measurable.

In the present paper we are specialized in the so-called Solovay model (SM)(see, [9]) which is the following system of axioms:

$$(ZF)\&(DC)\&(\text{every subset of R is measurable in the Lebesgue sense})$$

and consider one algorithm which allows us to construct the spatial partition for a Haar null set in a finite-dimensional Euclidean vector space.

In order to formulate the main goal of the present paper we need some auxiliary definitions.

Let R^n be an n-dimensional Euclidean vector space and let $\mathcal{B}(R^n)$ be the class of all Borel subsets in $R^n (n \geq 1)$.

Let K be the class of all non-zero finite measures defined on the Borel σ-field $\mathcal{B}(R^n)$. We denote by $\mathcal{B}(R^n)^\mu$ the completion of $\mathcal{B}(R^n)$ with respect to the measure μ for $\mu \in K$.

Definition 1. A set $E \subset R^n$ is called universally measurable if $E \in \cap_{\mu \in K} \mathcal{B}(R^n)^\mu$.

We denote by $\mathcal{U}(R^n)$ the class of all universally measurable subsets in R^n.

Now, let $h_1, \cdots, h_k$ be any linearly independent family of vectors in R^n. We put

$$L(h_1, \cdots, h_k) = \text{span}(h_1, \cdots, h_k).$$

Let $\lambda_{L(h_1,\cdots,h_k)}$ be a k-dimensional Lebesgue measure defined on $L(h_1, \cdots, h_k)$ by the formula

$$\lambda_{L(h_1,\cdots,h_k)}(\{t_1 h_1 + \cdots + t_k h_k : (t_1, \cdots, t_k) \in [0,1]^k\}) = 1.$$

Further, let $L^\perp(h_1, \cdots, h_k)$ be a linear compliment of the $L(h_1, \cdots, h_k)$ in R^n. We set

$$(\forall X)(X \in \mathcal{U}(R^n) \to G^{(R^n)}_{\lambda_{L(h_1,\cdots,h_k)}}(X) = \sum_{g \in L^\perp(h_1,\cdots,h_k)} \lambda_{L(h_1,\cdots,h_k)}(X - g \cap L(h_1, \cdots, h_k))).$$

It is not difficult to see that the functional $G^{(R^n)}_{\lambda_{L(h_1,\cdots,h_k)}}$ is a quasi-finite translation-invariant measure on $\mathcal{U}(R^n)$ and $G^{(R^n)}_{\lambda_{L(h_1,\cdots,h_n)}} = \lambda_{L(h_1,\cdots,h_n)}$.

Definition 2.[6, Definition 3.2,p.10]. A universally measurable set $E \subset R^n$ is said to be s-null if there exist a countable family $(E_k)_{k \in N}$ of universally measurable sets and a countable family $(h_k)_{k \in N}$ of elements in R^n that E can be written as union of sets $\{E_k : k \in N\}$ such that $G^{(R^n)}_{\lambda_{L(h_k)}}(E_k) = 0$ for $k \in N$.

The class of all s-null sets in R^n we denote by $s(R^n)$.

Definition 3. Following Christensen [3], a universally measurable set E is said to be Haar null if there is a Borel probability measure μ on R^n that every translation of E has $\overline{\mu}$-measure zero, where $\overline{\mu}$ denotes an usual completion of the measure μ.

The class of all Haar null sets in R^n we denote by $H(R^n)$.

Hongjia Shi [5, Theorem 3.2.4, p.65] proved that

$$s(R^2) = H(R^2).$$

He has stated the following

Problem 1.[6, Remark, p.13].*Whether there exists a Haar null set that is not s-null in R^n for $n > 2$?*

Remark 1. The main difficulty arising regarding this problem in Zermelo-Fraenkel model ZFC(cf.[4])is the argument that no every subset E_0 of the universally measurable Lebesgue null set E in R^n is universally measurable. We easily left this difficulty if we will work in the Solovay model where the following diametrally different argument holds: every subset in R^n is universally measurable.

In context to notice above can be mentioned especially K. Gödel's AC-model

$$(ZFC) \ \& \ (CH)$$

$((CH)$ denotes the Continuum Hypothesis(cf. [4])) where an arbitrary subset $E \subset R^2$ (including universally measurable Lebesgue null sets) can be decomposed in to two sets E_1 and E_2 such that $card(E_1 \cap L) \leq \aleph_0$ and $card(E_2 \cap S) \leq \aleph_0$, where L and S are lines in R^2 parallel to vectors e_1 and e_2, respectively. The proof of this fact can be dune easily, if we set $E_1 = A \cap E$ and $E_2 = B \cap E$, where (A, B) denotes Sierpinski's partition of R^2 (cf. [4]). The problem here is *whether E_1 or E_2 can be chosen in the class of universally measurable subsets*, which has been justly solved by H.Shi [5, Theorem 3.2.4, p.65].

The goal of the present paper is to consider Problem 1 in theories (SM) and $(ZF) \ \& \ (DC)$, respectively.

We need some auxiliary propositions.

Lemma 1. *In Solovay model, every subset in R^n is universally measurable for $n \geq 1$.*

Lemma 2. *Let E_1 and E_2 be Polish topological spaces. Let μ_1 be a σ-finite diffused Borel measure on E_1 and let μ_2 be a σ-finite diffused Borel measure on E_2. Then, in Solovay model we have that:*
  (i) the domain of a measure $\overline{\mu_1 \times \mu_2}$ coincides with a powerset of $E_1 \times E_2$;
  (ii) for $X \subset E_1 \times E_2$ which $0 \leq \overline{\mu_1 \times \mu_2}(X) < \infty$, we have

$$\int_{E_1} d\mu_1(x)\mu_2(\{y : (x; y) \in X\}) = \int_{E_2} d\mu_2(y)\mu_1(\{x : (x; y) \in X\}).$$

The proof of Lemma 2 can be obtained by using the result of Lemma 1 and the well known Fubini theorem.

Lemma 3. *Let F_1 and F_2 be vector subspaces in R^n such that $F_1 \subset F_2$. Let λ_{F_1} be a Lebesgue measure concentrated on F_1. In Solovay model, we set*

$$(\forall X)(X \subset F_2 \rightarrow G_{F_1}^{F_2}(X) = \sum_{g \in F_2/F_1} \lambda_{F_1}(X - g \cap F_1)),$$

where F_2/F_1 denotes a factor group of F_2 with respect to F_1. Then $G_{F_1}^{F_2}$ is a quasi-finite translation-invariant measure in F_2.

The main result is formulated as follows.

Lemma 4. *Let F_1, F_2 and F_3 be vector subspaces in R^n such that $F_1 \subset F_2 \subset F_3$. Then, in Solovay model the following formula is valid*

$$(\forall X)(X \subset F_3 \rightarrow G_{F_1}^{F_3}(X) = \sum_{g \in F_3/F_2} \sum_{f \in F_2/F_1} \lambda_{F_1}(X - g - f \cap F_1)).$$

Definition 4. Let X be any subset in $L(e_1, \cdots, e_n)$. For $1 \leq k \leq n-1$, we set

$$F_{L(e_1,\cdots,e_k)}^{L(e_1,\cdots,e_n)}(X) = \cup_{t_1 e_1 + \cdots + t_k e_k \in \Theta_k(X)}(X \cap L(e_{k+1}, \cdots, e_n) + t_1 e_1 + \cdots + t_k e_k),$$

where

$$\Theta_k(X) = \{t_1 e_1 + \cdots + t_k e_k : \lambda_{L(e_{k+1},\cdots,e_n)}(X \cap L(e_{k+1}, \cdots, e_n) + t_1 e_1 + \cdots + t_k e_k) > 0\},$$

and

$$F_{L(e_1,\cdots,e_n)}^{L(e_1,\cdots,e_n)}(X) = X.$$

Below, in Solovay model, we consider an algorithm which gives the partition $\{E_k : 1 \leq k \leq n\}$ of the Lebesgue null set $E \subset R^n$ such that e_k is transverse to E_k for $1 \leq k \leq n$(i.e., $G_{L(e_k)}^{L(e_1,\cdots,e_n)}(E_k) = 0$ for $1 \leq k \leq n$.)

Theorem 1. *Let $(e_k)_{1 \leq k \leq n}$ be a basis in R^n. Then, in Solovay model, every Lebesgue null set E can be written as a union of sets $\{E_k : 1 \leq k \leq n\}$ such that*

$$E_n = F_{L(e_n)}^{L(e_1,\cdots,e_n)}(E),$$

$$E_{n-1} = F_{L(e_{n-1},e_n)}^{L(e_1,\cdots,e_n)}(E \setminus E_n),$$

$$E_{n-2} = F_{L(e_{n-2},e_{n-1},e_n)}^{L(e_1,\cdots,e_n)}(E \setminus (E_n \cup E_{n-1})), \cdots$$

$$E_{n-k} = F_{L(e_{n-k},\cdots,e_n)}^{L(e_1,\cdots,e_n)}(E \setminus (E_n \cup E_{n-1} \cup \cdots \cup E_{n-k+1})), \cdots$$

$$E_1 = F_{L(e_1,\cdots,e_n)}^{L(e_1,\cdots,e_n)}(E \setminus (E_2 \cup E_3 \cup \cdots E_n))$$

and $G_{L(e_k)}^{L(e_1,\cdots,e_n)}(E_k) = 0$ for $1 \leq k \leq n$.

Proof. Let prove Theorem 1 by the method of mathematical induction. The result of Theorem 1 is valid for $n = 1$.

Let us show that the result of Theorem 1 is valid for $n = 2$. Indeed, assume that $E \subset R^2$ and $\lambda_{L(e_1,e_2)}(X) = 0$. We have that E can be writhed as a union of sets $\{E_k : 1 \leq k \leq 2\}$ such that

$$E_2 = F_{L(e_2)}^{L(e_1,e_2)}(E),$$

$$E_1 = F_{L(e_1,e_2)}^{L(e_1,e_2)}(E \setminus E_2).$$

By using the result of Lemma 2, we easily conclude that $G_{L(e_2)}^{L(e_1,e_2)}(E_2) = 0$ and $G_{L(e_1)}^{L(e_1,e_2)}(E_1) = 0$.

Thus the result of Theorem 1 is proved for $n = 2$ in Solovay model.

Now assume that, in Solovay model, the result of Theorem 1 is valid for every Christensen's null set Y in R^m for $1 \le m \le n$.

Let show the validity of Theorem 1 for every Lebesgue null set E in R^{n+1}.

We set

$$\Theta_1 = \{h : h \in L(e_{n+1}) \,\&\, \lambda_{L(e_1,\cdots,e_n)}(E \cap L(e_1,\cdots,e_n) + h) > 0\}$$

and

$$\Theta_2 = \{h : h \in L(e_{n+1}) \,\&\, \lambda_{L(e_1,\cdots,e_n)}(E \cap L(e_1,\cdots,e_n) + h) = 0\}.$$

By using the result of Lemma 2, we have

$$\lambda_{L(e_1,\cdots,e_{n+1})}(E) = \int_{L(e_{n+1})} d\lambda_{L(e_{n+1})}(h)\lambda_{L(e_1,\cdots,e_n)}(\{g : h + g \in E\}) = 0,$$

which follows that

$$\lambda_{L(e_{n+1})}(\Theta_1) = 0.$$

We have

$$E = \cup_{h \in \Theta_1}(X \cap (L(e_1,\cdots,e_n) + h)) + \cup_{h \in \Theta_2}(X \cap (L(e_1,\cdots,e_n) + h)).$$

We set

$$\cup_{h \in \Theta_1}(E \cap (L(e_1,\cdots,e_n) + h)) = E_{n+1}.$$

Let consider a set

$$\cup_{h \in \Theta_2}(E \cap (L(e_1,\cdots,e_n) + h)).$$

Since $E^{(h)} = (E \cap (L(e_1,\cdots,e_n) + h) - h$ is $\lambda_{L(e_1,\cdots,e_n)}$-null set for $h \in \Theta_2$, by the assumption of the mathematical induction, $E^{(h)}$ can be writhed as a union of sets $\{E_k^{(h)} : 1 \le k \le n\}$ such that

$$E_n^{(h)} = F_{L(e_n)}^{L(e_1,\cdots,e_n)}(E^{(h)}),$$

$$E_{n-1}^{(h)} = F_{L(e_{n-1},e_n)}^{L(e_1,\cdots,e_n)}(E^{(h)} \setminus E_n^{(h)}),$$

$$E_{n-2}^{(h)} = F_{L(e_{n-2},e_{n-1},e_n)}^{L(e_1,\cdots,e_n)}(E^{(h)} \setminus (E_n^{(h)} \cup E_{n-1}^{(h)})), \cdots$$

$$E_{n-k}^{(h)} = F_{L(e_{n-k},\cdots,e_n)}^{L(e_1,\cdots,e_n)}(E^{(h)} \setminus (E_n^{(h)} \cup E_{n-1}^{(h)} \cup \cdots \cup E_{n-k+1}^{(h)})), \cdots$$

$$E_1^{(h)} = F_{L(e_1,\cdots,e_n)}^{L(e_1,\cdots,e_n)}(E^{(h)} \setminus (E_2^{(h)} \cup E_3^{(h)} \cup \cdots E_n^{(h)}))$$

and $G_{L(e_k)}^{L(e_1,\cdots,e_n)}(E_k^{(h)}) = 0$ for $1 \le k \le n$ and $h \in \Theta_2$.

We set $E_k = \cup_{h \in \Theta_2}(E_k^{(h)} + h)$ for $1 \le k \le n$.

We have

$$E_{n+1} = \cup_{h \in \Theta_1}(E \cap (L(e_1,\cdots,e_n) + h)) = F_{L(e_{n+1})}^{L(e_1,\cdots,e_{n+1})}(E).$$

For $1 \leq k \leq n$ we have

$$E_k = \cup_{h \in \Theta_2}(E_k^{(h)} + h) = \cup_{h \in \Theta_2}(F_{L(e_{k+1},\cdots,e_n)}^{L(e_1,\cdots,e_n)}(E^{(h)} \setminus (E_n^{(h)} \cup E_{n-1}^{(h)} \cup \cdots \cup E_{k+1}^{(h)})) + h) =$$

$$F_{L(e_{k+1},\cdots,e_{n+i})}^{L(e_1,\cdots,e_{n+1})}(\cup_{h \in \Theta_2}(E^{(h)} + h) \setminus \cup_{h \in \Theta_2}(E_n^{(h)} + h) \cup \cdots \cup \cup_{h \in \Theta_2}(E_{k+1}^{(h)} + h)) =$$

$$F_{L(e_{k+1},\cdots,e_{n+1})}^{L(e_1,\cdots,e_{n+1})}((E \setminus E_{n+1}) \setminus (E_n \cup \cdots \cup E_{k+1})) =$$

$$F_{L(e_{k+1},\cdots,e_{n+1})}^{L(e_1,\cdots,e_{n+1})}((E \setminus (E_{n+1} \cup E_n \cup \cdots \cup E_{k+1})).$$

Also, we have

$$G_{L(e_{n+1})}^{L(e_1,\cdots,e_{n+1})}(\cup_{h \in \Theta_1}(X \cap (L(e_1,\cdots,e_n) + h)) =$$

$$\sum_{g \in L(e_1,\cdots,e_n)} \lambda_{L(e_{n+1})}([\cup_{h \in \Theta_1}(X \cap (L(e_1,\cdots,e_n) + h)] - g \cap L(e_{n+1}))) \leq$$

$$\sum_{g \in L(e_1,\cdots,e_n)} \lambda_{L(e_{n+1})}(\Theta_1) = 0.$$

By using the main assumption of the mathematical induction, we have that

$$G_{L(e_k)}^{L(e_1,\cdots,e_n)}(E_k^{(h)}) = 0$$

for all $h \in \Theta_2$ and $1 \leq k \leq n$. Last relation follows that

$$G_{L(e_k)}^{L(e_1,\cdots,e_n)}(E_k^{(h)}) = \sum_{f \in L(e_1,\cdots,e_n)/L(e_k)} \lambda_{L(e_k)}(E_k^{(h)} - f \cap L(e_k)) = 0$$

for all $h \in \Theta_2$ and $1 \leq k \leq n$.

We have

$$G_{L(e_k)}^{L(e_1,\cdots,e_{n+1})}(\cup_{h \in \Theta_2}(E_k^{(h)} + h)) =$$

$$\sum_{f \in L(e_1,\cdots,e_{n+1})/L(e_k)} \lambda_{L(e_k)}((\cup_{h \in \Theta_2}(E_k^{(h)} + h)) - g \cap L(e_k)) =$$

$$\sum_{g \in L(e_1,\cdots,e_{n+1})/L(e_1,\cdots,e_n)} \sum_{f \in L(e_1,\cdots,e_n)/L(e_k)} \lambda_{L(e_k)}((\cup_{h \in \Theta_2}(E_k^{(h)} + h)) - g - f \cap L(e_k)) =$$

$$\sum_{f \in L(e_1,\cdots,e_n)/L(e_k)} \sum_{g \in L(e_{n+1})} \lambda_{L(e_k)}((\cup_{h \in \Theta_2}(E_k^{(h)} + h)) - g - f \cap L(e_k))) =$$

$$\sum_{f \in L(e_1,\cdots,e_n)/L(e_k)} \sum_{h \in \Theta_2} \lambda_{L(e_k)}(E_k^{(h)} - f \cap L(e_k)) =$$

$$\sum_{h \in \Theta_2} \sum_{f \in L(e_1,\cdots,e_n)/L(e_k)} \lambda_{L(e_k)}(E_k^{(h)} - f \cap L(e_k)) = \sum_{h \in \Theta_2} G_{L(e_k)}^{L(e_1,\cdots,e_n)}(E_k^{(h)}) = 0.$$

By the principle of the mathematical induction, the proof of Theorem 1 is completed.

$\square$

Since Haar null sets and Lebesgue null sets in R^n coincide in Solovay model, as consequence of the Theorem 1, we get

Corollary 1. *In Solovay model the following equality*

$$s(R^n) = H(R^n)$$

holds for every $n \geq 1$.

Definition 5. A Haar null set $X \subset R^n (n > 2)$ is said to be m-dimensional Haar null if there exists m-dimensional vector space $L_m \subset R^n$ such that $G_{L_m}^{R^n}(X) = 0$ and $G_L^{R^n}(X) > 0$ for every k-dimensional vector subspace $L \subset R^n$, where $1 \leq k \leq m - 1$.

Remark 2. In 1928 Besicovitch [2]constructed a set of $E \subset R^2$ of Lebesgue measure zero which includes line segments of length 1 in every orientation. Thus any Borel set containing the set E and having Lebesgue measure zero is a 2-dimensionally Haar null set.

Remark 3. Let F be a n-dimensional Lebesgue measure zero set of R^n. K.J.Falconer[5, Theorem 7.13,pp.106] proved that there is a 2-dimensional subspace P of R^n such that every translate of P intersects F in a set of 2-dimensional measure zero. By using this result H.Shi[6,Theorem 3.2.3,pp.65] proved that in $R^n(n > 2)$ any Haar null set is at most 2-dimensional.

Now we consider the following

Problem 2. [6,Problem 7, p.66]. In $R^n(n > 2)$ can any 2-dimensional Haar null set be decomposed into two at most 1-dimensionally haar null set?

The answer on this problem is contained in the following proposition.

Theorem 2. *In Solovay model, every 2-dimensional Haar null set in $R^n(n > 2)$ can be decomposed into two at most 1-dimensionally Haar null set.*

Proof. Without loss of generality, we can assume that X is a Haar null set which is defined by $L(e_1, e_2)$, i.e.,

$$G_{L(e_1,e_2)}^{L(e_1,\cdots,e_n)}(X) = 0.$$

Last relation means that

$$\sum_{h \in L(e_1,e_2)^\perp} \lambda_{L(e_1,e_2)}(X - h \cap L(e_1, e_2)) = 0,$$

equivalently, $\lambda_{L(e_1,e_2)}(X - h \cap L(e_1, e_2)) = 0$ for $h \in L(e_1, e_2)^\perp$. By using the result of Theorem 1, we have $X - h \cap L(e_1, e_2) = X_1^{(h)} + X_2^{(h)}$, such that

$$G_{L(e_1)}^{L(e_1,e_2)}(X_1^{(h)}) = 0$$

and

$$G_{L(e_2)}^{L(e_1,e_2)}(X_2^{(h)}) = 0.$$

We set

$X_1 = \cup_{h \in L(e_1,e_2)^\perp}(X_1^{(h)} + h)$ and $X_2 = \cup_{h \in L(e_1,e_2)^\perp}(X_2^{(h)} + h)$.

We have

$$G_{L(e_1)}^{L(e_1,\cdots,e_n)}(X_1) = \sum_{g \in L(e_1)^\perp} \lambda_{L(e_1)}(X_1 - g \cap L(e_1)) =$$

$$\sum_{g_1 \in L(e_2)} \sum_{g_2 \in L(e_3,\cdots,e_n)} \lambda_{L(e_1)}((\cup_{h \in L(e_1,e_2)^\perp}(X_1^{(h)} + h)) - g_1 - g_2 \cap L(e_1)) =$$

$$\sum_{g_1 \in L(e_2)} \sum_{g_2 \in L(e_3,\cdots,e_n)} \lambda_{L(e_1)}(X_1^{(g_2)} - g_1 \cap L(e_1)) =$$

$$\sum_{g_2 \in L(e_3,\cdots,e_n)} \sum_{g_1 \in L(e_2)} \lambda_{L(e_1)}(X_1^{(g_2)} - g_1 \cap L(e_1)) =$$

$$\sum_{g_2 \in L(e_3,\cdots,e_n)} 0 = 0.$$

Analogously, we can show that

$$G_{L(e_2)}^{L(e_1,\cdots,e_n)}(X_1) = 0.$$

This ends the proof of Theorem 2. □

By using the result of Theorem 2 and Remark 3, we get the following

Corollary 2. In $R^n(n > 2)$ any Haar null set can be decomposed into two at most 1-dimensionally Haar null sets.

The next assertion belongs to J. Mycielski [11].

Theorem 3. *In the theory (ZF) & (DC), for every universally measurable set X of Lebesgue measure 0 in $\mathbb{R}^n$ there exists a partition of X into n universally measurable subsets sets $A_1, \cdots, A_n$, such that the intersection of X with A_i and any line parallel to the i-th axis of coordinates is of linear measure 0.*

Proof Let us prove Theorem 3 for $n = 2$. By Fubini theorem we claim that for every measure zero set N in R^2 there is a G_delta set A in the horizontal axis X such that A is a G_δ of measure 0 and very vertical line that does not meet A meets N in a set of linear measure 0. Thus the cylinder C over A is a G_δ and meets every horizontal line in a set of measure 0 and the intersection of every vertical line with N and the complement of C is of measure 0.

Now, let us prove theorem for $n = 3$. Thus let X be a subset of measure 0 in $R^3 = R \times \dot{R} \times R$. By Fubini, we have a G_delta set A of measure 0 in the plane $P = R \times R \times \{0\}$ such that X intersected with every line perpendicular to P that does not meet A is of linear measure 0. Thus let the first Borel set B_1 be the complement of $A \times R$. Now, it suffices to split $A \times R$ into two Borel sets B_2 and B_3 such that the intersection of every line $R \times \{y\} \times \{z\}$ and of B_2 is of measure 0 and the intersection of every line $\{x\} \times R \times \{z\}$ and of B_3 is of measure 0. But of course the theorem for $n = 2$ gives such sets B_2 and B_3.

Now one can get the generalization of this fact for $n > 3$.

Remark 3. Theorem 3 demonstrates that the answer to Problem 1 is no in the system of axioms (ZF) & (DC). This extends the result of Theorem 1.

References

[1] N. Aronszajn, Differentiability of Lipschitzian mappings between Frchet spaces, Studia Math., **57** (1976),147-190.

[2] A. B. Besicovitch, On the fundamental geometrical properties of linearly rneasurable plane sets of points, *Math. Annalen,* **98** (1928), 422-464.

[3] J. R. Christensen, Topology and Borel Structure, Amsterdam: *North-Holland Publishing Company.*(1974)

[4] J. Cichon, A. Kharazishvili and B. Weglorz, *Subsets of the real line, Wydawnictwo Uniwersytetu Lodzkiego,* Lodz (1995).

[5] K. J. Falconer, The geometry of fractal sets, *Cambridge University Press* (1985).

[6] Hongjia Shi, *Measure-Theoretic Notions of Prevalence,* Ph.D.Dissertation (under Brian S.Thomson), Simon Fraser University, October (1997), $ix+165$ pages.

[7] B. R. Hunt, T. Sauer and J. A. Yorke, Prevalence: a translation-invariant almost every on infinite-dimensional spaces, *Bull. Amer. Math. Soc.* (N.S.) **27(2)** (1992), 217238.

[8] G. R. Pantsulaia, Relations between Shy sets and Sets of ν_p-Measure Zero in Solovay Model, *Bull. Polish Acad. Sci.,* **52 (1)** (2004), 63–69

[9] Solovay R.M., A model of set theory in which every set of reals is Lebesgue measurable, *Ann. Math.,* **92** (1970), 156.

[10] http://en.wikipedia.org/wiki/Non-measurable set.

[11] Jan Mycielski, Personal communications, June 21 (2008).

In: Focus on Science and Technology...
Editor: Sergo Gotsiridze, pp. 119-124

ISBN 978-1-61209-970-5
© 2011 Nova Science Publishers, Inc.

Chapter 11

ON A CERTAIN APPLICATION
OF PREISS-TIšER GENERATORS

*Gogi Pantsulaia**

Department of Mathematics, Georgian Technical University
0175 Kostava Str. 77, Tbilisi, Georgia
I.Vekua Institute of Applied Mathematics
Tbilisi State University
0143 University Str. 2, Tbilisi, Georgia

Abstract

For $n \in \mathbf{N}$, let $G^n_{P\&T}$ be an n-dimensional Preiss-Tišer generator in an infinite-dimensional topological vector space V (cf. [7]). For $D \subseteq V$, we establish the validity of the following formula

$$\dim(D) = \inf\{m - 1 : G^{(m)}_{P\&T}(\overline{D}) = 0\},$$

where $\overline{D}$ denotes a closed convex hull of the set D in V.

2010 Mathematics Subject Classification: Primary 28Axx, 28Cxx, 28Dxx; Secondary 28C20, 28D10, 28D99

Keywords and phrases: Convex set, Dimension of a set, Preiss-Tišer generator

1. Introduction

For an arbitrary set in an infinite-dimensional topological vector space, we apply a technique of Preiss-Tišer generators [7] for a calculation its natural dimension.

The paper is organized as follows. In Section 2 we present preliminary notions and facts that we need for our investigation. In Section 3 we prove the main result. In Section 4 we consider some examples.

---

*E-mail address: gogi_pantsulaia@hotmail.com
The designated project has been fulfilled by financial support of the Georgia National Science Foundation (Grants: # GNSF /STO 7/3-178, # GNSF /STO 8/3-105).

2. Preliminary Notions and Facts

Let V be a Polish topological vector space, by which we mean a vector space with a complete metric for which the addition and the scalar multiplication are continuous. Let $B(V)$ be the σ-algebra of Borel subsets of V and μ be a non-zero non-negative measure defined on $B(V)$.

Definition 2.1. ([5], Definition 1, p. 221p) A set $X \subset V$ is called shy if it is a subset of a Borel set X' for which $\mu(X' + v) = 0$ for every $v \in V$ and some Borel probability measure μ such that $\mu(K) = \mu(B)$ for some compact K. In this case we say that the measure μ is transverse to S, or that the measure μ is a testing measure for the X.

A compliment of a shy set in V is called prevalent.

The class of all shy sets in V we denote by $\mathcal{S}.\mathcal{S}.(V)$.

Definition 2.2. ([4], Definition 2.3.5, p. 11) A Borel set $S \subset V$ is said to be an n-dimensional Preiss-Tišer null set if every Lebesgue measure μ_i concentrated on any n-dimensional vector subspace L_i in V is transverse to S.

We denote by $\mathcal{P}.\mathcal{T}.\mathcal{N}.\mathcal{S}.(V, n)$ the class of all n-dimensional Preiss-Tišer null sets in V.

It can be shown that $\cup_{n \in \mathbf{N}} \mathcal{P}.\mathcal{T}.\mathcal{N}.\mathcal{S}.(V, n) \subset \mathcal{S}.\mathcal{S}.(V)$.

As Christensen (cf.[2], p.119) notes, there is no σ-finite (equivalently, probability) measure μ such that S being shy is equivalent to $\mu(S) = 0$. Slightly more can be said that any σ-finite measure μ must assign 0 to a prevalent set of points. On these ground, in [7] we have introduced the following

Definition 2.3. ([5], Definition 15.1.1) A Borel measure μ in V is called a generator (of shy sets) in V, if

$$(\forall X)(\overline{\mu}(X) = 0 \rightarrow X \in \mathcal{S}.\mathcal{S}.(V)),$$

where $\overline{\mu}$ denotes a usual completion of the Borel measure μ.

Various examples of Generators of shy sets (for, example, Mankiewicz, Kharazisvili, Baker and etc) can be found in [6],[7].

A notion of an n-dimensional Preiss-Tišer generator in V ($n \in \mathbf{N}$) is given as follows.

Definition 2.4.   For $n \in \mathbf{N}$, a quasi-finite[1] translation-invariant Borel measure μ in V is called an n-dimensional Preiss-Tišer generator of shy sets (or shortly, an n-dimensional Preiss-Tišer generator), if the following equality

$$\mathcal{P}.\mathcal{T}.\mathcal{N}.\mathcal{S}.(V, n) = \mathcal{N}.\mathcal{S}.(\mu)$$

holds, where $\mathcal{N}.\mathcal{S}.(\mu)$ denotes a class of all sets of μ measure zero.

An existence of an n-dimensional Preiss-Tišer generator of shy sets in V is guaranteed by the following lemma.

Lemma 2.1.([6], Theorem 15.3.3) *Let* $(\Gamma_i)_{i \in \mathbf{N}}$ *be a family of all an n-dimensional vector subspaces in V and μ_i be an n-dimensional Lebesgue measure concentrated on Γ_i for $i \in I$*

---

[1]A measure μ is called quasi-finite if there exists a measurable subset A with $0 < \mu(A) < +\infty$.

($n \in \mathbf{N}$). Let $\Gamma_i^{\perp}$ be any linear complement of Γ_i in V for $i \in I$. We put

$$(\forall X)(X \in \mathcal{B}(V) \to G_{P\&T}^{(n)}(X) = \sum_{i \in I} \sum_{g \in \Gamma_i^{\perp}} \mu_i((X - g) \cap \Gamma_i)).$$

Then the functional $G_{P\&T}^{(n)}$ is a quasi-finite translation-invariant n-dimensional Preiss-Tišer generator in V.

Remark 2.1. Preiss-Tišer generators have many interesting geometric properties (see, for example, [6], Theorems 15.3.1 –15.3.2).

Definition 2.5. A subset $D \subset V$ is said to be n-dimensional if there exists n-dimensional vector subspace $L_n \subset V$ and an element $g_n \in V$ such that $D - g_n \subset L_n$ and , for $0 \leq m < n$ there is no an m-dimensional vector subspace $L_m \subset V$ and an element $g_m \in V$ such that $D - g_m \subset L_m$.
  In this case we write that $\dim(D) = n$.

Definition 2.6. A subset $B \subseteq V$ is called infinite-dimensional if it is not n-dimensional for any $n \in \mathbf{N}$. In this case we write that $\dim(D) = +\infty$.

Remark 2.2. We can give an equivalent definition of the notion of a dimension for sets in V as follows: let D be a subset in V and $x \in D$. A dimension of the vector space generated by vectors $\{y - x : y \in D\}$ is called a dimension of the set D.
  Note that this definition does not depend on the choice of a point $x \in D$.

Lemma 2.2. Let $D \subset V$. Then D is n-dimensional if and only if its closed convex hull $\overline{D}$ is n-dimensional.

3. Main Result

Theorem Let V be an infinite-dimensional topological vector space. For $D \subset V$, the following formula
$$dim(D) = \inf\{m - 1 : G_{P\&T}^{(m)}(\overline{D}) = 0\}$$
is valid, where $\overline{D}$ denotes a closed convex hull of the set D in V. The $\inf$ taken over an empty set is assumed to be equal to $+\infty$.

Proof. Here are possible the following two cases:

Case 1. $\dim(D) = n$, where $n \in \mathbf{N}$.
  By Lemma 2.2, we have that $\overline{D}$ is n-dimensional. This means that there exists an n-dimensional vector subspace Γ_i and $g \in \Gamma_i^{\perp}$ such that $\overline{D} - g \subseteq \Gamma_i$. Let us show that $G_{P\&T}^{(n)}(\overline{D}) > 0$. Indeed, if we assume the contrary then we obtain that $\mu_i(\overline{D} - g \cap \Gamma_i) = 0$. Since $\overline{D}$ is convex and $\mu_i(\overline{D} - g \cap \Gamma_i) = 0$, by Fubini theorem we claim that there exists an $n - 1$ dimensional subspace Γ in Γ_i such that $\overline{D} - g \subseteq \Gamma$. This means that $\overline{D}$,

equivalently D is not n-dimensional and we get a required contradiction. Now it is obvious that $G_{P\&T}^{(m)}(\overline{D}) = 0$ for every $m > n$.

Case 2. $dim(D) = +\infty$.

Assume the contrary and let $G_{P\&T}^{(m)}(\overline{D}) = 0$ for any $m \in \mathbf{N}$. Let k be nearest from below for m natural number such that $G_{P\&T}^{(k)}(\overline{D}) > 0$. If such natural number does not exist then we get that $G_{P\&T}^{(0)}(\overline{D}) = 0$ which implies that $\overline{D}$ is a singleton. The later relation contradicts to the condition $dim(D) = +\infty$.

Since $G_{P\&T}^{(k)}(\overline{D}) > 0$, there exist $i_0 \in I$ and $g_0 \in \Gamma_{i_0}^{\perp}$ such that $\mu_{i_0}(\overline{D} - g_0 \cap \Gamma_{i_0}) > 0$, where Γ_{i_0} is any k-dimensional vector subspace in V and μ_{i_0} is an k-dimensional Lebesgue measure on Γ_{i_0}.

Let us show that $\overline{D} - g_0 \subseteq \Gamma_{i_0}$. Indeed, assume the contrary and let $g_1 \in \Gamma_{i_0}^{\perp}$ be such a non-zero element that $\overline{D} - g_0 \cap \Gamma_{i_1} \neq \emptyset$, where $\Gamma_{i_1} = \Gamma_{i_0} + g_1$. We set $y \in \overline{D} - g_1 \cap \Gamma_{i_1}$.

Let Γ be an $k+1$-dimensional vector subspace of V generated by Γ_{i_0} and the element y. Further, let μ be an $k+1$-dimensional Lebesgue measure on Γ.

Since $\overline{D}$ is closed(hence, measurable), convex and $\mu_{i_0}(\overline{D} - g_0 \cap \Gamma_{i_0}) > 0$, by Fubini theorem we claim that $k+1$-dimensional Lebesgue measure of the closed convex hull of the set $F = (\overline{D} - g_0 \cap \Gamma_{i_0}) \cup \{y\}$ (which is in Γ) is positive. Since $F \subseteq \overline{D}$, we claim that $0 < \mu(F) \leq G_{P\&T}^{(k+1)}(F) \leq G_{P\&T}^{(k+1)}(\overline{D})$. It contradicts to the definition of the natural number k and Theorem 2.1 is proved. □

4.   Some Examples

Example 4.1. *Let* $(H, < \cdot, \cdot >)$ *be an infinite-dimensional separable Hilbert space. There exists a set* E *in* H *such that*

(i) *For every isometric transformation* g *of* H *we have* $G_{P \& T}^{(1)}(g(E)) = +\infty$;

(ii) E *is shy in* H;

(iii) $dim(E) = +\infty$.

Proof. Let $(e_k)_{k \in \mathbf{N}}$ be an orthonormal basis in H. Let $H(e_1, e_2)$ be a vector subspace of H generated by vectors e_1 and e_2.

Let E_0 be a Besicovitch set in $H(e_1, e_2)$(cf. [1])

We set $E = E_0 + H(e_1, e_2)^{\perp}$, where $H(e_1, e_2)^{\perp}$ denotes a linear compliment of $H(e_1, e_2)$ in H.

Since E_0 is a set of segments of length 1 of all directions in $H(e_1, e_2)$, we claim that E contains segments with positive length in all directions in H. Hence, $G_{P \& T}^{(1)}(E) = +\infty$. Since under isometric transformation $g : H \to H$ different lines perform in to different lines, we claim that $g(E)$ contains segments with positive length in all directions in H.

Hence, $G_{P \& T}^{(1)}(g(E)) = +\infty$.

Since Jordan measure of E_0 is zero, we claim that 2-dimensional Lebesgue measure μ_2 concentrated on $H(e_1, e_2)$ is a testing measure for a set E. Hence, E is a shy set in H.

For every natural number $n \in \mathbf{N}$, there exist $n + 1$ points $x_1, \cdots, x_{n+1}$ in E which do not lie on any n-dimensional vector subspace. Hence, we get

$$G^{(n)}_{P \& T}(\overline{E}) \geq G^{(n)}_{P \& T}(\overline{\{x_1, \cdots, x_{n+1}\}}) > 0.$$

By Theorem 2.1 we deduce that

$$\dim(E) = +\infty.$$

Example 4.2. *Let H be an infinite-dimensional Hilbert space. Let S be a unite sphere in H. Then we have*
(i) $G^{(n)}_{P \& T}(S) = 0$ for every $n \in \mathbf{N}$;
(ii) $\dim(S) = +\infty$.

Proof. For every n-dimensional vector subspace Γ_i in H and for $g \in H$, an intersection $\Gamma_i - g \cap S$ is an $n-1$-dimensional sphere in Γ_i, a point or an empty set. Hence, $G^{(n)}_{P \& T}(S) = 0$ for every $n \in \mathbf{N}$.

Note that a closed convex hull of the S is the unite closed ball $B[\mathbf{0}, 1]$ in H. Hence, for every n-dimensional vector subspace Γ_i in H, an intersection $\Gamma_i \cap B[\mathbf{0}, 1]$ is an n-dimensional unite ball in Γ_i. Hence, for $n \in \mathbf{N}$, we have

$$G^{(n)}_{P \& T}(B[\mathbf{0}, 1]) = \sum_{\mathbf{i} \in \mathbf{I}} \sum_{\mathbf{g} \in \Gamma_{\mathbf{i}}^{\perp}} \mu_{\mathbf{i}}(\mathbf{B}[\mathbf{0}, \mathbf{1}] - \mathbf{g} \cap \Gamma_{\mathbf{i}}) \geq \sum_{\mathbf{i} \in \mathbf{I}} \mu_{\mathbf{i}}(\mathbf{B}[\mathbf{0}, \mathbf{1}] \cap \Gamma_{\mathbf{i}}) = +\infty.$$

By Theorem 2.1 we deduce that

$$\dim(S) = \inf\{n - 1 : G^{(n)}_{P \& T}(B[\mathbf{0}, 1]) = \mathbf{0}\} = +\infty.$$

References

[1] Besicovitch A. S., On Kakeya's problem and a similar one. *Math. Z.* **27** (1928), no. 1, 312–320

[2] Christensen J. R., *Topology and Borel Structure.* Amsterdam : North-Holland Publishing Company (1974).

[3] Halmos P. R., *Measure theory, Princeton, Van Nostrand* (1950).

[4] Shi H., Measure-Theoretic Notions of Prevalence, *Ph. D. Dissertation (under Brian S. Thomson), Simon Fraser University,* October 1997, ix + 165 pages

[5] Hurt B. R., Sauer T., Yorke J. A., Prevalence : A Translation-Invariant "Almost Every" On Infinite-Dimensional Spaces, *Bulletin (New Series) of the American Mathematical Society,* **27**(2), 10 (1992), 217–238.

[6] Pantsulaia G. R., *Invariant and Quasiinvariant Measures in Infinite-Dimensional Topological Vector Spaces,* Nova Science Publishers, 2007, xii+231 pp.

[7] Pantsulaia G. R., On generators of shy sets on Polish topological vector spaces, *New York J. Math.,* **14** (2008), 235 - 261 .

Received: October 5, 2009

In: Focus on Science and Technology...
Editor: Sergo Gotsiridze, pp. 125-129

ISBN 978-1-61209-970-5
© 2011 Nova Science Publishers, Inc.

Chapter 12

ON MEASURABILITY OF UNIONS OF PLANE DISKS

Gogi Pantsulaia[*]
Department of Mathematics, Georgian Technical University
0175 Kostava Str. 77, Tbilisi, Georgia
I.Vekua Institute of Applied Mathematics
Tbilisi State University
0143 University Str. 2, Tbilisi, Georgia

Abstract

Using N.Lusin's arithmetic example of a non-Borel analytic set [4], an example of an uncountable family of polygons is constructed in the Euclidean plane $\mathbb{R}^2$ such that their union is non-Borel. In the system of axioms $ZF + DC$ it is proved that for every uncountable family of plane disks their union is ℓ_2-measurable iff every subset of the real axis $\mathbb{R}$ is ℓ_1-measurable.

2010 Mathematics Subject Classification: Primary 28A05, 54H05, 03E65; Secondary 03E15, 03E60

Keywords and phrases: Plane disk, Lebesgue measurable set, Axiom of Determinacy, Axiom of Choice

Let b_2 and ℓ_2 denote two-dimensional classical Borel measure and two-dimensional Lebesgue measure defined on the Euclidean plane $\mathbb{R}^2$, respectively. Let $(X_i)_{i\in\mathbb{I}}$ be an arbitrary family of polygons. Then we can prove that the set $\cup_{i\in I} X_i$ is measurable with respect to the Lebesgue measure ℓ_2. The proof of this fact employs classical Vitali's covering theorem (see, for example, [2]).

Here naturally arises a question asking whether the same union is measurable with respect to the Borel measure b_2. Below we construct an example of an uncountable family of polygons in $\mathbb{R}^2$ such that their union is not measurable in the Borel sense.

We need some notions and auxiliary propositions.

[*]E-mail address: gogi_pantsulaia@hotmail.com
The designated project has been fulfilled by financial support of the Georgia National Science Foundation (Grants: # GNSF /STO 7/3-178, # GNSF /STO 8/3-105).

A set is called analytic if it is the image of a Borel function defined on $[0,1]$. Analytic sets form a class larger than Borel, but all analytic sets are universally measurable (see, for example [5]).

Let us consider a space of all natural-valued sequences $\mathbb{N}^{\mathbb{N}}$. We say that a sequence $\alpha = (\alpha_n)_{n \in \mathbb{N}} \in \mathbb{N}^{\mathbb{N}}$ is composite if "among of its members there are infinitely many numbers such that one is divided by the another"(see [4]).

The following assertion plays a key role in our future investigating.

Lemma 1.(Luzin [5]) *The set E is a non-Borel analytic set in $\mathbb{N}^{\mathbb{N}}$.*

Remark 1. In 1977 M.A.Lunina [6] observed that Luzin's proof is not correct. Later, V.Kanovei [8] proved that, in spite of Luzin's uncorrectable, his "arithmetic example" satisfies all required properties(i.e., E is an example of analytic set in $\mathbb{N}^{\mathbb{N}}$ that is not Borel).

Let $\mathcal{J}$ denotes a set of all irrational numbers in $[0,1]$. It is well known that a function $f : \mathbb{N}^{\mathbb{N}} \to \mathcal{J}$, defined by

$$f((\alpha_k)_{k \in \mathbb{N}}) = \cfrac{1}{\alpha_0 + \cfrac{1}{\alpha_1 + \cfrac{\ddots}{\ddots + \cfrac{1}{\alpha_n + \ddots}}}}$$

defines a continuous one-to-one mapping between $\mathbb{N}^{\mathbb{N}}$ and $\mathcal{J}$.

Since every homeomorphism sends analytic and non-Borel sets again in analytic and non-Borel sets, we claim that $D = f(E)$ is an example of an analytic set in $[0,1]$ that is not Borel.

Proposition 1. *Let X be any polygon in $\mathbb{R}^2$. Then there exists an uncountable family of translations $(h_i)_{i \in I}$ such that a set $\cup_{i \in I}(X + h_i)$ is a non-Borel subset in $\mathbb{R}^2$.*

Proof Let L be such a line in $\mathbb{R}^2$ that $L \cap X$ consists only one point. Without loss of generality we can assume that L coincides with the real axis $\{0\} \times \mathbb{R}$ and $(\{0\} \times \mathbb{R}) \cap X = \{(0,0)\}$. We set $I = \{0\} \times D$ and $h_i(x) = x + i$ for $x \in \mathbb{R}^2$ and $i \in I$. A set $\cup_{i \in I}(X + h_i)$ is a non-Borel subset in $\mathbb{R}^2$. Indeed, if we assume the contrary then an intersection $L \cap \cup_{i \in I}(X + h_i) = I$ must be Borel which is a required contradiction.

We have more strict result in infinite-dimensional Polish topological vector spaces.

Proposition 2. *Let F be any compact set in infinite-dimensional Polish topological vector space V. Then there exists an uncountable family of translations $(h_i)_{i \in I}$ such that a set $\cup_{i \in I}(F + h_i)$ is a non-Borel subset in V.*

Proof Let us show that there exists a line M in V that $L \cap F$ consists only one point. Let us consider a function $f : \mathbf{R} \times F \times F \to V$ defined by the following formula $f(\alpha, x, y) = \alpha(x - y)$. It is clear that $f(\mathbf{R} \times F \times F) = \cup_{n \in N} f([-n, n] \times F \times F)$. Clearly, $f([-n, n] \times F \times F)$ is the compact subset of V for every $n \in N$. Since V is the second category set we have $V \setminus f(\mathbf{R} \times F \times F) \neq \emptyset$. Let $v \in V \setminus f(\mathbf{R} \times F \times F)$. Let us show that the vector v spans a line L such that every translation of L meets F in at most one point. Indeed, let us assume the contrary and let $y_1, y_2 \in F$ and $y_2 = y_1 + \alpha v$, where $\alpha \neq 0$. Then for the element

$v = \frac{1}{\alpha}(y_2 - y_1)$ we can indicate such a natural number n_0, that $v \in f([-n_0, n_0] \times F \times F)$ and we obtain a contradiction.

Now let M be such a line in V which is parallel to the line L and $M \cap F \neq \emptyset$. The later relation gives that $M \cap F$ consists only one point. We set $I = D$ and $h_i = iv$ for $i \in I$. Then for a family $(F + h_i)_{i \in I}$ we have that the set $\cup_{i \in I}(F + h_i)$ is a non-Borel subset in V.

Recall that a subset of $\mathbb{R}^\infty$ is called a plane disk if it is an image of the closed unite ball under any homeomorphism of $\mathbb{R}^\infty$.

Following [9], there exists a family $(P_i)_{i \in I}$ of plane disks such that $\cup_{i \in I} P_i$ is not measurable in the Lebesgue sense. In this context A.Kharazisvili stated the following

Question 1. ([9], p. 467) Let $(P_i)_{i \in I}$ be any uncountable family of plane disks. Whether the union $\cup_{i \in I} P_i$ is ℓ_2-measurable?

We denote by $ZF + DC$ the system of axioms ZF and DC, where ZF denotes the Zermelo-Fraenkel set theory and DC denotes the axiom of Dependent Choices(see, for example, [1])

Lemma 2.([3], p. 271) *In the system of axioms $ZF + DC$ there exists a plane disk whose boundary has a positive ℓ_2 measure.*

Proposition 3. *In the system of axioms $ZF + DC$ the following two conditions are equivalent*

(i) For every uncountable family of plane disks their union is ℓ_2-measurable;

(ii) Every subset of the real axis $\mathbb{R}$ is ℓ_1-measurable.

Proof Proof of the implication $(i) \rightarrow (ii)$**.** Assume the contrary and let there exists a subset Y of the real axis $\mathbb{R}$ which is not ℓ_1-measurable. The later relation implies that for every subset $X \subset \mathbb{R}$ with $\ell_1(X) > 0$ there exists a non-measurable (in the Lebesgue sense) subset of X. Indeed, assume the contrary. Then for any set X with $\ell_1(X) > 0$ its every subset will be measurable in the Lebesgue sense. Since ℓ_1 is metrically transitive measure, there exists a family of elements $(h_k)_{k \in \mathbb{N}}$ of the real axis such that $\ell_1(Y \setminus \cup_{k \in \mathbb{N}}(X + h_k)) = 0$. The following representation

$$Y = (Y \setminus \cup_{k \in \mathbb{N}}(X + h_k)) \cup (Y \setminus \cap_{k \in \mathbb{N}}(X + h_k))$$

implies that Y is measurable in the Lebesgue sense and we get a required contradiction. Now using Fubini theorem, one can extend this fact for every subset $X \subseteq \mathbb{R}^2$ with $\ell_2(X) > 0$.

For $x \in \mathbb{R}^2$ and $r > 0$, we denote by $S(x, r)$ and $B(x, r)$ a circle and a closed ball with center at $x \in \mathbb{R}^2$ and radius r, respectively.

By Lemma 2 there exists a homeomorphism $f : \mathbb{R}^2 \rightarrow \mathbb{R}^2$ such that $\ell_2(f(S(O, 1))) > 0$. Let Γ be a non-measurable (with respect to Lebesgue measure ℓ_2) subset of the set $f(S(0, 1))$. For $x \in \Gamma, y \in f(B(0, 1)) \setminus f(S(0, 1))$ we denote by $D(x, y)$ a plane disk which has the following properties:

1) $f^{-1}(x), f^{-1}(y) \in D(x, y)$;

2) $(D(x, y) \setminus \{f^{-1}(x)\}) \subseteq (B(0, 1) \setminus S(0, 1))$;

3) $D(x, y) \cap S(0, 1) = \{f^{-1}(x)\}$.

Let us consider the following family

$$(f(D(x, y)))_{x \in \Gamma, y \in f(B(0,1)) \setminus f(S(0,1))},$$

which again is a family of plane disks, and let us consider their union. It is obvious to see that this union is presented as follows

$$\Gamma \cup int(f(B(0, 1))),$$

where $int(f(B(0, 1)))$ denotes a set of all inner points of the set $f(B(0, 1))$(It is easy to see that $int(f(B(0, 1))) = f(B(0, 1)) \setminus f(S(0, 1))$). The later representation implies that the union of the above considered plane disks is not ℓ_2-measurable. This is a required contradiction and the validity of the implication $(i) \rightarrow (ii)$ is proved.

The proof of the implication $(ii) \rightarrow (i)$ is obvious.

Remark 2. Let us denote by $(P_i)_{i \in I}$ the family of plane disks described in [9]. Then Question 1 asking whether the set $\cup_{i \in I} P_i$ is ℓ_2-measurable is not solvable in the theory $ZF + DC$. Indeed, on the one hand, in the system of axioms $ZF + DC + AD$(here AD denotes the axiom of Determinacy(cf. [1])), following Mycielski and Swierczkowski well known classical result [7], every subset of the real axis $\mathbb{R}$ is ℓ_1-measurable. This fact and Proposition 3 imply that $\cup_{i \in I} P_i$ is ℓ_2-measurable. On the other hand, in the system of axioms $ZF + AC = ZF + DC + AC$, following A.B.Kharazishvili result [9], the set $\cup_{i \in I} P_i$ is not ℓ_2-measurable.

References

[1] Cichon J., Kharazishvili A., Weglorz B., *Subsets of the real line*, Wydawnictwo Uniwersytetu Lodzkiego, Lodz (1995).

[2] Natanson, I. P.. *Teoriya funktsii veshchestvennogo peremennogo.* (Russian) [Theory of functions of real variable] Second edition, revised. Gosudarstv. Izdat. Tehn.-Teoret. Lit., Moscow, 1957. 552 pp. (errata insert).

[3] Gelbaum, Bernard R.; Olmsted, John M. H., *Counterexamples in analysis, Corrected reprint of the second* (1965) edition. Dover Publications, Inc., Mineola, NY, 2003. xxiv+195 pp.

[4] Luzin, N. N., Sobranie sochineni. Tom II: Deskriptivnaya teoriya mnozhestv. (Russian) [Collected works. Vol. II: Descriptive set theory] *Izdat. Akad. Nauk SSSR,* Moscow 1958 744 pp.

[5] Luzin, N. N., Lekcii ob analiticheskih mnoestvah i ih priloeniyah. (Russian) [Lectures on analytic sets and their applications.] *Gosudarstv. Izdat. Tehn.-Teor. Lit.,* Moscow, 1953. 359 pp.

[6] Lunina, M. A., Luzin's arithmetic example of an analytic set that is not a Borel set. (Russian) *Mat. Zametki* **22** (1977), no. 4, 525–534.

[7] Mycielski J., Swierczkowski S., On the Lebesgue measurability and the axiom of determinateness, *Fund.*,**54**,(1964),67-71.

[8] Kanovei, V. G., A proof of a theorem of Luzin. (Russian) *Mat. Zametki* **23** (1978), no. 1, 61–66.

[9] Kharazishvili, A. B., Quasipolygons and their uncountable unions. (Russian)*Soobshch. Akad. Nauk Gruzin. SSR* **124** (1986), no. 3, 465–468.

[10] Solovay R.M., A model of set theory in which every set of reals is Lebesgue measurable,*Ann.Math.*,**92**, 1970, 1–56.

In: Focus on Science and Technology...
Editor: Sergo Gotsiridze, pp. 131-142

ISBN 978-1-61209-970-5
© 2011 Nova Science Publishers, Inc.

Chapter 13

ON RIESZ-THORIN THEOREMS
IN THE LIZORKIN-TRIEBEL-MORREY TYPE
SPACES WITH DOMINANT MIXED DERIVATIVES

*Alik M. Najafov**

Department of Supreme Mathematics,
Azerbaijan Architectural and Civil Engineering University,
5, T.Shahbazi str., AZ1073 Baku, Azerbaijan

Abstract

In this paper, is studied some differential properties of functions $f(x)$ belonging to the intersection of Lizorkin-Triebel-Morrey type spaces with dominant mixed derivatives $S^{l^{\mu}}_{p_{\mu},\theta_{\mu},a,æ,\tau}F(G)$ $(\mu = 1, 2, \ldots, N)$.

2000 Mathematics Subject Classification: 26A33, 46E30, 42B35, 46E35.

Keywords and phrases: Lizorkin-Triebel-Morrey type space, dominant mixed derivatives, embedding theorems, Holder condition.

Spaces with dominant mixed derivatives Sobolev and Nikolskii's type spaces $S^l_p W$ and $S^l_p H$ have been introduced and studied by S.M.Nikolskii [18] and Besov type spaces $S^l_{p,\theta} B$ with different methods by A.D.Djabrailov [2] and T.I.Amanov [1] and Sobolev-Liuville type spaces by P.I.Lizorkin and S.M.Nikolskii [4] and Lizorkin-Triebel type spaces $S^l_{p,\theta} F$ by H.Triebel [20]. Introduced by C.Morrey [8,9.10]parameter spaces $L_{p,\lambda}(R^n)$ later on Sobolev-Morrey space $W^l_{p,a,æ}(G)$ studied by V.P.Il'yin [7] and Nikolskii-Morrey spaces $H^l_{p,\lambda}(R^n)$ studied by I.Ross [19] and Besov-Morrey space $B_{p,\theta,a,æ}l(G)$ studied by Y.V.Netrusov [17].The spaces of Besov-Morrey type $B^l_{p,\theta,a,æ,\tau}(G)$ and of Lizorkin-Triebel-Morrey type $F^l_{p,\theta,a,æ,\tau}(G)$ were introduced and studied in [5, 6]. The spaces of Besov-Morrey type $S^l_{p,\theta,a,æ,\tau}B(G)$ and Lizorkin-Triebel-Morrey type with dominant mixed

---

*E-mail address:nadjafov@rambler.ru

derivatives $S^l_{p,\theta,a,\ae,\tau}F(G)$ were introduced and studied in [11-13]. In the paper [15] applying the Riesz-Thorin type theorems in the $S^l_{p,\theta,a,\ae,\tau}B(G)$ (theorems like that gives opportunity to increase the number of vectors in the Holder's spaces C_{ν,σ_1}) obtained in [13] was studied the local smoothness of solutions of some class of hypoelliptic equations.

In this paper as like as are studied Riesz-Thorin type theorems in the Lizorkin-Triebel-Morrey type spaces $S^l_{p,\theta,a,\ae,\tau}F(G)$ ($p,\ \theta\ \in\ (1,\infty)$, $a\ \in\ [0,1]^n$, $\ae,l\ \in\ (0,\infty)^n$, $\tau\ \in\ [1,\infty]$) with dominant mixed derivatives and Holder's condition of functions belonging to the intersection of these types of spaces are valid too.

Let $e_n = \{1,2,\ldots,n\}$, $e \subseteq e_n$, $m = (m_1,m_2,\ldots,m_n)$, m_j be natural, $k = (k_1,k_2,\ldots,k_n)$, k_j be integer non-negative numbers, $j \in e_n$; $k^e = (k^e_1,k^e_2,\ldots,k^e_n)$, $k^e_j = k_j$ for $j \in e$, $k^e_j = 0$ for $j \in e_n\backslash e = e'$; $[t_j]_1 = \min\{1,t_j\}$, for $j \in e_n$, and let $h_0,t_0 \in (0,\infty)^n$ be fixed vectors and

$$\int_{a^e}^{b^e} f(x)dx^e = \left(\prod_{j\in e}\int_{a_j}^{b_j}dx_j\right)f(x),$$

that, is the integral involves only the variable x_j, and indices belong to set e.

We say that the open set $G \subset R^n$ satisfies the condition (A_1), if for any $x \in G$ and $T \in (0,\infty)^n$ there exists the vector-function

$$\rho(t,x) = (\rho_1(t_1,x),\rho_2(t_2,x),\ldots,\rho_n(t_n,x)),\ 0 \le t_j \le T_j,\ j \in e_n$$

with the following properties:

1) for all $j \in e_n$, the functions $\rho_j(u_j,x)$ are absolutely continuous with respect to u_j on $[0,T_j]$, and $\left|\rho'_j(u_j,x)\right| \le 1$ for almost all $u_j \in [0,T_j]$, where $\rho'_j(u_j,x) = \frac{\partial}{\partial u_j}\rho_j(u_j,x)$;

2) $\rho_j(0,x) = 0$ for all $j \in e_n$,

$$x + V(x,\omega) = x + \bigcup_{0\le t_j\le T_j,\ j\in e_n}[\rho(t,x)+t\omega I] \subset G,$$

where $\omega = (\omega_1,\omega_2,\ldots,\omega_n)$, $\omega_j \in (0,1]$, $j \in e_n$; $I = [-1,1]^n$, $t\omega I = \{(t_1\omega_1 y_1,\ldots,t_n\omega_n y_n) : y \in I\}$. If $t_1 = t^{\lambda_1},\cdots,t_n = t^{\lambda_n}$, $\rho(t,x) = \rho(t^\lambda,x)$, $\lambda = (\lambda_1,\lambda_2,\ldots,\lambda_n)$, $\omega = (\omega^{\lambda_1},\omega^{\lambda_2},\ldots,\omega^{\lambda_n})$, $\omega \in (0,1]$, then $V(x,\omega) = \bigcup_{0\le t\le T}[\rho(t^\lambda,x)+t^\lambda\omega^\lambda I]$ is a flexible $\lambda-$ horn introduced by O.V.Besov [3].

Definition 1. *By the space of Lizorkin-Triebel-Morrey-$S^l_{p,\theta,a,\ae,\tau}F(G)$ is meant the Banach space of localy summable on G function f with finite norm* $(m_j > l_j > k_j \ge 0,\ j \in e_n)$:

$$\|f\|_{S^l_{p,\theta,a,\ae,\tau}F(G)} =$$

$$= \sum_{e\subseteq e_n}\left\|\left\{\int_0^{h^e_0}\left[\frac{\delta^{m^e}(h)D^{k^e}f(\cdot)}{\prod_{j\in e}h_j^{l_j-k_j}}\right]^\theta\frac{dh^e}{\prod_{j\in e}h_j}\right\}^{\frac{1}{\theta}}\right\|_{p,a,\ae,\tau}, \tag{1}$$

where

$$\delta^{m^e}(h) = \left(\prod_{j\in e}\delta_j^{m_j}(h_j)\right)f(x), \; \delta_j^{m_j}(h_j)f(x) = \int_{-1}^{1}\left|\Delta_j^{m_j}(h_ju_j, G_h)f(x)\right|du,$$

$$\Delta_j^{m_j}(h_j) = \sum_{i=0}^{m_j}(-1)^{m_j-i}C_{m_j}^i f(x+ih_j), \; G_h = \{x : x+hI \in G\},$$

$$\|f\|_{p,a,\text{æ},\tau;G} = \|f\|_{L_{p,a,\text{æ},\tau}(G)} =$$

$$= \sup_{x\in G}\left\{\int_0^{t_{01}}\cdots\int_0^{t_{0n}}\left[\prod_{j\in e_n}[t_j]_1^{-\frac{\text{æ}_ja_j}{p}}\|f\|_{p,G_{t\text{æ}}(x)}\right]^{\tau}\prod_{j\in e_n}\frac{dt_j}{t_j}\right\}^{\frac{1}{\tau}}, 1 \le \tau < \infty, \quad (2)$$

$$\|f\|_{p,a,\text{æ},\infty;G} = \sup_{x\in G, t>0}\left(\prod_{j\in e_n}[t_j]_1^{\frac{\text{æ}_ja_j}{p}}\|f\|_{p,G_{t\text{æ}}(x)}\right)^{\frac{1}{\tau}}.$$

Note that the space $L_{p,a,\text{æ},\infty}(G)$ large then space of $\mathcal{L}_{\sqrt{},\dashv,\text{æ}}(\mathcal{G})$ introduced and studied by V.P.Il'yin [7]. Here we point out some properties of the spaces $S_{p,\theta,a,\text{æ},\tau}^l F(G)$.

1. The following embedding holds for arbitrary $\text{æ}_j > 0$ and $0 \le a_j \le \le 1$ for $j \in e_n$

$$\|f\|_{S_{p,\theta}^l F(G')} \le \|f\|_{S_{p,\theta,a,\text{æ}}^l F(G)} \le C\|f\|_{S_{p,\theta,a,\text{æ},\tau}^l F(G)}, \qquad \overline{G'} \subset G. \quad (3)$$

2. For all real $c > 0$ the expression

$$\|f\|_{S_{p,\theta,a,c\text{æ},\tau}^l F(G)} \sim \|f\|_{S_{p,\theta,a,\text{æ},\tau}^l F(G)}. \quad (4)$$

3. For every $\text{æ}_j > 0$ for $j \in e_n$

$$\|f\|_{S_{p,\theta,a,\text{æ},\infty}^l F(G)} = \|f\|_{S_{p,\theta}^l F(G)}. \quad (5)$$

4. Let $1 < \theta \le r \le s \le \sigma < \infty$, $\theta \le p \le \sigma$, then

$$S_{p,\theta,a,\text{æ},\tau}^l B(G) \hookrightarrow S_{p,r,a,\text{æ},\tau}^l F(G) \hookrightarrow S_{p,s,a,\text{æ},\tau}^l F(G) \hookrightarrow S_{p,\sigma,a,\text{æ},\tau}^l B(G).$$

In particular, for $l \in N^n$, $r = s = 2$, then

$$S_{p,\theta,a,\text{æ},\tau}^l B(G) \hookrightarrow S_{p,a,\text{æ},\tau}^l W(G) = S_{p,2,a,\text{æ},\tau}^l F(G) \hookrightarrow S_{p,\sigma,a,\text{æ},\tau}^l B(G).$$

Note that the space type of Sobolev -Morrey $S_{p,a,\text{æ},\tau}^l W(G)$ was defined and studied in [16].

Let $\beta_\mu \geq 0$ $(\mu = 1, 2, \ldots, N)$, $\sum_{\mu=1}^{N} \beta_\mu = 1$, $\frac{1}{p} = \sum_{\mu=1}^{N} \frac{\beta_\mu}{p_\mu}$, $\frac{1}{\theta} = \sum_{\mu=1}^{N} \frac{\beta_\mu}{\theta_\mu}$, $\frac{1}{r} = \sum_{\mu=1}^{N} \frac{\beta_\mu}{r_\mu}$,

$l = \sum_{\mu=1}^{N} \beta_\mu l^\mu$, $\varepsilon_j = \sum_{\mu=1}^{N} \beta_\mu l_j^\mu - \nu_j - (1 - æ_j a_j)\left(\frac{1}{p} - \frac{1}{q}\right)$,

$$\varepsilon_{j,0} = \sum_{\mu=1}^{N} \beta_\mu l_j^\mu - \nu_j - (1 - æ_j a_j)\frac{1}{p},$$

and let $L_e \in C^\infty(R^n \times R^n)$ be informly finite with respect to z from an arbitrary compact, i.e.,

$$S(L_{e,i}) = \sup p(L_{e,i}) \subset I_1 = \{x : |x_j| < 1, j \in e_n\},$$

and let $0 \leq T_j \leq 1$, $j \in e_n$. Suppose

$$V = \bigcup_{0 \leq t_j \leq T_j, j \in e_n} \left\{ y : \left(\frac{y}{t^e + T^{e'}}\right) \in S(L_e) \right\}.$$

Clearly, $V \subset I_T = \{x : |x_j| < T_j, j \in e_n\}$. U we denote an open set contained in the domain G, and henceforth it will be always assumed that $U + V \subset G$. Further, let $G_{tæ}(U) = \bigcup_{x \in U} G_{tæ}(x)$; note that if $0 < æ_j \leq 1, 0 \leq T_j \leq 1$ for $j \in e_n$, then $I_T \subset I_{Tæ}$, since $U + V \subset G$ and $U + V \subset G_{tæ}(U) = Z$.

Lemma 1. *Let* $1 \leq p_\mu \leq q_\mu \leq r_\mu \leq \infty$ $(\mu = 1, 2, \ldots, N)$; $0 < æ_j \leq 1, 0 < t_j \leq T_j \leq 1$ *for* $j \in e_n$; $1 \leq \tau \leq \infty$; $\eta = (\eta_1, \eta_2, \ldots, \eta_n)$, $0 \leq \eta_j \leq \leq T_j, j \in e_n$; $\nu = (\nu_1, \nu_2, \ldots, \nu_n)$, $\nu_j \geq 0-$*are integers* $j \in e_n$; $\Delta^{m^e}(h)f \in L_{p_\mu, a, æ, \tau}(Z)$, $\mu = 1, 2, \ldots$, $\ldots, N$ *and let*

$$F_\eta^e(x) = \prod_{j \in e'} T_j^{-1-\nu_j} \int_{0^e}^{\eta^e} \frac{\varphi_e(x, t^e + T^{e'})dt^e}{\prod_{j \in e} t_j^{2+\nu_j}}, \tag{6}$$

$$F_{\eta,T}^e(x) = \prod_{j \in e'} T_j^{-1-\nu_j} \int_{\eta^e}^{T^e} \frac{\varphi_e(x, t^e + T^{e'})dt^e}{\prod_{j \in e} t_j^{2+\nu_j}}, \tag{7}$$

where

$$\varphi_e(x, t^e + T^{e'}) = \int_{R^n} L_e^{(\nu)}\left(\frac{y}{t^e + T^{e'}}, \frac{\rho(t^e + T^{e'}, x)}{t^e + T^{e'}}\right) f_e(x + y, t^e)dy,$$

$$|f_e(x, t^e)| \leq C \int_{-1^e}^{1^e} \delta^{m^e}(\delta t) f(x + t_1^e u_1^e + \cdots + t_n^e u_n^e)du^e.$$

Then the following inequalities hold:

$$
\sup_{\overline{x}\in U}\left\|F_\eta^e\right\|_{q,U_{\gamma^{\mathrm{æ}}}(\overline{x})} \le C_1 \prod_{\mu=1}^{N}\left\{\left\|\prod_{j\in e} t_j^{-l_j^\mu}\,\delta^{m^e}(\delta u)f\right\|_{p_\mu,a,\mathrm{æ},\tau}\right\}^{\beta_\mu} \times
$$

$$
\times \prod_{j\in e'} T_j^{-\nu_j-(1-\mathrm{æ}_j a_j)(\frac{1}{p}-\frac{1}{q})} \prod_{j\in e_n}[\gamma_j]_1^{\frac{\mathrm{æ}_j a_j}{q}} \prod_{j\in e}\eta_j^{\varepsilon_j}\ (\varepsilon_j>0), \tag{8}
$$

$$
\sup_{\overline{x}\in U}\left\|F_{\eta,T}^e\right\|_{q,U_{\gamma^{\mathrm{æ}}}(\overline{x})} \le C_2 \prod_{\mu=1}^{N}\left\{\left\|\prod_{j\in e} t_j^{-l_j^\mu}\,\delta^{m^e}(\delta u)f\right\|_{p_\mu,a,\mathrm{æ},\tau}\right\}^{\beta_\mu} \times
$$

$$
\times \prod_{j\in e'} T_j^{-\nu_j-(1-\mathrm{æ}_j a_j)(\frac{1}{p}-\frac{1}{q})} \prod_{j\in e_n}[\gamma_j]_1^{\frac{\mathrm{æ}_j a_j}{q}}
\begin{cases}
\prod\limits_{j\in e} T_j^{\varepsilon_j}, & \varepsilon_j>0, \\[4pt]
\prod\limits_{j\in e} \ln\frac{T_j}{\eta_j}, & \varepsilon_j=0, \\[4pt]
\prod\limits_{j\in e} \eta_j^{\varepsilon_j}, & \varepsilon_j<0,
\end{cases} \tag{9}
$$

where $\gamma\in(0,\infty)^n$, C_1 and C_2 are the constans independent of f,γ,η and T.

Proof of Lemma 1. For the given $\overline{x}\in U$, applying the generalized Minkovskii inequality, we obtain

$$
\left\|F_\eta^e\right\|_{q,U_{\gamma^{\mathrm{æ}}}(\overline{x})} \le
$$

$$
\le C_1 \prod_{j\in e'} T_j^{-1-\nu_j} \int_{0^e}^{\eta^e} \frac{\left\|\varphi_e(\cdot,t^e+T^{e'})\right\|_{q,U_{\gamma^{\mathrm{æ}}}(\overline{x})}\,dt^e}{\prod\limits_{j\in e} t_j^{2+\nu_j}}, \tag{10}
$$

Estimate the norm $\left\|\varphi_e(\cdot,t^e+T^{e'})\right\|_{q,U_{\gamma^{\mathrm{æ}}}(\overline{x})}$. Applying the Holder inequality with the exponents

$$
\alpha_\mu=\frac{q_\mu}{\beta_\mu q},\ \mu=1,2,\ldots,N\ \left(\sum_{\mu=1}^{N}\frac{1}{\alpha_\mu}=q\sum_{\mu=1}^{N}\frac{\beta_\mu}{q_\mu}=1,\right)
$$

for $\left|\varphi_e(x,t^e+T^{e'})\right|$ we obtain

$$
\left\|\varphi_e(\cdot,t^e+T^{e'})\right\|_{q,U_{\gamma^{\mathrm{æ}}}(\overline{x})} \le C_2 \prod_{\mu=1}^{N}\left\{\left\|\varphi_e(\cdot,t^e+T^{e'})\right\|_{q_\mu,U_{\gamma^{\mathrm{æ}}}(\overline{x})}\right\}^{\beta_\mu} \tag{11}
$$

By virtue of the Holder's inequality, with regard for $q_\mu\le r_\mu$, $\mu=1,2,\ldots,N$ we have

$$
\left\|\varphi_e(\cdot,t^e+T^{e'})\right\|_{q_\mu,U_{\gamma^{\mathrm{æ}}}(\overline{x})} \le \prod_{j\in e_n}\gamma_j^{(\frac{1}{q_\mu}-\frac{1}{r_\mu})}\left\|\varphi_e(\cdot,t^e+T^{e'})\right\|_{r_\mu,U_{\gamma^{\mathrm{æ}}}(\overline{x})}. \tag{12}
$$

Let χ be the charasteristic function of the set $S(L_e)$. Note that $1 \leq p_\mu \leq r_\mu$, $s_\mu \leq r_\mu$ $\left(\frac{1}{s_\mu} = 1 - \frac{1}{p_\mu} + \frac{1}{r_\mu}\right)$, $\mu = 1, 2, \ldots, N$, and and apply Holder's inequality for $\left|\varphi_e(x, t^e + T^{e'})\right|$ (by virtue of $\frac{1}{r_\mu} + \left(\frac{1}{p_\mu} - \frac{1}{r_\mu}\right) + \left(\frac{1}{s_\mu} - \frac{1}{r_\mu}\right) = 1$, $\mu = 1, 2, \ldots, N$), we obtain

$$\left\|\varphi_e(\cdot, t^e + T^{e'})\right\|_{r_\mu, U_{\gamma \text{æ}}(\overline{x})} \leq$$

$$\leq \sup_{x \in U_{\gamma \text{æ}}(\overline{x})} \left(\int_{R^n} |f_e(x + y, t^e)|^{p_\mu} \chi\left(\frac{y}{t^e + T^{e'}}\right) dy\right)^{\frac{1}{p_\mu} - \frac{1}{r_\mu}} \times$$

$$\times \sup_{y \in V} \left(\int_{U_{\gamma \text{æ}}(\overline{x})} |f_e(x + y, t^e)|^{p_\mu} dx\right)^{\frac{1}{r_\mu}} \times$$

$$\times \left(\int_{R^n} \left|L_e^{(\nu)}\left(\frac{y}{t^e + T^{e'}}, \frac{\rho(t^e + T^{e'}, x)}{t^e + T^{e'}}\right)\right|^{s_\mu} dy\right)^{\frac{1}{s_\mu}} \tag{13}$$

Since $U + V \subset Z$, $Z_{t^e + T^{e'}}(x) \subset Z_{(t^{\text{æ}})^e + (T^{\text{æ}})^{e'}}(x)$ for arbitrary $0 < \text{æ}_j \leq 1, 0 < t_j \leq T_j \leq 1$ for $j \in e_n$ and $x \in U$, we have

$$\int_{R^n} |f_e(x + y, t^e)|^{p_\mu} \chi\left(\frac{y}{t^e + T^{e'}}\right) dy \leq$$

$$\leq \int_{Z_{(t^{\text{æ}})^e + (T^{\text{æ}})^{e'}}(x)} |f_e(x + y, t^e)|^{p_\mu} \chi\left(\frac{y}{t^e + T^{e'}}\right) dy \leq$$

$$\leq \prod_{j \in e} t_j^{l_j^\mu p_\mu} \left\|\prod_{j \in e} t_j^{-l_j^\mu} \delta^{m^e}(\delta u) f\right\|_{p_\mu, Z_{(t^{\text{æ}})^e + (T^{\text{æ}})^{e'}}(x)}^{p_\mu} \leq$$

$$\leq \prod_{j \in e} t_j^{l_j^\mu p_\mu} \left\|\prod_{j \in e} t_j^{-l_j^\mu} \delta^{m^e}(\delta u) f\right\|_{p_\mu, a, \text{æ}}^{p_\mu} \prod_{j \in e} t_j^{\text{æ}_j a_j} \prod_{j \in e'} T_j^{\text{æ}_j a_j}. \tag{14}$$

At $y \in V$

$$\int_{U_{\gamma \text{æ}}(\overline{x})} |f_e(x + y, t^e)|^{p_\mu} dx \leq \int_{U_{\gamma \text{æ}}(\overline{x} + y)} |f_e(x, t^e)|^{p_\mu} dx \leq$$

$$\leq \int_{Z_{(t^{\text{æ}})^e + (T^{\text{æ}})^{e'}}(x)} |f_e(x, t^e)|^{p_\mu} dx \leq$$

$$\leq \prod_{j\in e} t_j^{l_j^\mu p_\mu} \left\| \prod_{j\in e} t_j^{-l_j^\mu} \delta^{m^e}(\delta u)f \right\|_{p_\mu,a,\ae}^{p_\mu} \prod_{j\in e_n} [\gamma_j]_1^{\ae_j a_j}, \tag{15}$$

$$\int_{R^n} \left| L_e^{(\nu)} \left(\frac{y}{t^e + T^{e'}}, \frac{\rho(t^e + T^{e'}, x)}{t^e + T^{e'}} \right) \right|^{s_\mu} dy =$$

$$= \prod_{j\in e} t_j \prod_{j\in e'} T_j \left\| L_e^{(\nu)} \right\|_{s_\mu}^{s_\mu} \tag{16}$$

It follows from (11)-(16) that

$$\left\| \varphi_e(\cdot, t^e + T^{e'}) \right\|_{q,U_{\gamma\ae}(\overline{x})} \leq C_3 \left\{ \left\| \prod_{j\in e} t_j^{-l_j^\mu} \delta^{m^e}(\delta u)f \right\|_{p_\mu,a,\ae} \right\}^{\beta_\mu} \times$$

$$\times \prod_{j\in e'} T_j^{-\nu_j - (1-\ae_j a_j)(\frac{1}{p}-\frac{1}{r})} \prod_{j\in e_n} [\gamma_j]_1^{\frac{\ae_j a_j}{r}} \prod_{j\in e_n} \gamma_j^{\frac{1}{q}-\frac{1}{r}} \prod_{j\in e} t_j^{1-(1-\ae_j a_j)(\frac{1}{p}-\frac{1}{r})}. \tag{17}$$

Taking into account the inequality (3) and substituting (17) in (10) for $r = q$ we arrive at (8). In a similar way we can prove inequality (9).

Lemma 2. *Let* $1 \leq p_\mu \leq q_\mu \leq \infty$, $\mu = 1, 2, \ldots, N$; $0 < \ae_j \leq 1, 0 \leq T_j \leq 1, j \in e_n$; $1 \leq \tau_1 \leq \tau_2 \leq \infty$; *and let* $\varepsilon_j > 0$. *Then for every function* $B_T^{e,i}$ *defined by (6) the estimate*

$$\|F_T^e\|_{q,b,\ae,\tau_2;U} \leq C \prod_{\mu=1}^N \left\{ \left\| \prod_{j\in e} t_j^{-l_j^\mu} \delta^{m^e}(\delta u)f \right\|_{p_\mu,a,\ae,\tau_1} \right\}^{\beta_\mu}, \tag{18}$$

isvalid, where C *is the constant, independent of* f, *and* $b = (b_1, b_2, \ldots, , \ldots, b_n)$, b_j *is an arbitrary number satisfying the inequalities*
  $0 \leq b_j \leq 1$, *if* $\varepsilon_{j,0} > 0$ *for* $j \in e$;
  $0 \leq b_j < 1$, *if* $\varepsilon_{j,0} = 0$ *for* $j \in e$; $0 \leq b_j \leq a_j$ *for* $j \in e'$;
  $0 \leq b_j < 1 + \frac{\varepsilon_{j,0} q(1-a_j)}{1-\ae_j a_j}$, *if* $\varepsilon_{j,0} < 0$ *for* $j \in e$.

Theorem 1. *Assume that an open set* G *satisfies the condition* (A_1), $1 \leq p_\mu \leq q_\mu < \infty$, $1 \leq \theta_\mu \leq \infty$, $\mu = 1, 2, \ldots, N$; $\overline{\ae} = c\ae$, $\frac{1}{c} = \max_{j\in e_n} l_j \ae_j$; $1 \leq \tau_1 \leq \tau_2 \leq \infty$;

$\nu = (\nu_1, \nu_2, \ldots, \nu_n)$, $\nu_j \geq 0 -$ *are integers* $j \in e_n$; $f \in \bigcap_{\mu=1}^N S_{p_\mu,\theta_\mu,a,\ae,\tau_1}^{l_\mu} F(G)$ *and let*

$\varepsilon_j > 0$ *for* $j \in e_n$.
  Then there exists the generalized derivative $D^\nu f$ *in the domain* G' $(\overline{G'} \subset G)$ *and the following inequalities are valid:*

$$\|D^\nu f\|_{q,G'} \leq C_1 F(T) \prod_{\mu=1}^N \left\{ \|f\|_{S_{p_\mu,\theta_\mu,a,\ae,\tau_1}^{l_\mu} F(G)} \right\}^{\beta_\mu}, \tag{19}$$

$$\|D^\nu f\|_{q,b,\ae,\tau_2;G} \le C_2 \prod_{\mu=1}^{N} \left\{ \|f\|_{S_{p_\mu,\theta_\mu,a,\ae,\tau_1}^{l^\mu} F(G)} \right\}^{\beta_\mu}, \tag{20}$$

where $F(T) = \sum\limits_{e \subseteq e_n} \prod\limits_{j \in e_n} T_j^{s_{e,j}}, s_{e,j} = \left\{ \begin{array}{ll} \varepsilon_j & for\, j \in e, \\ -\nu_j - (1 - \ae_j a_j)(\frac{1}{p} - \frac{1}{q}) & for\, j \in e'. \end{array} \right.$
In particular, if $\varepsilon_{j,0} > 0$ *for* $j \in e_n$, *then* $D^\nu f$ *is continuous*

$$\sup_{x \in G} |D^\nu f(x)| \le C_1 F(T^0) \prod_{\mu=1}^{N} \left\{ \|f\|_{S_{p_\mu,\theta_\mu,a,\ae,\tau_1}^{l^\mu} F(G)} \right\}^{\beta_\mu}, \tag{21}$$

where $F(T^0) = \sum\limits_{e \subseteq e_n} \prod\limits_{j \in e_n} T_j^{s_{e,j}^0}, s_{e,j}^0 = \left\{ \begin{array}{ll} \varepsilon_{j,0} & for\, j \in e, \\ -\nu_j - (1 - \ae_j a_j)\frac{1}{p} & for\, j \in e', \end{array} \right.$
$T_j \in (0, \min(1, t_{0,j})]$ *for* $j \in e_n$, C_1, C_2 *are the cons*tan*s independent of* f, *and* C_1
is independent of T *as well.*

Proof of Theorem 1. First of all, it should be noted that since $\overline{\ae} = c\ae$, $c > 0$, using property 2), we can assume that $f \in \bigcap\limits_{\mu=1}^{N} S_{p_\mu,\theta_\mu,a,\ae,\tau_1}^{l^\mu} F(G)$ and in inequalities (19)-(21) and in the expression for ε_j we can replace $\ae_j$ by $\overline{\ae_j}$ for $j \in e_n$. Here we will prove these very inequalities (the greater $\ae_j$ the the greater ε_j, $j \in e_n$). Let $f \in \bigcap\limits_{\mu=1}^{N} S_{p_\mu,\theta_\mu,a,\overline{\ae},\tau_1}^{l^\mu} F(G)$ then $f \in S_{p_\mu,\theta_\mu,a,\overline{\ae},\tau_1}^{l^\mu} F(G)$ for all $\mu = 1, 2, \ldots, N$. The existence of the generalized derivative $D^\nu f$ under the conditions of the theorem follows from Theorem 1 of [12]. Then for almost every point $x \in G'$ the following equality holds:

$$D^\nu f(x) = \sum_{e \subseteq e_n} (-1)^{|\nu|} \prod_{j \in e'} T_j^{-1-\nu_j} \int_{0^e}^{T^e} \frac{dt^e}{\prod\limits_{j \in e} t_j^{2+\nu_j}} \times$$

$$\times \int_{R^n} L_e^{(\nu)} \left(\frac{y}{t^e + T^{e'}}, \frac{\rho(t^e + T^{e'}, x)}{t^e + T^{e'}} \right) \chi(G') f_e(x + y, t^e) dy, \tag{22}$$

obtained in [14]. The support of representation (22) is contained in the set $x + V \subset G'$, and the parameter of the representation $\delta > 0$ is assumed to be sufficiently small. Therefore $\Delta^{m^e}(\delta u; G'_{\delta u})f = \Delta^{m^e}(\delta u)f$. On the basis on Minkovskii's inequality, we have

$$\|D^\nu f\|_{q,G} \le C_1 \sum_{e \subseteq e_n} \|F_T^e\|_{q,G} \tag{23}$$

By means of (8), for $U = G'$, $\eta = T$, as $\gamma \to \infty$, we find that

$$\|F_T^e\|_{q,G'} \le C_2 \prod_{j \in e_n} T_j^{s_{e,j}} \prod_{\mu=1}^{N} \left\{ \left\| \prod_{j \in e} t_j^{-l_j^\mu} \delta^{m^e}(\delta u)f \right\|_{p_\mu,a,\overline{\ae},\tau_1} \right\}^{\beta_\mu}. \tag{24}$$

Then from inequality (23) we have

$$\|D^\nu f\|_{q,G} \le C_1 \sum_{e \subseteq e_n} \prod_{j \in e_n} T_j^{s_{e,j}} \prod_{\mu=1}^N \left\{ \left\| \prod_{j \in e} t_j^{-l_j^\mu} \delta^{m^e}(\delta u) f \right\|_{p_\mu, a, \overline{\ae}, \tau_1} \right\}^{\beta_\mu}. \qquad (25)$$

whence with regard for $1 \le \theta_\mu \le \infty$ and $p_\mu \le \theta_\mu$, $\mu = 1, 2, \ldots, N$ we get inequality (19). Similarly, using (18), we obtain estimate (20).

Assume now that $\varepsilon_{j,0} > 0$ for $j \in e_n$. Let us show that $D^\nu f$ is continuous in G'. On the basis identity (22) and inequality (25) for $q = \infty$, we have

$$\|D^\nu f - D^\nu f_T\|_{\infty,G} \le \sum_{e \subseteq e_n} \prod_{j \in e_n} T_j^{s_{e,j}^0} \prod_{\mu=1}^N \left\{ \left\| \prod_{j \in e} t_j^{-l_j^\mu} \delta^{m^e}(\delta u) f \right\|_{p_\mu, a, \overline{\ae}, \tau_1} \right\}^{\beta_\mu},$$

whence it follows that the left hand side of the inequality tends to zero as $T_j \to 0$, $j \in e$. Since $D^\nu f_T$ is continuous in G', in this case the convergence of $L_\infty(G')$ coincides with the uniform convergence and, concequently, the limit function $D^\nu f$ is continuous in G'.

Theorem 1 is proved.

Let ξ be an n−dimensional vector.

Theorem 2. *Let the domain G, the vector $\overline{\ae}$, and, parametres p_μ, q_μ, θ_μ ($\mu = 1, 2, \ldots, N$), τ, satisfy the conditions of Theorem 1. If $\varepsilon_j > 0$ for $j \in e_n$, then the derivative $D^\nu f$ satisfies the Holder condition in G' in the metric of L_q with exponent σ_j^1, or more exactly,*

$$\|\Delta(\xi, G) D^\nu f\|_{q,G'} \le C \prod_{\mu=1}^N \left\{ \|f\|_{S_{p_\mu, \theta_\mu, a, \ae, \tau_1}^{l^\mu} F(G)} \right\}^{\beta_\mu} \prod_{j \in e_n} |\xi_j|^{\sigma_j^1}, \qquad (26)$$

where σ_j^1 is an arbitrary number satisfying the inequalities:

$0 \le \sigma_j^1 \le 1$, *if $\varepsilon_j' > 1$ for $j \in e$;*
$0 \le \sigma_j^1 < 1$, *if $\varepsilon_j' = 1$ for $j \in e$; $0 \le \sigma_j^1 \le 1$ for $j \in e'$;*
$0 \le \sigma_j^1 \le \varepsilon_j'$, *if $\varepsilon_j' = 1$ for $j \in e$, where $\varepsilon_j' = \varepsilon_j + \alpha_j$, $0 < \alpha_j < 1/\tau$ for $j \in e_n$.*
If $\varepsilon_{j,0} > 0$ for $j \in e_n$, then

$$\sup_{x \in G'} |\Delta(\xi, G) D^\nu f(x)| \le C \prod_{\mu=1}^N \left\{ \|f\|_{S_{p_\mu, \theta_\mu, a, \ae, \tau_1}^{l^\mu} F(G)} \right\}^{\beta_\mu} \prod_{j \in e_n} |\xi_j|^{\sigma_j^{1,0}} \qquad (27)$$

where $\sigma_j^{1,0} (j \in e_n)$ satisfy the same conditions as σ_j^1 ($j \in e_n$), but with the substitution of ε_j by $\varepsilon_{j,0}$ ($j \in e_n$).

Proof of Theorem 2. Just as in the proof of Theorem 1, below we can replace the vector $\ae_j$ by $\overline{\ae}_j$, $j \in e_n$. By Lemma 8.6 [3] there is a domain $G_z \subset G'$ ($z = (z_1, z_2, \ldots, z_n)$, $z_j = \alpha_j r(x)$, $\alpha_j > 0$, $j \in e_n$; $r(x) = dist(x, \partial G')$, $x \in G'$). Suppose that $|\xi_j| < z_j$ ($j \in e_n$), then for every $x \in G_z$ the segment with endpoints x and $x + \xi$ lies in G'.

Consequently, (22) is valid for all points of that segment with the same kernels. After several transformations we obtain

$$|\Delta(\xi, G)D^\nu f(x)| \le C_1 \sum_{e \subseteq e_n} \prod_{j \in e'} T_j^{-1-\nu_j} \int_{0^e}^{|\xi^e|} \frac{dt^{e \vee i}}{\prod_{j \in e} t_j^{2+\nu_j}} \times$$

$$\times \int_{R^n} \left| L_e^{(\nu)} \left(\frac{y}{t^e + T^{e'}}, \frac{\rho(t^e + T^{e'}, x)}{t^e + T^{e'}} \right) \right| |\Delta(\xi, G)f_e(x+y, t^e)|\, dy +$$

$$+ C_2 \sum_{e \subseteq e_n} \prod_{j \in e'} T_j^{-1-\nu_j} \prod_{j \in e_n} |\xi_j| \int_{|\xi^e|}^{T^e} \frac{dt^{e \vee i}}{\prod_{j \in e} t_j^{3+\nu_j}} \int_{R^n} \left| L_e^{(\nu+1)} \left(\frac{y}{t^e + T^{e'}}, \frac{\rho(t^e + T^{e'}, x)}{t^e + T^{e'}} \right) \right| \times$$

$$\times \int_0^1 |f_e(x + y + u_1\xi_1 + \cdots + u_n\xi_n, t^e)|\, du\, dy =$$

$$= C_1 \sum_{e \subseteq e_n} F_1^e(x, \xi) + C_2 \sum_{e \subseteq e_n} F_2^e(x, \xi), \tag{28}$$

where $|\xi^e| = (|\xi_1^e|, |\xi_2^e|, \ldots, |\xi_n^e|)$, $\left|\xi_j^e\right| = \left|\xi_j\right|$ for $j \in e$, $\left|\xi_j^e\right| = 0$ for $j \in e'$, and $0 < T_j \le \min(1, t_{0,j})$, $j \in e_n$. We can consider, that $|\xi_j| < T_j$ for $j \in e_n$, consequently, $|\xi_j| < \min(z_j, T_j)$ for $j \in e_n$. If $x \in G' \backslash G_z$, then by definition,

$$\Delta(\xi, G')D^\nu f(x) = 0.$$

By (28) we have

$$\left\| \Delta(\xi, G')D^\nu f \right\|_{q,G'} = \left\| \Delta(\xi, G')D^\nu f \right\|_{q,G_z} \le$$

$$\le C_1 \sum_{e \subseteq e_n} \| F_1^e(\cdot, \xi) \|_{q,G_z} + C_2 \sum_{e \subseteq e_n} \| F_2^e(\cdot, \xi) \|_{q,G_z}. \tag{29}$$

Using inequality (8) for $U = G'$, $\eta_j = |\xi_j|$, as $\gamma \to \infty$, we obtain

$$\| F_1^e(\cdot, \xi) \|_{q,G_z} \le C_3 \prod_{j \in e_n} |\xi_j|^{\varepsilon_j} \times$$

$$\times \prod_{\mu=1}^N \left\{ \left\| \prod_{j \in e} t_j^{-l_j^\mu} \delta^{m^e}(\delta u) f \right\|_{p_\mu, a, \overline{\mathstrut x}, \tau} \right\}^{\beta_\mu}, \tag{30}$$

and by inequality (9) for $U = G'$, $\eta_j = |\xi_j|$, as $\gamma \to \infty$, we find that

$$\| F_2^e(\cdot, \xi) \|_{q,G_z} \le C_4 \prod_{j \in e_n} |\xi_j|^{\varepsilon_j'} \prod_{j \in e} |\xi_j|^{\varepsilon_j'-1} \times$$

$$
\times \prod_{\mu=1}^{N} \left\{ \left\| \prod_{j\in e} t_j^{-l_j^{\mu}} \delta^{m^e}(\delta u) f \right\|_{p_\mu, a, \overline{\overline{\mathrm{æ}}}, \tau} \right\}^{\beta_\mu} \leq
$$

$$
\leq C_6 \prod_{j\in e_n} |\xi_j|^{\sigma_j^1} \prod_{\mu=1}^{N} \left\{ \left\| \prod_{j\in e} t_j^{-l_j^{\mu}} \delta^{m^e}(\delta u) f \right\|_{p_\mu, a, \overline{\overline{\mathrm{æ}}}, \tau} \right\}^{\beta_\mu} . \tag{31}
$$

From inequalities (29)-(31), under the condition $1 < \theta_\mu < \infty$ and $p_\mu \leq \theta_\mu$ ($\mu = 1, 2, \ldots, N$), we obtain inequality (26).

Suppose now that $|\xi_j| \geq \min(z_j, T_j)$ for $j \in e_n$. Then

$$
\left\| \Delta(\xi, G') D^\nu f \right\|_{q, G'} \leq 2 \left\| D^\nu f \right\|_{q, G'} \leq C(z, T) \left\| D^\nu f \right\|_{q, G'} \prod_{j\in e_n} |\xi_j|^{\sigma_j^1} .
$$

In this case, estimating $\left\| D^\nu f \right\|_{q, G}$ by means of (19), we again obtain the desired inequality. Theorem 2 is proved.

References

[1] T.I.Amanov, Theorems representations and imbeddings functions spaces SB. *Tr. MIAN SSSR*, 1965, v.77, 5-34 (Russian).

[2] A.D.Djabrailov, On some functions of spaces, *DAN SSSR*, 1964, v.159, No 2, 254-257 (Russian).

[3] O.V.Besov, V.P.Il'yin, S.M.Nikolskii, Integral representati- ons of functions and imbeddings theorems. M. *Nauka*, 1996, 480p.

[4] P.I.Lizorkin and S.M.Nikolskii, Classification of differenti- able functions on the basis of spaces with dominant mixed derivatives. *Trud. Math. Inst. Steklov*, 1965,v.77, 143-167 (Russian).

[5] V.S.Guliev and A.M.Najafov, On some imbeddings theorems of Besov-Morrey and Lizorkin-Triebel-Morrey type spaces. *Voronejjsk. Zimn. math. shkola*, 1999, 71 (Russian).

[6] V.S.Guliev and A.M.Najafov, The imbedding theorems on the Lizorkin-Triebel-Morrey type spaces. Progress in Analysis. *Proceeding of the 3rd International ISAAC Congress*, Berlin, 2001, v.1, 23-30.

[7] V.P.Il'yin, On some properties of the functions of spaces $W_{p, a, \mathrm{æ}}^l(G)$. *Zap. Nauch. sem. LOMI AN SSSR*, 1971, v.2, 33-40 (Russian).

[8] C.B.Morrey, *Multiple integral problems in the calculus of variations and related topics*. Univ. California Puble., 1943, v.1, 34-63.

[9] C.B.Morrey, Second order elliptic of differential equations. *Ann. Math. studies*, 1954, v. 33, 101-159.

[10] C.B.Morrey, Second order elliptic equation in several and Holder continuity, *Math. Zet.*, 1959, v. 72, No 2, 146-164.

[11] A.M.Najafov, The imbedding theorems for the space of Besov-Morrey type with dominant mixed derivatives. *Proceedings of IMM of NAS Azerbaijan*, 2000, v.XII(XX), 97-104.

[12] A.M.Najafov, The imbedding theorems in the Lizorkin- Triebel-Morrey type space with dominant mixed derivatives. *Proceedings of IMM of NAS Azerbaijan*, 2001, v.XV (XXIII), 121-131.

[13] A.M.Najafov, Some properties of functions from the intersection of Besov-Morrey type spaces with dominant mixed derivatives. *Proceedings of A.Razmadze Math. Institute*, 2005, v.139, 71-82.

[14] A.M.Najafov, On integral representations functions from spaces with dominant mixed derivatives. Vest. Bakin. Univer., *physic-mathematical science series*, 2005, No 3, 31-39 (Russian).

[15] A.M.Najafov, Problem on the smoothness of solutions of one class of hypoelliptic equations. *Proceedings of A.Razmadze Math. Institute*, 2006, v.140, 131-139.

[16] A.M.Najafov, Embedding theorems in the Sobolev-Morrey type spaces $S_{p,a,æ,\tau}^{l} W(G)$ with dominant mixed derivatives. *Siberian Math. Jour.*, 2006, v.47, No 3, 613-625.

[17] Yu.V.Netrusov, On some imbedding theorems of Besov-Morrey type spaces. *Zap. Nauch. sem. LOMI AN SSSR,* 1984, v.139, 139-147 (Russian).

[18] S.M.Nikolskii, The functions with dominating mixed derivatives satysfying the multiple Holder condition. *Siberian Mathematical Journal*, 1963, v.4, No 6, 1342-1364 (Russian).

[19] J.Ross, A Morrey-Nicolskii inequality. *Proceed. Amer. Soc.*, 1980, v.78, 97-102.

[20] H.Triebel, Interpolation theory. Functions spaces. *Differential operators*. Berlin, 1978.

In: Focus on Science and Technology...
Editor: Sergo Gotsiridze, pp. 143-147

ISBN: 978-1-61209-970-5
© 2011 Nova Science Publishers, Inc.

Chapter 14

SIGNIFICANCE AND TENDENCIES OF SMALL BUSINESS DEVELOPMENT

T. Shengelia and Kh. Berishvili*
Tbilisi State University, Georgia

Before the 60s of the XXI century small business was considered to have no prospects and to be the most weakly developed sector of economy. It started development after the end of the II World War, when developed countries orientated at full employment of the population and mass production of goods. After that small business has become one of the indispensable and dynamically developing parts in the economy of developed countries. Due to its special place it plays an essential role not only in socio-economic, but in public-political life of these countries as well. Through small business these countries achieve common national wellbeing, social security and stability of wide circles of the society. Quite important part of GDP and national income is being formed in the sphere of small business. Diverse produce of both production designation and wide consumption is being prepared here in large volumes. However, more important is the situation that a considerable part of the population able to work is engaged just in small business. For example, in 6 countries of EU (Spain, the Netherlands, Belgium, Germany, France, and Italy) a specific share of employed in small and medium enterprises fluctuates within 40-70%. Small and medium enterprises of these countries make up 90% of their general amount and they produce 35-71% of newly formed value of the produce (1, p 8).

A good example of small business development is the USA. Still in 1953 the law "on small business" was adopted there, and special state body – "administration of small business issues" – was formed with its regional department and 4000 staff members. One of the speeches of renowned President Ronald Reagan in 1984 says: "Good health and force of small enterprises of America is an important key of health and power of our economy. In reality, small business is America". Therefore, support to small business activities is most vital for the American politicians. This is just that very sector, in which along with medium-sized enterprises, there is annually produced more than 50% of output, and more than 2/3 of

---

* E-mail address: shengelia.temur@gmail.com

increase in new work places. 20 million work places were formed in the USA through small entrepreneurship only in the last ten years. In 1999 in the USA there were registered and successfully functioned 4.4 million independent small companies and 5.5 million small enterprises. In 2006 their amount reached 19 millions, and their specific share in GDP was 40% (2, p 60). This number does not involve small companies working in agriculture. Small companies in this sector make up 75% in general amount of farms. 30% of the land plots, 40% of labor force and 50% of agricultural machines come on them. In 1989-1999 a general amount of enterprises increased by 15, 6% in the entire economy of the USA, among them of small enterprises – by 29, 3% (3, p 17).

To adopt measures for small business development is the subject of special care of the central government in the USA. 4 billion USD is being annually allocated from the USA state budget for subsidies designated for small and medium business, and 300 million USD – to fulfill the grant programs as well as privileged taxes. As for technical-technological, management and marketing assistance and information software for small enterprises, this is fully done by the administration for small business issues within the programs to be elaborated by it.

Japan follows the USA with high level of small business development. Small business sector in Japan played a key role in rapid development of the country in the post-II World War ten years. In this period small and medium-sized enterprises were demanded to be very adaptive and flexible. They succeeded in this and managed to improve their technical and managerial potentials. In result, the difference in efficiency of workers and wages considerably reduced in them as compared with big enterprises. Broad network of small and medium-sized enterprises makes firm basis of world-renowned Japanese concerns. It plays a key role in the economic life of this country. By the data of 2008, 55% of industrial produce of Japan comes on small business, and 40% of export of Japan is conducted by small and medium-sized enterprises (4, p 74). To develop small business, cooperation between big and small companies is very important for Japan along with the accord of their interests. It is especially characteristic to the sphere of high technologies. The government of Japan turns special attention and renders intensive support to such symbioses and through legislation regulates the relations between big (contractors) and small producers (subcontractors). This legislation is elaborated so that it gives certain guarantees and profit to small entrepreneurs. Such beneficial terms cause the situation that 99, 2% of the enterprises in Japanese economy are small and medium-sized. They provide with work 80, 6% of employed in this sector (5, p 135). Small business development in Japan is promoted by tax privileges and cheap sources of getting credits. There exist a state establishment in the country, the so-called "small business financial corporation", which gives privileged credits to the small enterprises and also special soft loans for modernization of production, for import of deficit raw material, and for leaving the home market for the world market, etc.

Leading positions are occupied by small business in the economy of France as well, where in the 90s small and medium-sized companies occupied 99, 4% in the total amount of enterprises (6, p 75). Small entrepreneurship here is most developed in engineering industry, food industry, production of electric appliances, wood processing and furniture production. 50% of labor resources of France are employed in small entrepreneurship; small companies produce 45% of industry in this country (6, p 73). Like all the developed countries the government of France conducts numerous measures to support small entrepreneurship. Privileged taxation and crediting should be specially mentioned. To credit small entrepreneurship there are formed in France state and semi-state finance institutions (e.g. national cash department for crediting the agriculture, deposit cash department, etc), and credits given by them are far cheaper. In addition, small enterprises in France have privileges in taxing their incomes and VAT.

In Great Britain in 2000-2008 the amount of small companies increased from 200 000 to 1.5 million, and the amount of employed in them – from 2 million to 2.9 million (4, p 73). In total amount of processing industry enterprises in Great Britain a specific share of small enterprises makes up 90% (2, p 62).

Small entrepreneurship is largely developed in Germany, Italy, Spain and other countries. It was always of great importance in Germany to form a stratum of entrepreneurs, which had great support. The government of Germany simplified the terms of small enterprises formation. There also acts here the "the program of filling the authorized capital". Through this program 25 billion Deutsche marks were allocated for the beginner-entrepreneurs. There also exists an investment grant designated for entrepreneurs to fill their private starting capital in business. The government of Germany also sponsors the so-called "European revival program (ERP)", which assists formation of enterprises through the "Marshal Assistance Partnership Fund". Soft loans are given in great amount automatically for those small enterprises, which do not satisfy the investment grant demands. Tax privileges also exist in Germany for small companies. When annual income of a company is less than 2 millions, its rate of tax on income reduces by half. As for other services for small business – information, consulting, engineering, etc, this is headed by Trade-Industrial Chamber of Germany, which assists also financially the small companies, working on export. About 500 small and medium-sized enterprises of Germany conducted export of goods and services at about 50 billion Deutsche marks in 2005, i.e. 7% of the entire German export (9, p 6).

The policy of small business support in Italy started long ago. Since 1952 the law has existed, on the basis of which the institute was founded for allocating the credits to small and medium-sized enterprises. In 1976 "the national fund for privileged crediting" was founded. In 1977 "the conservation and reconstruction fund" was formed for giving loans to the enterprises, which were being reconstructed and modernized. A special fund for technical renovation functions in the Ministry of Industry in Italy. Its 20% is given to small and medium-sized enterprises for the term of 15 years and more (8, p 56).

In Spain the amount of small and medium companies reaches 2 millions. 65% of labor force is employed in them and they produce 64% of the output (8, p 56). A 5-year program was elaborated in Spain for 2000-2005 for full assistance to the small and medium companies.

Small business development played an important role in upsurge of economy in other countries as well. It made an economic miracle in new industrial countries of Asia – Korea, Taiwan, Singapore and others.

Small business becomes one of the most important sectors of economy in the countries of East Europe. For example, in Hungary 14, 8% of those, working in industry, are engaged in small entrepreneurship, and 9, 8% - in Bulgaria. Promotion to small economy development is regulated in Hungary at the highest governmental level. In this country, organizational-legal issues for small business promotion are charged on Economic Chamber of the country. Special organizations for consulting services are formed in it, where the representatives of small enterprises can get conjecture survey on goods of certain kinds, suppliers of produce, prices, etc. There is formed here a special scheme for privileged crediting of enterprises. In

Bulgaria, still in 1984 a governmental program was adopted – "on direction of accelerated development of small and medium enterprises in economy till 1990", which stimulated formation of such enterprises on the developed technological basis. Such a law was adopted in Romania in 1980.

The same approaches are to small business development in Poland, Czechia, Slovenia and other countries.

Small business is properly developed in Latin America and Africa. In African countries the amount of employed in small enterprises does not exceed 10, in Asian countries – 50, and in Latin America – average 75. The foundation capital is determined respectively within 100, 190 and 350 thousand USD. An efficient mechanism of state support to small and medium entrepreneurship is formed everywhere, which assists them annually in producing the output worth 200-250 thousands (9, p 108).

Table 1. Specific share of small business in economic indicators of entrepreneurial sphere of Georgia in 2000-2008

	2000	2008	Change in 2000-2008 (+, -)
1.Total amount of enterprises,	28 547	23 137	5 410
among them the amount of small enterprises,	4 726 150	18 576	7 574
specific share of small enterprises	91. 6%	80. 3%	11. 3
2. Total production (mln GEL),	2 382. 6	9 645. 4	+7 262. 8
among them production in small enterprises,	493. 7	392. 0	-101. 7
specific share of production in small enterprises	20. 7%	4. 1%	-16. 6
3. Goods turnover (mln GEL), total,	3 781. 4	17 544. 4	+13 763
among them goods turnover in small enterprises (mln GEL),	1 265. 5	932. 9	-332. 6
specific share of small enterprises in goods turnover	33. 5%	5. 3%	-28. 2
4. Amount of employed, total (man),	378 055	337 765	-40 290
among them amount of employed in small enterprises	125 807	46 435	-79 372
specific share of employment in small enterprises	33. 2%	13. 8%	-19. 4

Source: Indicators of the Georgian Statistics Department, Tb., 2009.

In this background of world experience small business development in Georgia seems too modest. As of 1 January 2008 in total amount of enterprises in Georgia small enterprises occupy 80, 3%. This indicator is equal to corresponding indicators of EU countries, but is behind them in output production and employment by specific indicators. Here it occupies respectively 4, 1 and 5, 3% (see Table 1). In EU countries these indicators fluctuate within 35-71% and 40-70% (1, p 8). The situation is the same in Japan, the small enterprises of which employ 80, 6% of work-able population and form 59% of industrial produce. In France, in small enterprises of which 45% of industrial produce of the country is produced and 50% of work-able population are engaged. In Spain, these indicators correspondingly make up 64% and 65%, etc.

The figures, given in Table 1, witness that no terms have been formed in Georgia yet, necessary for promoting small business development. Unfortunately, dynamics of economic indicators of small enterprises within recent years does not give any hope. The same Table shows that in 2000-2008 in small sector of Georgian economy production of output reduced by 16, 6% item, the amount of small enterprises – by 7 574 units, goods turnover – by 28, 2% item, and employment – by 79 237 men. Consequently, in 2000 there were better terms for small business development in Georgia than today. Otherwise, what we can do with the facts that in 2000 small sector occupied 33, 2% in entire employment (in 2008 – 13, 8%), in goods turnover – 33, 5% (in 2008 – 5, 3%), in output production – 20, 7% (in 2008 – 4, 1%), and in the amount of enterprises – 91, 6% (in 2008 – 80, 3%). In the background of reduction in the amount of taxes and in the rates of some taxes in Georgia after 2005, such recession in small sector of entrepreneurship should not have happened. However, it did happen. This was caused by wrong measures. Reduction started with poor protection of entrepreneurs' rights of ownership on the part of the state. Obviously, in such a situation the entrepreneurs, mostly small entrepreneurs, abstained from starting business, and already functioning small enterprises stopped their activities for an indefinite period of time.

In Georgia improvement of the situation in small business depends on direct foreign investments in conditions of scanty budget, political will of the local government directed at overcoming the poverty, formation of middle stratum, and settlement of the problems, which are still urgent for the development of Georgia.

References

[1]   Policy of small and medium entrepreneurs development in Georgia, Tb., *"Smeda"*, 1997.
[2]   Shengelia, T. *Fundamentals of business administration.* Tb., 2008.
[3]   Shurghaia, O. Place of small business in economy. *Jrnl. "Economics"*, 2006, Nos. 6-8.
[4]   Razumnov, I. I. Small companies in the USA, *economy and management. M.*, Nauka, 2008.
[5]   Kevkhishvili, N.   Small business in economy of developed countries. *Jrnl. "Economics"*, Nos. 4-5, 1994.
[6]   Vilenski, A.Y. State policy of Japan in regard to small and medium enterprises. *Jrnl. "ECO'*, No. 6, 2006.
[7]   Gvelesiani, R.. Strategy and culture of success in small and medium entrepreneurship. Tb., *"Samshoblo"*, 1999.
[8]   A. Grishikashvili. *Countries of transitional economy.* Tb., 2009.
[9]   Ulyakhin, V. K. *Small capitalistic entrepreneurship in countries of the East.* M., 2009.
[10] *Statistical collection* – Entrepreneurship in Georgia. Tb., 2009.

In: Focus on Science and Technology…
Editor: Sergo Gotsiridze, pp. 149-154

ISBN: 978-1-61209-970-5
© 2011 Nova Science Publishers, Inc.

Chapter 15

THE TRIBOLOGICAL EFFICIENCY AND THE MECHANISM OF HIGHLY DISPERSED AMORPHOUS CARBON AS A DOPE TO PLASTIC LUBRICANTS

D.S. Iosebidze, E.R. Kutelia, G.S. Abramishvili, T.M. Apakidze, A.P. Chkheidze and M.T. Khvedelidze

Georgian Technical University, Kostava Str. 77, 0175 Tbilisi, Georgia

Nowadays chemically active antiwear and antiscoring dopes and dispersive additives (graphite, molybdenum disulfide, powder-like lead, copper, babbit etc.) are used for improving the tribological properties of plastic lubricants.

Currently technical carbon (carbon black), which is characterized by high thermal and chemical stability and also by low cost, is widely used as a lubricant additive. At the sume time, commercial carbon black does not manifest high antiscoring properties. Moreover, in some cases, by catalyzing the oxidotiun processes, it makes worse the thermooxidation stability of dispersion medium and, hence, the rheological and protective properties of lubricants. Consequently, we have modified the commercial carbon black for improving its efficiency.

Commercial carbon black PM-100 was modified with the fractions of polycyclic hydrocarbons with 15-21 carbon atoms in side chains, which were isolated by thermal-diffusion separation of highly aromatic oil, in the absorbing column.The spindle oil thickened with 20 % modified carbon black PM-100 (i.e. PM-100m) has better tribological and other properties than the plastic lubricant wich the same amount of commercial carbon black PM-100 (Table 1). At the same time, doping of 5 % PM-100m to lubricant AC-1 increases its ultimate strength, colloidal stability, drop point and lubricating properties. The composition of lubricant AC-1 with 5 % PM-10m (Table 1) has undergone standard tests in hinged joints of field binoculares 7x35 and appeared to fit well the specifications.

As is known, the crystallites of carbon black particles form a so-called turbostratic (disordered layer) structure with the disorder of mutual arrangement of crystallites increasing from the surface of the particle to its centre while theim density decreases (Figure 1). It should be emphasized that the order of crystallites also decreases with reduction in particle sizes and the degree of their graphitization [1-4], This regularity is of great importance for increasing the degree of amorphism of carbon black by increasing the degree of its dispersion and, thus, of non-equilibrium of carbon black particles, which is the basis of highly efficient multifunctional mechanism of action of oils with highly dispersed amorphous carbon [5-6].

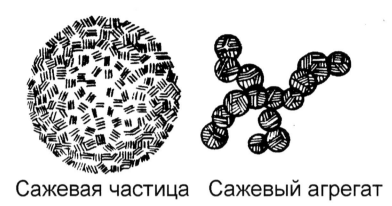

Сажевая частица   Сажевый агрегат

Figure 1. The structure of carbon black.

In this regard the most promising are the additives of highly dispersed amorphous carbon (HDAC) produced at electrocracking of liquid hydrocarbons, e.g. benzene (CEKb) and oxidizing pyrolysis of natural gas, e.g. methane (CPm). Their comparison with higly dispersed materials PM-100 and DG-100 (GOST 7885-68) which are furnace and channel blacks, the most widespread types of industrial carbon black at present, showed that the fundamental difference between additives CEKb and CPm and serial carbon black additives is the higher dispersion of the first ones. So, the mean diameters of particles of CEKb and CPm make up 250 Å and 290 Å, respectively, while those of serial additives PM-100 and D-100 make up 320 Å and 353 Å, respectively. At the same time, CEKb and CPm contain a class of very fine particles which are close by their sizes and have the diameter less than 200 Å. The higher degree of dispersion of CEKb and CPm is also attested by the fact that their geometric specific area is greater than that of the abovementioned industrial blacks (Table 2).

In view of the abovementioned and the fact that the degree of order of crystallites in carbon black particles decreases with the decreasing size of the latter, additives CEKb and CPm are more non-equilibrium than commercial carbon black additives PM-100 and DG-100. Due to high non-equilibrium, additives CEKb and CPm are capable of undergoing polymorphic conversions at friction with formation of secondary structures on rubbing surfaces possessing better tribological and other functional properties as compared to technical carbon black additives [5-6].

We have established that the main mechanism of formation of tribologically efficient surface structures is the synthesis of heterophase thin-film structure, containing the phases of amorphous carbon, graphite, and hexagonal (lonsdaleite) and cubic diamond (Figure 2) at friction of metallic parts in media of oils and plastic lubricants with HDAC (CEKb and CPm). Those phases are dispersion-ordered in the form of regular graphite-carbon cells and a network with statistically uniform distribution of tribosynthesized diamond nanoparticles [5].

Table 1. Properties of lubricants and their compositions with CPm and modified technical carbon PM-100 (PM-100m)

Index	Spindle oil AV		Lubricant AC-1		Dope Acor-1		Lubricant PFMS-4s	Oil PFMS-4s		Conventional antifriction lubricant [7]
	+20 % PM-100m	+20 % PM-100m	-	+5% PM-100m	+30% PM-100m	+30% CPm		+30% PM-100m	+30% CPm	
Ultimate stength, Pa, at °C:										
20	450	720	860	1140	420	610	100-200	320	470	40-2000
50	260	440	-	-	-	280	-	-	-	100-700
Colloidal stability, weight %	10.5	13.0	9.6	4.0	plasti pressed out	6.5	2	2.5	3.1	1-35
Wash-off index, %										
5 min.	3.8	6.0	-		2.5	2.5	-	-	-	0-100
30 min..	15.5	12.4	-		6.1	5.9	-	-	-	0-100
Thermal strength, %	41.6	11.5	-		15.2	7.5	8.2	6.8	4.3	0-100
Drop point, °C	250	250	130	240	240	240	251	253	258	60-260
Metal corrosion:										
steel	prssed out.*	prssed out.*	-		prssed out.	prssed out.	-	prssed out.	prssed out.	prssed out.
bronze	did not press out.**	prssed out.**	-		prssed out.	prssed out.	-	prssed out.	prssed out.	prssed out.
Lubricanting properties:										
Дн, мм	0.55	0.53	0.60	0.62	0.55	0.55	0.72	0.68	0.63	-
Рк, Н	800	940	750	1000	840	1000	450	500	750	550-1120
Рс, Н	2500	2820	2000	2660	3550	4000	1880	2500	3550	1750-2250

*Withstands **Does not withstand

Table 2. Characteristics of Highly Dispersed Amorphous Carbon Dopes

Composition and properties	Dope			
	CEKb	CPm	PM-100m	DG-100
Mean-arithmetic diameter, Å	250	290	320	353
Specific area, m²/g geometric adsorption	114 124	90-120 125	95-105 115	99-100 140
Size of crystallites, Å La Lc d_{002}	13-18 12-13 3.59	15-17 11-12 3.59	24-27 10-11 3.68	20-23 11-12 3.67

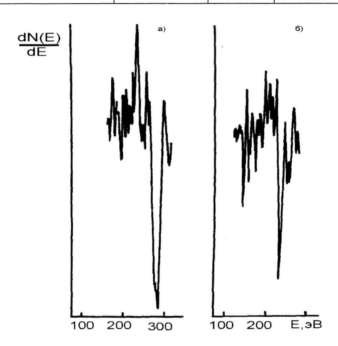

Figure 2. Differential K L L – Auger spectra of proportional carbon crystallites of natural diamond (a) and the diamond synthesized on the friction surface in the medium of plastic lubricant with highly dispersed amorphous carbon HDAC (b).

Apparently, one of the basic characteristics determining the opportunity of tribosynthesis of diamonds from highly dispersed amorphous carbon HDAC is the structure of its near order. According to the results of calculation of the radial function of atomic density distribution, HDAC (CEKb and CPm) has closer packing in the first three coordination spheres in comparison with commercial carbon PM-100 (Table 2). HDAC is characterized by the radius of the first coordination sphere practically coinciding with that of diamond and by a positive difference between the radii of the second and the third spheres. At the same time, the graphite cluster packing differs significantly from the diamond in all spheres with a negative difference between the corresponding radii. Based on the data listed in Table 3, it could be inferred that the "direct" transition of HDAC (CEKb and CPm) into diamonds under compression with a shift at friction is possible.

**Table 3. The Results of Calculation of the Radial Function
of Atomic Density Distribution of Carbon Materials**

Carbon material	The closest interatomic spacing, E						
	R_1	R_2	R_3	R_4	R_5	R_6	R_7
HDAC (CEKb, CPm)	1.55	2.75	4.10				
PM-100	1.65	2.90	4.20				
Graphite	1.42	2.45	2.83	3.34	3.63	3.75	4.2
Diamond	1.54	2.52	3.57	3.89			

According to the results of laboratory investigation, used as thickeners or dopes commercial carbon blacks PM-100 and DG-100, as well as modified black PM-100, are exceeded by highly dispersed carbon When CEKb and CPm (Tables 1 and 4). In particular, as is obvioous from Table 1, the sample of lubricant made by thickening of nitrated oil Acor-1 with 30% CPm by its structural-mechanical and lubricating properties and thermal strength exceeds the lubricant containing 30% modified black PM-100. At the same time, at thickening of silicone oil PFMS-4 (GOST 15866-70) with 30% CPm (Table 1), such indices of lubricating properties of the basic oil as critical load, welding load and a scoring index are significantly higher as well (750 H, 3550 H and 40.3, respectively) than in the case of thickening with 30 % commercial black DG-100 (450 H, 2500 H and 24.2, respectively).Thus, sample plastic lubricants Acor-1 with 30 % CPm and PFMS-4 with 30% CPm are quite efficient and promising as compared with the convetional antifriction lubricant (Table 1).

Highly dispersed amorphous carbon (CEKb and CPm) also manifest the properties of an efficient dope (Table 4) which significantly improves the tribological properties of various commercial antifriction lubricants (e.g. Russian-produced lubricants: Solidol S - of general purpose; Litol-24-multipurpose; PFMS-4s-heat-resistant; CIATIM-201-instrumental).

**Table 4. Lubricating Properties of Plastic Lubricants
with Highly Dispersed Carbon Dopes**

Name	Lubricating properties (GOST 9490-75)		
	Pw, H	Pc, H	I_s
Solidol S	560	2000	31.6
The same with 10% CEKb	750	2660	40.7
10% CPm	840	2500	45.6
10% CPm, 3% DF-11	1000	2820	47.9
10% PM-100	670	2370	35.6
Litol-24	630	2000	35.5
The same with 10% CEKb	840	2500	41.8
10% CPm	1000	2500	39.4
CIATIM-201	400	1500	34.2
The same with 10% CEKb	500	1880	37.5
10% CPm	560	2000	38.7
PFMS-4s	450	1880	30.7
The same with 10% CPm	670	3000	40.3
10% DG-100	500	2370	33.2

Thus, at addition of 10% of CEKb and CPm (Table 3), the increase in the indices of antiscoring properties such as critical load, welding load, scoring index etc. makes up 31% and 40%, respectively, for Solidol S; 30% and 38%, respectively, for CIATIM–201; 46% (at addition of 10% of CPm) for PFMS-4S. At the same time, the addition of 10% of commercial carbon black PM-100 and DG-100 to Solidol S and PFMS 4S increases the abovementioned indices by less than 17% and 15%, respectively. Dope CPm in the lubricant composition shows good compatibility with chemically active antiwear additive DF-11 which, in its turn, shows synergism in relation to antiscoring efficiency of CPm.

References

[1] Shulzhenko, I. V., Kobzon, R. I., Nikanorov, E. M. et al. *Black Plastic Lubricants. Topical Review.* –Moscow, TNIITENeftechim, 1983,44p.

[2] Gyulmisaryan, T. G. *Production Process of Technical Carbon (Black).* –Moscow, MINK and I. M. Gubkin GP, 1979, 78 p.

[3] Pechkovskaya, K. A. *Carbon Black as a Rubber Strengthener.* –Moscow, Chimia, 1968, 215 p.

[4] Smirnov, B. N. and Fialkov, A. S. The Structure of Carbon Black Particles according to the Data of Electron Microscopic Analyses. - Proceedings of the All-Union Research and Design Institute of Electrocarbon Items. –Moscow, *Energy,* 1972, pp. 4-10.

[5] Iosebidze, D. S., Kutelia, E. R., Bershadski, L. I. et al. On the Phenomenon of Tribosynthesis of Dissipative-Ordered Heterophase Structures on the Base of Carbon. -Reports of Georgian Academy of Sciences, *Tbilisi,* 1987, No, 1, 128, pp. 29-32.

[6] Iosebidze, D. S. *Research and Development of Automobile Transmission Lubricants Based on Highly Dispersed Carbon.* – Competitor's Report for the Degree of Doctor of Technical Sciences. –Ufa, Russia, 1989, 50p.

[7] Fuel, Lubricants, Technical Liquids: Assortment and Applications, -Moscow, *Techinform,* 1999, pp, 307-364.

In: Focus on Science and Technology…
Editor: Sergo Gotsiridze, pp. 155-166

ISBN: 978-1-61209-970-5
© 2011 Nova Science Publishers, Inc.

Chapter 16

MODELLING OF ELASTIC WAVES GENERATED BY A POINT EXPLOSION

Z. Kereselidze[1,], T. Gegechkori[1], N. Tsereteli[1] and V. Kirtskhalia[2]*

[1]Institute of Geophysics, M. Aleksidze str. 1, 0193 Tbilisi, Georgia
[2]Sukhumi Physical and Technical Institute, A. Kazbegi Ave. 12.,
0168 Tbilisi, Georgia

Abstract

On the point explosion area boundary, the spectrum of elastic oscillations, which generate volumetric standing waves, and also the index of its discreteness (ratio between individual frequencies) should not differ essentially from the frequency spectrum of volumetric seismic waves recorded at a far distance from the point explosion center. Therefore, using the proposed model, after defining the basic frequencies of natural oscillations of the point explosion area boundary by the spectral analysis of seismograms, we can obtain a sufficiently correct analytic solution of the inverse seismological problem which consists in defining the linear characteristics of the point explosion area and estimating the energy released.

Introduction

A point explosion is followed by an avalanche-like release of energy resulting in the generation of shock waves and plastic deformations. Formally, the explosion center will go on expanding till the moment of time at which a state of balance establishes between the pressure force causing plastic deformations and the elastic forces of earth rocks. It is obvious that that this balance will take place at distances where the explosion energy density will be commensurable with the density of elastic deformation energy. In a homogeneous disturbance medium, waves produced by a point explosion will propagate in a practically radial manner. Even in the non-homogeneous case, the disturbance of the shock wave sphericity front near its is less pronounced than at far distances where the explosion energy density drops essentially. There, the spherical symmetry may be disturbed, but this disturbance cannot be so

---

[*] E-mail address: sipt@sipt.org

essential that the explosion region boundary could not be approximated by some regular bodies, among which the spheroid seems to be the best suitable one. Below, we will consider in terms of approximation of an incompressible medium the problem of mechanical natural oscillations of the spheroid which has similarity not only with the point explosion region, but also with the focus of most earthquakes. Therefore we think that ellipsoid surface oscillations may model with sufficient reliability the picture of generation of seismic waves produced by elastic forces on the point explosion focus boundary.

Statement of the Problem

According to the model, the point explosion region is identified with a hollow sphere or with an elongated hollow spheroid that has a discrete spectrum of mechanical natural oscillations and is defined by the elastic properties of a medium and the linear characteristics of the spheroid. It is believed that oscillations can be generated only on the boundary subjected to periodic radial displacements of small amplitudes as compared with the linear dimensions of spheroids. Usually, only the case of volumetric deformations is considered, which is sufficiently correct only for a point explosion when, as different from earthquakes, the wave effects of shear deformations can be neglected [Robin, 1972]. For an earthquake we have to deal with the problem of defining a complete wave spectrum containing both volumetric and surface waves. In the case of spherical symmetry the solution of this problem was obtained by representing the shear modulus as a sum of normal oscillation modes [Stein, et al., 2003]. Analogously to an earthquake, the problem of defining the natural oscillation frequency spectrum of the point explosion boundary is less difficult in the case of a spherical cavity, for which we can use a physical analogy with hydromechanical natural oscillations of a liquid drop of spherical configuration. Such an analogy enables us to apply the classical mathematical methods worked out for a spherical drop preserving its shape because of the action of capillary forces in the case of small disturbances of its surface [Landau, Lifshitz, 1954]. Using these methods, in [Gvelesiani, et al., 1993], the solution was obtained, which is suitable for describing the natural oscillation spectrum of a drop of ellipsoidal configuration. Thus it was shown that natural oscillations can be modeled for both a relatively simple spherical cavity and a more complex ellipsoidal one.

We will follow the scheme of solution for an ellipsoid given in [Gvelesiani, et al., 1993] and require the fulfillment of several necessary conditions. In the first place, the oscillation amplitude (the value of radial displacements of the ellipsoid boundary) must be essentially smaller as compared with the characteristic linear dimension of the problem, which is, for example, a large semi-axis of the ellipsoid: $A \ll c$. Besides, it should be postulated that the method used is valid only in that region of the space, in which the condition

$$V \ll V_0 ,$$

(1)

is fulfilled, where V is the velocity of ellipsoid surface oscillations, V_0 the sound velocity. This condition means that the problem under consideration is valid only for a hollow body whose inner surface limits the plastic deformation volume, i.e. the core region where shock waves accompanying a point explosion propagate. Thus condition (1) excludes the generation of shock waves able to overlap the elasticity effect leading to small disturbances of the ellipsoid boundary.

It is obvious that in the point explosion region a medium is incompressible, which is characteristic of most earth rocks. Therefore it can be assumed that after a point explosion a medium also remains homogeneous and the earth mass motion obeys the Laplace equation

$$\Delta\psi = 0,$$

(2)

where $\vec{V} = grad\psi$, i.e. the motion is potential.

Let us introduce the ellipsoidal system and define the dimensionless coordinates σ, τ, φ, which correspond to the elongated elipsoid. This system is related with the Cartesian system by

$$x^2 = a^2(\sigma^2 - 1)(1 - \tau^2)\cos^2\varphi, \ y^2 = a^2(\sigma^2 - 1)(1 - \tau^2)\sin^2\varphi, \ z = a\sigma\tau,$$

(3)

where a is half the interfocal distance of the ellipsoid.

Equation (2) with relation (3) taken into account is reduced to the form

$$a^2\left\{\frac{\partial}{\partial\sigma}\left[(\sigma^2 - 1)\frac{\partial\psi}{\partial\sigma}\right] + \frac{\partial}{\partial\tau}\left[(1 - \tau^2)\frac{\partial\psi}{\partial\tau}\right] + \frac{\sigma^2 - \tau^2}{(\sigma^2 - 1)(1 - \tau^2)}\frac{\partial^2\psi}{\partial\varphi^2}\right\} = 0$$

(4)

According to the model, the equilibrium condition for the ellipsoid boundary can be obtained by modifying the Laplace formula for a liquid medium, in which the coefficient of capillary surface tension giving the elasticity effect is replaced by the product of the uniform compression modulus K by the characteristic linear dimension of the ellipsoid boundary c:

$$cK = \frac{Ec}{3(1 - 2\sigma_p)},$$

(5)

where E is the tension modulus (Young's modulus) and σ_p the Poisson coefficient.

Thus, on the surface of the modeled ellipsoid the equilibrium condition is given by an expression analogous to the Laplace formula for a liquid drop

$$P_1 - P_2 = Kc\left(\frac{1}{R_1} + \frac{1}{R_2}\right) = K \cdot \frac{c}{a}\left(\frac{1}{\sigma_1} + \frac{1}{\sigma_2}\right),$$

(6)

where P_1 and P_2 are the pressures respectively inside and outside the ellipsoid, $R1$ and $R2$ are the principal curvature radii of the oscillating surface, which for the spheroid are $R_1 \rightarrow a\sigma_1$ and $R_2 = a\sigma_2$. Since the medium density is invariable, the contribution of elastic forces to pressure equilibrium, i.e. the difference between inner and outer pressures on the ellipsoid boundary is a small correction to these values. Therefore the difference between the pressures can be defined by means of the expression [Landau, et all., 1954].

$$\Delta P = P_1 - P_2 = -\rho g u - \rho \frac{d\psi}{dt} \,, \tag{7}$$

where u is the radial displacement producing the oscillation of the ellipsoidal surface. In the Cartesian system, the displacement velocity is related with the motion potential by the expression

$$\frac{\partial u}{\partial t} = \frac{\partial \psi}{\partial z} \,, \tag{8}$$

Hence the aim of this paper is to define analytically the natural oscillation frequency spectrum of an ellipsoidal hollow body. But first we should define the value in terms of ellipsoidal coordinates $\frac{1}{R_1} + \frac{1}{R_2}$ by varying the equilibrium surface. For this, we use the relationship defined by (8). Since the gravity force does not influence the elastic deformation effect without loss of generality in (7), we can neglect the first term on the right-hand side, i.e. in the sequel it will be assumed that $\Delta P = \rho \frac{\partial \psi}{\partial t}$, which coincides in form with the first motion integral.

The surface of the elongated spheroid is defined by the integral [Korn et. all., 1968]

$$f = \int\int \left[1 + \frac{g_{\sigma\sigma}}{g_{\tau\tau}} \left(\frac{\partial \sigma}{\partial \tau}\right)^2 + \frac{g_{\sigma\sigma}}{g_{\varphi\varphi}} \left(\frac{\partial \sigma}{\partial \varphi}\right)^2 \right]^{1/2} \left(g_{\tau\tau} g_{\varphi\varphi}\right)^{1/2} d\tau d\varphi \,, \tag{9}$$

where $g_{\sigma\sigma} = a^2 \frac{\sigma^2 - \tau^2}{\sigma^2 - 1}$, $g_{\tau\tau} = a^2 \frac{\sigma^2 - \tau^2}{1 - \tau^2}$, $g_{\varphi\varphi} = a^2 \left(\sigma^2 - 1\right)\left(1 - \tau^2\right)$ are the metric tensor components.

Let us define a dimensionless small disturbance of the ellipsoid surface $u^1 \to \frac{u}{a}$ (in the sequel, the stroke will be omitted). Thus $\sigma = \sigma_0 + u$, where $\sigma_0 = const$. If we take into account the smallness of the second and third terms in the square brackets in expression (9) and use expansion into a series, then we obtain

$$f \approx a^2 \int\int \left\{ \left(\sigma^2 - 1\right) + \frac{1}{2} \left[\left(1 - \tau^2\right)\left(\frac{\partial u}{\partial t}\right)^2 + \frac{\sigma^2 - \tau^2}{\left(\sigma^2 - 1\right)\left(1 - \tau^2\right)} \left(\frac{\partial u}{\partial \varphi}\right)^2 \right] \right\} \cdot \left[\left(\sigma^2 - \tau^2\right)\left(\sigma^2 - 1\right)^{-1} \right]^{1/2} d\tau d\varphi \,, \tag{10}$$

Hence, neglecting the variation of the fractional terms, for the surface variation f with respect to the variable u we obtain

$$\delta f \approx a^2 \int\int \left\{ 2\sigma\delta u + \left[\left(1 - \tau^2\right)\frac{\partial u}{\partial \tau}\frac{\partial \delta u}{\partial \tau} + \frac{\sigma^2 - \tau^2}{\left(\sigma^2 - 1\right)\left(1 - \tau^2\right)}\frac{\partial u}{\partial \varphi}\frac{\partial \delta u}{\partial \varphi} \right] \right\} \left[\frac{\sigma^2 - \tau^2}{\sigma^2 - 1} \right]^{1/2} d\tau d\varphi \,, \tag{11}$$

where $|\tau| \leq 1, 0 \leq \varphi \leq 2\pi$. For partial integration with respect to the variables τ and φ in expression (11) we can neglect the term $\dfrac{\partial u}{\partial t} \tau \left(\dfrac{\sigma^2 - 1}{\sigma^2 - \tau^2} \right)^{\frac{1}{2}}$ having a higher degree of smallness as compared with the term $\dfrac{\partial^2 u}{\partial \tau^2} \left(\dfrac{\sigma^2 - \tau^2}{\sigma^1 - 1} \right)^{\frac{1}{2}}$. As a result we have

$$\delta f \approx a^2 \int\int \left\{ 2\delta - \left[\frac{\partial}{\partial t}\left(1 - \tau^2\right)\frac{\partial u}{\partial t} + \frac{\sigma^2 - \tau^2}{\left(\sigma^2 - 1\right)\left(1 - \tau^2\right)}\frac{\partial^2 u}{\partial \varphi^2} \right] \right\} \left[\left(\sigma^2 - \tau^2\right)\left(\sigma^2 - 1\right)\right]^{\frac{1}{2}} \left(\sigma^2 - 1\right)^{-1} d\tau d\varphi \delta u \tag{12}$$

Taking the expression $df = \left(g_{\tau\tau} g_{\varphi\varphi}\right)^{\frac{1}{2}} d\tau d\varphi$ in formula (12), performing the operations of multiplication and division by $\sigma_0\left(\sigma_0 + 2u\right)$ and expanding the first term into a series with respect to u, we obtain

$$\delta f = \int\int \left\{ \frac{2}{\sigma_0} - \frac{2u}{\sigma_0^2} - \frac{1}{\sigma_0^2}\left[\frac{\partial}{\partial \tau}\left(\left(1 - \tau^2\right)\frac{\partial u}{\partial \tau}\right) + \frac{\sigma^2 - \tau^2}{\left(\sigma^2 - 1\right)\left(1 - \tau^2\right)}\frac{\partial^2 u}{\partial \varphi^2} \right] \right\} \sigma_0^2\left(\sigma_0^2 - 1\right)^{-1} df \delta u \tag{13}$$

At the same time, expression (6) implies

$$\delta f = \int\int \left(\frac{1}{\sigma_1} + \frac{1}{\sigma_2} \right) df \delta u \tag{14}$$

using which and taking into account expression (7), we obtain $\left(\dfrac{\partial \psi}{\partial t} \to a^2 \dfrac{\partial \psi}{\partial t} \right)$

$$\rho a^2 \frac{\partial \psi}{\partial t} - \frac{K \cdot c}{a\left(\sigma_0^2 - 1\right)} \left\{ 2u - 2\sigma_0 + \frac{\partial}{\partial \tau}\left(1 - \tau^2\right)\frac{\partial u}{\partial \tau} + \frac{\sigma^2 - \tau^2}{\left(\sigma^2 - 1\right)\left(1 - \tau^2\right)}\frac{\partial^2 \psi}{\partial \varphi^2} \right\} = 0 \tag{15}$$

Let us differentiate expression (15) with respect to time and, by analogy with expression (8), define the radial velocity $\dfrac{\partial u}{\partial t} = \dfrac{a}{g_{\sigma\sigma}^{\frac{1}{2}}}\Big|_{\tau=\tau_0, \sigma=\sigma_0} \dfrac{\partial \psi}{\partial \sigma}$, where $\tau = \tau_0$ and $\sigma = \sigma_0$ are fixed. Thus we have

$$\frac{\partial^2 \psi}{\partial t^2}\Big|_{\sigma=\sigma_0} - \frac{K \cdot c}{\rho a^3\left[\left(\sigma_0^2 - 1\right)\left(\sigma_0^2 - \tau_0^2\right)\right]^{\frac{1}{2}}} \left\{ 2\frac{\partial \psi}{\partial \sigma} + \frac{\partial}{\partial \sigma}\left[\frac{\partial}{\partial \tau}\left(1 - \tau^2\right)\frac{\partial \psi}{\partial \tau} + \frac{\sigma^2 - \tau^2}{\left(\sigma^2 - 1\right)\left(1 - \tau^2\right)}\frac{\partial^2 \psi}{\partial \varphi^2} \right] \right\}\Big|_{\sigma=\sigma_0} = 0 \tag{16}$$

from which, using formula (5), we eventually obtain

$$\frac{\partial^2 \psi}{\partial t^2}\bigg|_{\sigma=\sigma_0} - \frac{K \cdot c}{\rho a^3 \left[\left(\sigma_0^2 - 1\right)\left(\sigma_0^2 - \tau_0^2\right)\right]^{1/2}} \left\{2\frac{\partial \psi}{\partial \sigma} - \frac{\partial}{\partial \sigma}\left[\frac{\partial}{\partial \sigma}\left(\left(\sigma^2 - 1\right)\frac{\partial \psi}{\partial \sigma}\right)\right]\right\}\bigg|_{\sigma=\sigma_0} = 0 \tag{17}$$

Following (3), a solution for (17) will be sought in the form of a standing wave

$$\psi = A \cdot F(\sigma, \tau, \varphi)\ell^{iwt}, \; A = const, \tag{18}$$

where the function $F(\sigma, \tau, \varphi)$ satisfies the Laplace equation (4). As is well known, any solution of this equation can be represented as a linear combination of volumetric spherical functions: $F(\sigma, \tau, \varphi) = X(\sigma)Y(\tau)Z(\varphi)$, where $X(\sigma), Y(\tau)$ and $Z(\varphi)$ satisfy the following equations

$$\frac{d}{d\sigma}\left[\left(1 - \sigma^2\right)\frac{dX}{d\sigma}\right] + n(n+1)X - \frac{m^2}{1 - \sigma^2}X = 0 \tag{19}$$

$$\frac{d}{d\tau}\left[\left(1 - \tau^2\right)\frac{dY}{d\tau}\right] + n(n+1)Y - \frac{m^2}{1 - \tau^2}Y = 0 \tag{20}$$

$$\frac{d^2 Z}{d\varphi^2} + m^2 Z = 0 \tag{21}$$

The general solutions of equations of form (19) and (20) are the Legendre spherical functions P_n^m and Q_n^m. In the case of an spheroid, when there is a tendency to a sphere, i.e. when $a \to 0$, we have $a\sigma \to r, \; \tau \to \cos\theta, \; \varphi \to \varphi$, and equation (20) transforms to the usual Legendre equation

$$\frac{1}{\sin\theta}\frac{d}{d\theta}\left(\sin\theta\frac{\partial Y}{d\theta}\right) + n(n+1)Y - \frac{m^2}{\sin^2\theta}Y = 0 \tag{22}$$

while equation (19) reduces to the Euler radial equation

$$r^2\frac{d^2 X}{dr^2} + 2r\frac{dX}{dr} - n(n+1)X = 0 \tag{23}$$

whose solution has the form $X = A_n r^n + B_n r^{-(n+1)}$, where B and C are constants. It is also well known that the inner solution of equation (23) is $P_n^m(\sigma)$ which, for $a\delta \to r$, transforms to the expression r^n, while the outer solution gives, in the same limit, $Q_n^m(\delta) \to r^{-(n+1)}$

Thus, using (17), (18) and (19), the natural oscillation frequency spectrum of the surface of an elongated hollow spheroid is defined by the formula

$$\omega_n^2 \approx \frac{cK}{\rho a^3 \left[\left(\sigma_0^2 - \tau_0^2\right)\left(\sigma_0^2 - 1\right)\right]^{1/2}} \left[(n-1)(n+2) - \frac{m^2}{1-\sigma_0^2}\right] \frac{d}{d\sigma} \ln X(\sigma)$$

(24)

When the ellipsoid tends to a sphere, natural oscillation frequencies do not degenerate ($m = 0$). Therefore in that case, the adjoint Legendre functions should be replaced by usual ones.

From (24), as $a \to 0$, when $a\delta \to r, c \to r$, using the general solution $X(\sigma) \to A_n r^n + B_n r^{-(n+1)}$, for oscillations of a hollow sphere of finite thickness, for which $n \geq 2$, we obtain ($n = 0$ corresponds to a state at rest, $n = 1$ to translational motion)

$$\omega_n^2 \approx \frac{K}{\rho r^2} \left[(n-1)(n+2) \frac{nA_n r^{n-1} - (n+1)B_n r^{-(n+2)}}{A_n r^{n-1} + B_n r^{-(n+2)}}\right]$$

(25)

Expression (25) differs from the formula defining oscillations of a solid spherical drop, where only the inner solution of the Euler equation is used ($Bn = 0$) [Landau et. all., 1954]. The reason for such a restriction is obvious since the solution for a solid sphere must be valid everywhere, including the focal point at which a solution of the form $r^{-(n+1)}$ is divergent. For a hollow body this problem does not exist and therefore we can use a general solution. Such a situation is quite favorable when modeling the point explosion region whose inner surface must bound the volume in which shock waves propagate and plastic deformations take place. An example of a relatively simple formula for a hollow sphere of finite thickness clearly shows that it is only by means of a general solution of the Euler equation that we can define the size of the region where elastic forces may generate seismic waves. According to (25), using the radius R of the inner boundary we can connect the constants present in the general solution

$$nA_n R^{n-1} = (n+1)B_n R^{-(n+2)}$$

(26)

which implies

$$B_n = \frac{n}{n+1} A_n \cdot R^{2n+1}$$

The constant coefficients will be thus excluded and formula (25) will take the form

$$\omega_n^2 = \frac{K}{\rho r^2}\left[(n-1)(n+2)\frac{r^{n-1}-\left(\dfrac{R}{r}\right)^{n+2}R^{n-1}}{\dfrac{1}{n}r^{n-1}+\dfrac{1}{n+1}\left(\dfrac{R}{2}\right)^{n+2}R^{n-1}}\right], \tag{27}$$

In the framework of the direct geophysical problem, all the values contained in formula (27) are the given ones. However, if we solve the inverse problem, i.e. additionally to the physical parameters of the considered medium it is assume that the spectrum of point explosion frequencies is also the given one, then we can define the unknowns R and r, which frequently needed for various practical purposes, for example, for estimating the power of an explosion. This problem can be regarded as the first stage of the inverse seismological problem of a point explosion. It is obvious that the sought radii can be defined in terms of approximation of a hollow sphere by using two equations corresponding to some pair of frequencies, i.e. for any two n's. Naturally, for this the most convenient are the base frequency ω_2 ($n = 2$) and its next harmonic ω_3 ($n = 3$) which can be defined on a seismogram. Moreover, if in addition to them, we also give the frequency ω_4 ($n = 4$), then the unknown linear parameters of the point explosion region can be defined from the ratios $\dfrac{\omega_3}{\omega_2}$ and $\dfrac{\omega_4}{\omega_2}$ which do not depend on the error connected with the coefficient $\dfrac{K}{\rho}$. Thus we solve in principle the inverse problem of elastic oscillations of the point explosion region in terms of spherical approximation. This problem consists in defining the radius R of the elasticity boundary and the radius r of the effective surface of generation of stationary volumetric elastic waves.

The necessity of passage from a relatively simple case of a hollow sphere to a more difficult case of an elongated hollow spheroid depends of the result of spectral analysis of a seismogram. If we observe the peaks corresponding to the base frequencies (i.e. which are not blurred), and the frequencies correlate exactly enough with the base frequency, as is defined by formula (27), then this means that the spherical approximation is sufficiently exact. Otherwise, we should use the formula obtained by simplifying expression (25) and being, to a certain extent, analogous to formula (27). However, in that case the number of unknowns, the definition of which is the aim of the inverse problem, increases. In particular it is necessary to define the σ-coordinate of the oscillating outer boundary of the hollow ellipsoid, the σ_0-of the inner boundary and also the interfocal distance a. To define them, we need either three explicit values of the frequency ω_n and the coefficient K or four frequencies if we use frequencies ratios. The formulas

$$a\sigma = \left(a^2 + b^2\right)^{1/2}, \quad a = \left(b^2 - c^2\right)^{1/2}, \tag{28}$$

allow us to define the semi-axes b and c of the ellipsoid and also its eccentricity $\varepsilon = \dfrac{c}{b}$. As to the number m which appears only when we consider the ellipsoid, it may vary in the interval $[-n \div n]$, i.e. take $2n + 1$ values. However, like in the case of a sphere, without an essential limitation if ellipsoid oscillations, it can be assumed that $m = 0$, $a\sigma \to r$, $\sigma \to \infty$ are degenerate. Further, choosing values of the coordinate $\tau 0$ varying within $-1 \le \tau_0 \le 1$, we can define an extent of frequency peaks blur on seismograms. After such simplifications and taking into account that $\dfrac{\partial}{\partial \sigma} ln P_n = \dfrac{\partial}{\partial \sigma} ln Q_n = \sigma^{-1}$, we have

$$\omega_n = \frac{K}{\rho a^2 \left[\left(\sigma_0^2 - \tau_0^2 \right) \left(\sigma_0^2 - 1 \right) \right]^{1/2}} \left[(n-1)(n+2) \frac{\sigma^{n-1} - \left(\dfrac{\sigma_0}{\sigma} \right)^{n+2} \sigma_0^{n-1}}{\dfrac{1}{n} \sigma^{n-1} + \dfrac{1}{n+1} \left(\dfrac{\sigma_0}{\sigma} \right)^{n+2} \sigma_0^{n-1}} \right], \quad (29)$$

Thus on the outer boundary of the hollow sphere or the spheroid approximating the region affected by a point explosion, the elasticity factor causes small periodic displacements generating discrete stationary volumetric waves. Since we have to estimate theoretically the error in solving the inverse problem, it is obvious that we should take into consideration the possibility of transformation of these waves, the propagation of which beyond the explosion region is not limited at all. This statement is obvious if we assume that the periodically displacing points of the surface, i.e. the points performing oscillations relative to the state of equilibrium may be point sources of individual traveling waves. We may assume that a seismogram shows the presence of the results of superposition of such waves that traveled over various distances before they reached the point of their fixation. However, the first wave pulse can be regarded as arriving from the nearest point of the explosion region boundary. Formally, it cannot be excluded that the frequencies both of the first and of the next waves defined on a seismogram may turn out different from the frequencies of standing volumetric waves on the boundary. This difference must increase with an increase of the distance from the point of explosion. However, in fact, it is highly probable that there exists identity between the frequencies of standing volumetric waves on the explosion region boundary and the frequencies of traveling seismic waves. We think that the qualitative analysis given below and concerning the modeling of the propagation process of seismic waves can serve as an argumentation for this conjecture.

Traveling waves, the location of a source and the character of its change in time, which, by the conditions of the problem, are assumed to be given, must obey d'Alembert's wave equation. This equation is in fact a generalization of the ordinary equation for a wave potential, having the so-called anticipation and delay solutions. For seismological problems it is only a delay solution that is physically acceptable.

In spherical coordinates and under the assumption of radial symmetry, it has the form [Courant, et all., 1951]

$$\Delta\psi - \frac{1}{c^2}\frac{\partial^2\psi}{\partial t^2} = 4\pi\eta(r,t)$$

(30)

where c is the phase velocity of a wave, η is the source. For $t \rightarrow \infty$, we obtain the well known Poisson equation.

Let us assume that each wave source with its own η lying on the explosion region boundary has the form of the so-called Hertz vibrator but with a certain difference. In the first place, these sources give not an individual monochromatic wave, but a discrete unity (bundle) of monochromatic waves. In view of the classical interpretation of the Hertz vibrator as a result of periodic change of an electric dipole in time [Tamm, 1966], we can use its analogue, in which the dipole axis is understood as a small displacement u multiplied by a small mass which is used instead of a point charge. As different from the classical dipole where electromagnetic oscillations are produced by changing the poles of charges, in the above-mentioned analogue it is the dipole axis u that oscillates. According to the theory and in view of the connection $\frac{\partial u}{\partial t} = \frac{\partial\psi}{\partial R}$, the displacement also obeys, d'Alembert's equation (30).

Thus the spectrum of natural oscillation frequencies on the boundary of the explosion region can be regarded as formed due to a total effect of individual elementary (point) sources. For the inverse seismic problem to be well-posed it is important that this spectrum would not be subjected to transformation at the observation point. This condition can be fulfilled, for example, if we use such a type of the source η that at large distances from the generation region ($u \ll r$) gives waves whose spectral characteristics are analogous to electromagnetic waves generated by the Hertz vibrator [Tamm, 1966]. In that case, d'Alembert's equation splits in fact into a sum of wave equations corresponding to individual frequencies. For example, wave equations of delay type, each corresponding to natural frequencies of an individual vibrator, will have the form

$$\psi_n = \frac{A_n \cos\left[\omega_{0n}\left(t - \frac{r}{c}\right)\right]}{r}$$

(31)

where A_n is an amplitude, ω_{0n} is a natural frequency of some oscillation mode of the vibrator defined by formulas (27) or (30), r is the distance from the vibrator to the observation point, c is the phase velocity.

A wave propagating in a real medium may attenuate. Also, its phase, which at the fixed point depends on a mode and phase velocity, may also change. For homogeneous rocks it can be assumed that $c = const$. Therefore, at the second stage of the inverse problem, ω_{0n} and c will be the known values, and A_n and r the unknown ones (the initial moment of time t is not important). Thus the wave potential corresponding to an infinite unity (bundle) of elementary vibrators can be represented as a double sum

$$\psi = \sum_{i=1}^{\infty} \sum_{n=1}^{\infty} \frac{A_{in} \cos\left[\omega_{on}\left(t - \frac{r_i}{c}\right)\right]}{r_i} , \qquad (32)$$

from which it follows that for each source i, an individual wave generated by it, i.e. any oscillation mode may have an identical frequency. Since there may exist a phase shift between wave packages originating from various sources, even in the elementary vibrator approximation we cannot exclude amplification (or attenuation) of some oscillation mode. However, at large distances from the point explosion region surface will be hardly essential.

Conclusion

Thus the use of the uniform compression modulus as a parameter defining the elastic behavior of a medium makes this work singular from the physical standpoint. Before, for this purpose the transverse shift modulus was used, by means of which, for example, the expression was obtained only for the base oscillation frequency, but not for the entire oscillation spectrum of a point explosion [Rodin, 1974]

$$\omega_0 = \frac{2V_p}{R_p} , \qquad (33)$$

where R_p is the radius of the spherical surface of elastic wave generation, $V_p = \left(\frac{G}{\rho}\right)^{\frac{1}{2}}$ is the phase velocity of elastic wave propagation, $G = \frac{E}{2(1+\sigma_p)}$ is the transverse shift modulus (relation of shift stress to shift strain). It is obvious that for $n = 2$, expressions (27) and (29) give the values of the base oscillation mode of approximately the same order as the value defined by formula (34).

References

Courant, R. and Hilbert, D. (1951). *Methods of mathematical physics*. Part 2, Izd-vo Tekh. Lit., Moscow, p. 544 (transl. into Russian).

Gvelesiani, A., Kereselidze, Z. and Khantadze, A. (1993). On the natural oscillation frequency spectrum of the Earth's magnetosphere. Tbilisi State University Press, *Tbilisi*, 36-49 (in Russian).

Korn, G. and Korn, T. (1968). *Handbook of mathematics*. Nauka, Moscow, p. 720 (in Russian).

Landau, L. and Lifshitz, E.(1954). *Continua mechanics*. Izd-vo Tekh. Lit., Moscow, p. 795 (in Russian).

Rodin, G. (1974). *Seismology of nuclear explosions*. Mir, Moscow, p. 190

Stein, S. and Wysession, M. (2003). An introduction to seismology, earthquakes and earth structure. *Blackwell Publ.,* p 498.

Tamm, I. (1966). *Fundamental principles of electricity.* Nauka, Moscow, p. 624.

In: Focus on Science and Technology…
Editor: Sergo Gotsiridze, pp. 167-178

ISBN: 978-1-61209-970-5
© 2011 Nova Science Publishers, Inc.

Chapter 17

MODEL OF GEOMAGNETIC FIELD PULSATIONS BEFORE EARTHQUAKES OCCUR

Z. Kereselidze[1,*], M. Kachakhidze[2], N. Kachakhidze[2] and V. Kirtskhalia[3]

[1]Institute of Geophysics, M. Aleksidze str. 1, 0193 Tbilisi, Georgia
[2]St. Andrew The First-Called Georgian University of the Patriarchy of Georgia I. Chavchavadze av. 53 A, 0162 Tbilisi, Georgia
[3]Sukhumi Physical and Technical Institute, A. Kazbegi Ave. 12., 0168 Tbilisi, Georgia

Abstract

Very low frequency (VLF) electromagnetic radiation (in diapason 1 kHz – 1MHz) in atmosphere, generated during earthquake preparation period, may be connected with linear size, characterizing incoming earthquake source. In order to argue this hypothesis very simple quasi-electrostatic model is used: local VLF radiation may be the manifestation of own electromagnetic oscillations of concrete seismoactive segments of lithosphere-atmosphere system. This model explains qualitatively well-known precursor effects of earthquakes. At the same time, it will be principally possible to forecast expected earthquake with certain precision if we use this model after diagnosing existed data.

On the base this model we consider modeling task according to which simultaneously of origination of the own electromagnetic oscillation of the LAI system some segment or certain delay it is possible to observe periodic perturbations of the geomagnetic fields in the incoming earthquakes epicentral area, frequencies of which will be rather less than characteristic frequencies of the VLF electromagnetic emission.

Introduction

In solid medium considerable accumulation of polarization charge may take place in such a place where heterogeneity, having definite scale lines, is already formed or is in the process

---

* E-mail address: sipt@sipt.org

of forming. Geological medium is more, or less equally stressed before earthquake preparation. Progressive increase of tectonic stress is accompanied by formation of inhomogeneous structural sources, or by qualitative change of medium. It is known that at final stage of earthquake preparation chaotically occurred microfractures may be formed as one – direction main fault. It is possible that maximum electropolarization effect which manifests itself at various times before earthquake occurrence due to structural peculiarities of geological medium corresponds to this very moment. Polarization effect is often accompanied by electromagnetic radiation. Formally this means that besides electrostatic effect, which forms capacity, polarization is also accompanied by induction effect. But while analyzing possibility of induction interaction in lithosphere-atmosphere system it should be taken into account that there are many possibilities of induction effect development. We can suppose that the source of this effect is always lithosphere in connection with seismic phenomena. So we can think that schematically we have to do with certain type electromagnetic circuit (contour), elements of which should be connected with lithosphere, as well as atmosphere. In particular, the fact that upper limit of VLF (recorded before an earthquake) is of MHz order, may indicate at that minimum size of the Earth heterogeneity cluster which can call forth electric induction effect in atmosphere [TakeoYoshino, 1991; Molchanov et al., 1993; Hayakawa et al., 2002]. Though there exists alternative version according to which it is not at all necessary that electric oscillations frequency variation in ionosphere is connected with seismic phenomena only. This means that induction source may be in the atmosphere, but the response to it -in the lithosphere. Especially original example of such alternative is the model of inductive prolongation of ionosphere SQ current system in the upper lithosphere [Duma et al., 2003].

As we consider below electromagnetic oscillations, generated in separate segments of lithosphere-atmosphere system in quasi-electrostatic approximation, we can operate only with atmospheric electric field, without taking into account atmospheric current. This makes easier the problem of mathematical modeling because it is simple in electrostatic approximation to connect polarization charges with atmospheric electric field which is broken at atmosphere-lithosphere boundary. That's why quasi-electrostatic model does not need presentation of the mechanism of electric conduction change in the atmosphere, in particular, assumption of radon emission from the lithosphere to the atmosphere. It should be noted that from the point of view of establishment of atmospheric current variation mechanism in seismoactive regions the foresight of this effect does not yield any universal result which would be equally true for regions with different geological structure. In particular, modification of so called "Frenkel's model of atmospheric capacitor (based on radon emission) in opinion of the authors, may be effective only for the Far East region and partially, for Middle Asia region [Liperovsky et al., 2008]. Though, according to the work [Smirnov, 2008] the mechanism of electric conduction variation of the atmosphere is vague even for Kamchatka region where volcanic earthquakes occur especially often and emanation effect of radon is much more probable than in regions which have geological structures different from Far East, for example in the Caucasus. That's why it is logical to assume that (except special cases) change of vertical electric current intensity in the atmosphere is chiefly connected with change of electric field stress. We can consider as special case, for example, sharp change of electric conduction of lithospheric medium before earthquake and its subsequent atmospheric effect caused by emanation of charged particles or ionization of medium. It seems that such phenomenon is very rare. Otherwise there would be considerable materials which would strengthen quantitatively, for example, qualitative model of ULF radiation, constructed on the principle of sharp break of electric conduction of the medium [TakeoYoshino 1991].

It is known that in the period of earthquake preparation piezo-electric effect, caused by mechanical stresses, is observed in rocks [Mognaschi, 2002; Triantis et al., 2008]. Generally, polarization charge should be distributed on a surface, which should be either limited by fault or formed along faults [TakeoYoshino 1991]. As the Earth surface has conditionally negative potential with respect to the atmosphere, that segment of the lithosphere where an earthquake is prepared can be considered as negatively charged before piezo-effect. As the result of tectonic stress increase heterogeneity will originate in this segment, or positive charge areas, which, like "Frenkel's generator", will call forth inductive polarization at certain height of the atmosphere.

According to the model, in source area of incoming earthquake, at final stage of its preparation, against a background of numerous fractures, definite, linear size main fault is being formed. So it can be represented as linear wire, the length of which considerably exceeds characteristic size of its section. Conductor of the same size but with opposite polarity should occur in the atmosphere by induction. It is obvious that such model is inverse or it can be assumed that initial conductor is in the atmosphere and secondary or induced one is in the lithosphere. Operation with linear conductors is noticeable enough because atmospheric discharges (lightning) are linear phenomena and not areal. Formally, if two, moved away from each other horizontal conductor with opposite polarity exist in the lithosphere and the atmosphere then a structure, resembling a capacitor should be formed which may be locked by vertical atmospheric electric field As electromagnetic induction is the reason of generation of such spatial formation, or it has certain inertia like usual oscillatory circuit, there should exist its own characteristic frequency of electromagnetic oscillation.

Thus, using physical analogy with linear conductors while explaining the mechanism of VLF atmospheric electromagnetic radiation connected with seismic activity is quite logical. Such analogy will not distort considerably quantitative results, for example, because of disregarding areal effect in model condenser capacity. Also, general picture should not change qualitatively even when in seismically active region the system is considered which is formed not by one, but several electromagnetic circuits.

Usually, in electromagnetic oscillatory circuit the system capacity C is concentrated in capacitor, and inductance L – in the coil. In such circuit capacity and inductance of connecting wires, as well as capacity of the coil, are disregarded. When electromagnetic dissipation is disregarded, circuit's own oscillation frequency is defined by well-known Tompson's formula

$$\omega^2 = \frac{1}{L \cdot C} \qquad (1)$$

which is more precise when capacity outside the condenser and inductance outside the coil are the lesser. It is obvious that oscillatory circuit's own (characterizing) frequency increases when capacity and inductance decrease. But now capacity and inductance of connecting wires become considerable. That's why within very high frequencies there is no necessity of condenser and coil because inter own capacity and inductance of connected wires (linear conductor) will be absolutely enough for oscillation generation. At the same time, it is not necessary that virtual wires were tied strictly in circuit frame. It means, the circuit can be presented in open state too The main thing is the existence of locking mechanism of wires, function of which perform components of atmospheric electric field in the given model. From the point of view of physical analogy, this means that if we charge two conductors with similar charge which have opposite signs, and then lock them, current and connected with it magnetic field will appear in the system.

As conductors have inductance, electromotive force of induction will also occur or by all parameters circuit will be established in which electromagnetic oscillations will be generated.

Thus, presented model explains qualitatively generation mechanism of very low frequency electromagnetic waves in previous periods of an earthquake and indicates at the source of disturbance of atmospheric vertical electric field. As this field has the function of circuit locking, we should envisage that it is disturbed by oscillation frequency of the circuit, as well as according to characterizing time of ohmic damping.

Thus, disturbance of atmospheric electric field should have high and low frequency components. At the same time, in spite of disregard of ohmic resistance effect in the circuit, there sure will be energy loss due to electromagnetic radiation, intensity and propagation direction of which will depend on the form and spatial size of the circuit.

Let's say that the length of horizontal, opposite polarity conductors is l, characteristic quantity conductor's section is a, distance between conductors is h. It is known that inter capacity of conductors, when $h >> a$, is (International unit system is used)

$$C \approx \frac{\pi \varepsilon_0}{\ln\left(\dfrac{h}{a}\right)} l$$

and mutual induction of conductors $L \approx \dfrac{\mu_0}{\pi} \ln\left(\dfrac{h}{a}\right) l$ (it is assumed that relative electric and magnetic constants $\varepsilon = \mu = 1$).

Postulation of the very same section is not strict limitation as if the wares have different a and b sections, we have [Landau, Lifshic, 1957]

$$L \sim \ln \frac{h^2}{ab}$$

Thus, because the product of absolute dielectric and magnetic constants is $\varepsilon_0 \mu_0 = \dfrac{1}{c^2}$, from (1) formula of circuit's own electromagnetic oscillations we'll have

$$\omega = \left(\varepsilon_0 \mu_0 l^2\right)^{-\frac{1}{2}} = \frac{c}{l} \tag{2}$$

where c is velocity of light. Let's assume that l changes in (1-100) km interval, which corresponds to change diapason of characteristic scale of earthquake source. From (1) we'll receive that change diapason of analogous circuit's own electromagnetic oscillation frequency is $\omega = 3 \ (10^3 - 10^5)$ Hz. So it is obvious quantitative agreement with often recorded very low frequency atmospheric electromagnetic radiation spectrum in earthquake preparation period.

As for atmospheric electric field, which is the locking of polarized lines, certain freedom of circuit form exists here: if there is vertical locking, then according to our result the height does not mean anything If we consider second variant of open circuit, then occured horizontal component of atmospheric field plays the role of locking mechanism, and conductor length may exceed considerable linear sizes of polarization area, depending on inhomogenaty scale of the atmosphere. In case of horizontal circuit, positively polarized conductor may be coupled with opposite sign conductor having any length.

Statement of the Problem

According to own electromagnetic oscillation model of LAI system segment very low frequency (VLF) electromagnetic emission is connected with the origination of linear conductor by induction in the Earth towards polarization charges in the various height of the atmosphere. Due to it is formed the ordinary system similar to electromagnetic contour, the oscillation of which will be relaxed in the real dissipation medium. It means that the intensity of the atmospheric current in contour must decrease by the time. Because of system inductivity this process perhaps have oscillation character, though it is clear that the frequency of the noted oscillations which depends on the medium macroscopic features, must be less than VLF emission frequency. But the magnetic field always is jointed with current. The magnetic field distribution in the space and its changing character by time may be determined by mathematical modeling. Because of it perturbations of the magnetic field are considered as one of indicators of the process taking place in the possible earthquake focus.

It is necessary underline that first indicator or initial element of the model is the effect of polarization and VLF electromagnetic emission connected with it, but the other effects, for instance perturbations of the geomagnetic field, thermal and radiation effects inside the earth, are secondary.

So we consider modeling task according to which simultaneously of origination of the own electromagnetic oscillation of the LAI system some segment or certain delay it is possible to observe periodic perturbations of the geomagnetic fields in the incoming earthquakes epicentral area, frequencies of which will be rather less than characteristic frequencies of the VLF electromagnetic emission.

There are numerous experimental data, according to these data geomagnetic field pulsations with characteristic frequencies $1 \div 0.01 Hz$ are fixed during the tectonic stress increasing [TakeoYoshino., 1991; Hayakawa et al., 2002; Liperovsky, et al., 2008]. We note that the spectrum of these pulsations factual coincides with the frequency of the short periodical Pc1-3 regular geomagnetic pulsations, however different from the regular pulsations, by their forms they are short term wavy train which characterizes irregular geomagnetic pulsations Pi1-2 [Nishida,1978]. We consider that the source of the geomagnetic field perturbations is the portion of the linear current with I intensity, the length of which is commensurable with LAI system segment's linear size. According to condition direction of the I in the atmosphere has no principal meaning because of it may be considered by vertical or horizontal atmospheric current portion but its length must be too more than cross section.

Let us use rectangular coordinate system with the centre in the Earth surface. Z axis is directed to vertical (up). Horizontal component of the geomagnetic field by y axis is H_{oy}, vertical component – H_{oz}. Magnetic declinations ignored.

We mean that the atmosphere practical is immovable before origination of the electromagnetic contour of LAI segment. It is not very strict limitation however (as it will be shown below) it is handy by the getting of the analytic solution.

Below it will be used the classic scheme of the perturbation theory, which is given in monograph [Sutton, Sherman, 1965]. Let us bring in a little perturbation V and h which are vectors of hydrodynamical velocity of the medium connected with I current and magnetic field. It is obvious that the first one is taken as perturbation of the immovable atmosphere, but the second one-toward geomagnetic field. According to the model the amplitudes of the perturbated values depend on x and y coordinates by linear law, but the wave feature is given by $\exp^{i(K_z Z - \omega t)}$, where K_z is the wavy number by z direction, and ω -oscillation frequency.

If we limit ourselves by members of first rate smallness in the system of MHD equations, we will have system of equations for the incompressible medium

$$\frac{\partial \vec{V}}{\partial t} = -\frac{1}{\rho_0}\nabla p + \lambda\Delta\vec{V} + \frac{1}{4\pi\rho_0}\left[rot\vec{h}\cdot\vec{H}_0\right] \quad , \tag{1}$$

$$\frac{\partial \vec{h}}{\partial t} = \left(\vec{H}_0\nabla\right)\vec{V} + \lambda_m\Delta\vec{h} \quad , \tag{2}$$

$$div\vec{h} = 0 \quad , \tag{3}$$

$$div\vec{V} = 0 \quad , \tag{4}$$

where λ and λ_m are the ordinary (kinematic) and magnetic viscosity of the medium (weaker ionized plasma), p is the pressure perturbation, ρ_0 is the initial density of the ionized medium. In order of model simplicity we consider only one component H_{0z} of H_0 geomagnetic field which is not the principal limitation. Using (3) and well known formula of the vector analyses, (1) will get following expression

$$\frac{\partial \vec{V}}{\partial t} = -\frac{1}{\rho_0}\nabla\left[p + \frac{\left(\vec{h}\cdot\vec{H}_0\right)}{4\pi}\right] + \frac{1}{4\pi\rho_0}\left(\vec{Z}_0\nabla\right)\vec{h} + \lambda\Delta\vec{V} \quad . \tag{5}$$

Let us admit that $h_z=0$ and both sides of (5) operate by div. In this case we will get Laplace's equation $\Delta p = 0$ for the pressure perturbation

$$\Delta p = 0 \quad . \tag{6}$$

But far from the source of perturbation we consider their values $\vec{V}$ and $\vec{h}$ are equal to zero on the certain ultimate distances. There p=0 too. It means that perturbation of the pressure equal to zero all over (3). So (1) will get following expression

$$\frac{\partial \vec{V}}{\partial t} = \frac{1}{4\pi\rho_0}\left(H_{0z}\nabla\right)\vec{h} + \lambda\Delta\vec{V} .$$ (7)

In monograph [Sutton, Sherman, 1965] the amplitudes of perturbated values are considered as constants. But it is probable that they even change by certain rules. For instance, we assume that amplitudes of the perturbated values linear decreases by x and y, i.e. we have

$$V_x = V_0\left(\frac{a-x}{a}\right)\exp^{i(K_z Z-\omega t)},$$

$$V_y = -V_0\left(\frac{b-y}{b}\right)\exp^{i(K_z Z-\omega t)},$$ (8)

$$h_x = h_0\frac{a-x}{a}\exp^{(K_z Z-\omega t)},$$

$$h_y = h_0\frac{b-y}{b}\exp^{i(K_z Z-\omega t)},$$

where a and b are linear scales (characteristic sizes of the earthquake area). It is obvious that such imagination satisfies equations (3) and (4), because by condition $V_z = h_z = 0$, which is not any special physical limitation). So (2) and (7) equations significantly simplifies:

$$\frac{\partial V_x}{\partial t} = \frac{H_{0z}}{4\pi\rho_0}\frac{\partial h_x}{\partial z} + \lambda\frac{\partial^2 V_x}{\partial z^2} .$$ (9)

$$\frac{\partial V_y}{\partial t} = \frac{H_{0z}}{4\pi\rho_0}\frac{\partial h_y}{\partial z} + \lambda\frac{\partial^2 V_y}{\partial z^2} .$$ (10)

$$\frac{\partial h_x}{\partial t} = H_{0z}\frac{\partial V_x}{\partial z} + \lambda_m\frac{\partial^2 h_x}{\partial z^2} .$$ (11)

$$\frac{\partial h_y}{\partial t} = H_{0z}\frac{\partial V_y}{\partial z} + \lambda_m\frac{\partial^2 h_y}{\partial z^2} .$$ (12)

It is obvious that (9) and (10), like (11) and (12) are identical by the form. Because of it in future we may use some pair of equations written by one of indexes, for instance by (9) and (11) equations. We product (9) by Z and (11) by t.

We will get:

$$\frac{\partial V_x}{\partial z \partial t} = \frac{H_{0z}}{4\pi\rho_0}\frac{\partial^2 h_x}{\partial z^2} + \lambda\frac{\partial^2}{\partial z^2}\left(\frac{\partial V_x}{\partial z}\right).$$
(13a)

$$\frac{\partial^2 h_x}{\partial t^2} = H_{0z}\frac{\partial^2 V_x}{\partial t \partial z} + \lambda_m\frac{\partial}{\partial t}\left(\frac{\partial^2 h_x}{\partial z^2}\right).$$
(13b)

Let us define $\dfrac{\partial V_x}{\partial z}$ from (9) and bring in (13). We will get:

$$\frac{\partial^2 V_x}{\partial z \partial t} = \frac{H_{0z}}{4\pi\rho_0}\frac{\partial^2 h_x}{\partial z^2} + \frac{\lambda}{H_{0z}}\frac{\partial^2}{\partial z^2}\left(\frac{\partial h_x}{\partial t} - \lambda_m\frac{\partial^2 h_x}{\partial z^2}\right).$$
(14)

Let us alter the left side of (14) by its significance from (13). We will get:

$$\frac{\partial^2 h_x}{\partial t^2} = V_a^2\frac{\partial^2 h_x}{\partial z^2} + \left(\lambda + \lambda_m\right)\frac{\partial^2}{\partial z^2}\left(\frac{\partial h_x}{\partial t}\right) - \lambda\lambda_m\frac{\partial^4 h_x}{\partial z^4}.$$
(15)

where $V_a = \dfrac{H_{0z}}{\sqrt{4\pi\rho_0}}$ is the Alven velocity, determined towards density of ionized component.

So we got the equation for the component of magnetic field perturbation, in which Alven wave velocity is included as parameter, ordinary and magnetic viscosity coefficients too. Velocity of Alven is connected with solid features of conductive medium, which are gotten by external magnetic field. Although the rate of Earth's atmosphere ionization in the ordinary conditions is smaller below of ionospheric levels, the magnetic solidity of the medium may play the certain role in the period before earthquakes when it is possible sharp increases of air ionization level: for instance because of radon induction in the atmosphere. It is obvious that the equation for h_y component is similar to (15). So we may us with vector equation in general :

$$\frac{\partial^2 \vec{h}}{\partial t^2} = V_a^2\frac{\partial^2 \vec{h}}{\partial z^2} + \left(\lambda + \lambda_m\right)\frac{\partial^2}{\partial z^2}\left(\frac{\partial \vec{h}}{\partial t}\right) - \lambda\lambda_m\frac{\partial^4 \vec{h}}{\partial z^4}.$$
(16)

Because our aim is to get dispersive equation which joints wave number and frequency, it is necessary to use (8) for the magnetic field i.e. $\vec{h} = h_0(x, y)\exp^{i(K_z z - \omega t)}$ following of which will get equation of forth rate

$$\omega^2 = V_a^2 K^2 - i\omega\left(\lambda + \lambda_m\right)K^2 - \lambda\lambda_m K^4.$$
(17)

Solving of the equation is difficult in general, however we know that it must consist of Alven wave spectrum, rapid and low MHD waves spectrums too. The Alven transverse waves have nondissipative nature and (17) is in accordance with $\lambda = \lambda_m = 0$. In general, in case of ionospheric medium the task simplifies because here, as a rule, $\lambda_m >> \lambda$. It is shown below by estimations which are carried out for the typical significances of parameters of ionosphere E layer.

Let us estimate magnetic viscosity $\lambda_m = \frac{c^2}{4\pi\sigma}$, where σ is the specific electrical conduction. It is clear that first of all we must reveal characteristic value of specific conductivity. There are different conductivities in the ionosphere. The largest one among these conductivities is the Pedersen conductivity i.e. conductivity in direction of magnetic field flux. It's formula is [Sutton, Sherman, 1965]

$$\sigma_0 = e^2 N \left[\frac{1}{m_e \nu_e} + \frac{1}{m_i \nu_i} \right],$$

(18)

where e is elementary charge, N – concentration of electrons, m_e, m_i are masses of electrons and ions, ν_e, ν_i are the number of collision with neutrals. In conditions $m_e << m_i$ and $\nu_e \approx \nu_i \approx 10^{4 \div 6}$ sec^{-1} it is obvious that we may ignore the second item in the brackets of (18) (CGSE unit system is used).

So because $e \approx 5 \bullet 10^{-10}$ CGSE, $m_e \approx 10^{-28}$ gr , N$\approx 10^6$ sm^{-3}, we will have $\lambda_m \approx 2,5 \bullet 10^{9 \div 11}$ sm^2.sec^{-1}. As for ordinary viscosity: it is known that typical significances in ionosphere are $\lambda \approx 10^{6 \div 7}$ sm^2.sec^{-1} but to ignore of last member in right side's of (17) is justified only in case if $V_a^2 K^2 >> \lambda \lambda_m K^4$. it is necessary to use the characteristic values of wave number for it's estimation for the first time: $K_z = \frac{2\pi}{l_0}$ where l_0 is characteristic height of E - layer of ionosphere. In our model $l_0 \approx 10^7$ sm, i.e. $K_z \approx 6 \bullet 10^{-7}$ sm^{-1}.

After that in order to study characteristic value of Alven wave velocity it is necessary to determine density of the ionization component on the above mentioned height. The characteristic value of common density here is $\rho \approx 10^{-9}$ gr sm^{-3} but due to quazineutrality of ionospheric medium here is approximately the same quantity of ions as electrons $n \approx 10^6$ sm^{-3}. It seems, that the quantity of ionized components i.e. density of ions in Alven velocity formula $\rho \approx 10^{-15}$ gr sm^{-3} . The characteristic value of geomagnetic field on the average latitudes is $H_{0z} \approx 3 \bullet 10^{-1}$ Gauss. Accordingly: $V_a = \frac{H_{0z}}{\sqrt{4\pi\rho_0}} \approx 3 \bullet 10^6$ sm sec^{-1}. So members

$V_a^2 \bullet K^2$ (17) by our estimation, is more for several rates than $\lambda\lambda_m K^4$ member. Because of it we may use dispersive equation of the second rate

$$\omega^2 = V_a^2 K_z^2 - i\omega\lambda_m K_z^2.$$ (19)

We are interested in low frequency MHD waves, which causes changing of geomagnetic field. Due to it we have to use condition $\dfrac{\omega\lambda_m}{V_a^2} \ll 1$. This demand is justified even when $V_a \approx 3\bullet10^6$ sm sec^{-1} and $K_z \approx 6\bullet10^{-7}$ sm^{-1} we will have characteristic (maximal) frequency $\omega = 1$ Hz for Alven waves ($\omega \approx K_z V_a$). It is obvious that the frequency of Alven wave satisfies above noted condition even for maximal λ_m. So if we use the condition $\dfrac{\omega\lambda_m}{V_a^2} \ll 1$ and series expansion, we will have (20) for the complex K_z.

$$K_z = \eta + iK \approx \frac{\omega}{V_a}\left(1 + \frac{i\omega\lambda_m}{2V_a^2}\right) = \frac{\omega}{V_a} + i\frac{\omega^2\lambda_m}{2V_a^3},$$ (20)

using of it for the geomagnetic field perturbation finally gets:

$$\vec{h} = \vec{h}_0(x,y)\exp^{-\frac{\omega^2\lambda_m}{2V_a^3}Z} \cdot \exp^{i(\eta Z - \omega t)}$$ (21)

According to (2) h has decrement of logarithmic relaxation by distance i.e. $z_0 = \dfrac{2V_a^3}{\omega^2\lambda_m}$.

If $\omega_o \approx 1$ Hz which factual corresponds to regular geomagnetic pulsation Pc1-2, and $\lambda_m \approx 10^{9\div11}$ sm^2 sec^{-1} and $V_a \approx 3\bullet10^6$ sm sec^{-1}, we will have $z_0 = 5 (10^{8\div10})$ sm. It means that in case of having such characteristic values for such magnetic viscosities values above of earthquake focus, perturbation of geomagnetic field factual does not relax. Following of it we have to say that if the beginning amplitude of the perturbation is enough large, it means that the perturbation may be fixed as far as distances. In addition to it relaxation is weaker as low frequency waves to consider.

Conclusion

According to the LAI model, first considered by us in this paper, i.e. in the frame of electromagnetic contour tied by linear wires different from MHD waves, relatively high frequency oscillations ($\omega \approx (1-100)$ KHz take place in the space. With these oscillations by of medium's inductivity (which is the medium feature) if may be generated low frequency waves too which causes perturbation of magnetic field in the frequency diapason of geomagnetic pulsation and variations. As for amplitudes of perturbation: as we consider linear model, characteristic amplitudes of perturbation v_0 and h_0 are considered as constants. But it is obvious that they must exist limited minimal significances for amplitudes less of which will not be fixed by measurements. We must note that as a rule, magnetic viscosity in real may decrease only in the vicinity near earthquake focus due to charges abundance caused polarization effect or due to charged particles emission from Earth to atmosphere.

So the length of attenuation of above given MHD waves correspond to the ideal case, when λ_0 is low all over. In reality this value inside of earthquake focus may be rather more than $\lambda_m \approx 10^{11}$ sm^2 sec^{-1}. So in reality $z_0 \ll 10^8$ sm, e.i. 10^3 km, which means that perturbation of magnetic field must fix only near of VLF radiation source. The cyclotron resonance effect may take place in this area, which may strengthen VLF emission. In particular as cyclotron frequencies of protons and electrons in E-layer are $\omega_p = \dfrac{eH}{m_p c} \approx 5 \bullet 10^3$ Hz and $\omega_e = \dfrac{eH}{m_e c} \approx 1.6 \bullet 10^7$ Hz, it means that if resonance effect takes place near, the 5 kHz and lower (corresponding to ions) and 16 MHz too, the reason will be lugging of electric field energy to particle energy, but not particle – particle and wave-particle type interactions which characterizes ionosphere in the ordinary conditions. It means that for instance perturbation of ionosphere must be fixed in the narrow area during earthquake preparation period indeed. [Mikhailov at all,2004]. It certainly must occur in constant frequencies the weight of which must be larger than other frequencies weights in spectrum of VLF emission.

References

Duma, G. and Ruzhin, Y. Diurnal changes of earthquake activity and geomagnetic Sq-variations. *Natural Hazards and Earth System Sciences*, **3**,171-177, 2003.

Hayakawa, M. and Molchanov, O. A. Seismo-Electromagnetics. Lithosphere-Atmosphere-Ionosphere Coupling. *TERRAPUB*, Tokyo, p. 1–477, 2002.

Landau, L. D. and Lifshic, E. M. Electrodynamics of continuum.M.,Tech.-teor. *Lit.*, 1957, 532 p.(on Russian).

Liperovsky, V. A., Meister, C. V., Liperovskaya, E. V. and Bogdanov, V. V. On the generation of electric field and infrared radiation in aerosol clouds due to radon emanation in the atmosphere before earthquakes. *Nat. Hazards Earth Syst. Sci.*, **8**, 1199–1205, 2008.

Mikhailov, Y. M., Mikhailova, G. A., Kapustina, O. V., Buzevich, A. V. and Smirnov, S. E. Power spectrum features of near-Earth atmospheric electric field in Kamchatka, *Ann. Geophys.*, **47**(1), 237–245, 2004.

Mognaschi, E. R., IW2GOO. On The Possible Origin, Propagation And Detectebility Of Electromagnetic Precursors of Eaerthquakes. *Atti Ticinensi di Scienze della Terra*, vol. 43, p. 111-118, 2002.

Molchanov, O. A., Mazhaeva, O. A., Golyavin, A. N. and Hayakawa, M. Observation by the Intercosmos-24 Satellite of ELF-VLF electromagnetic emissions associated with earthquakes, Ann. Geophys., Atmos. *Hydrospheres Space Sci.*, **11**, 5,431–440, 1993.

Nishida, A. *Geomagnetic Diagnosis of the magnetosphere.* New York-Heidelberg-Berlin, 1978.

Smirnov, S. Association of the negative anomalies of the quasistatic electric field in atmosphere with Kamchatka seismicity. *Nat. Hazards Earth Syst. Sci.,* **8,** 745–749, 2008.

Sutton George W., Sherman Artur. *Engineering magnetohydrodynamics.* McGraw-Hill Boock Company. New York- St.-Luis- San-Francisco-Toronto-London-Sydney.1965.

TakeoYoshino. Low-Frequency Seismogenic Electromagnetic Emissions as Precursors to Earthquakes and Volcanic Eruptions in Japan. *Journal of Scientific Exploration.* Vol. 5, No. I, 121-144, 1991.

Triantis, D., Anastasiadis, C. and Stavrakas, I. The correlation of electrical charge with strain on stressed rock samples. *Nat. Hazards Earth Syst. Sci.,* **8,** 1243-1248, 2008. www.nat-hazards-earth-syst- sci.net/8/1243/2008.

In: Focus on Science and Technology…
Editor: Sergo Gotsiridze, pp. 179-194

ISBN: 978-1-61209-970-5
© 2011 Nova Science Publishers, Inc.

Chapter 18

THE AMOUNT OF CD34+, CD 34-CD 133+ AND GPA+ IN PERIPHERAL BLOOD LIVING NUCLEAR CELLS IN PATIENTS WITH HEART FAILURE

M. Rogava[*], Z. Gurtskaia and G. Gigilashvili

Georgian Technical University, National Center of Therapy, Georgia

Abstract

The aim of the present study was to determine the percentage of CD34+, CD34-CD133+ and GP-A+ cells in peripheral blood nuclear cells of the study population and its relation to the severity of the condition.

The data obtained indicate that symptoms of HF are associated with the increase in peripheral blood stem cells and progenitors when compared to the control subjects. Increase in GP-A is specific only for relatively mild forms of CHF (I-IIfc, NYHA), while aggravation of the condition (HF III-IV fc, NYHA) is associated with significant drop in GP-A levels in peripheral blood. Stem cell and progenitor CD34+, CD34-CD133+) number increases in parallel with the progression of CHF.

Actuality of the Theme

Cardiovascular diseases are the most significant problem of public health all over the world. Among them diseases of cordial muscle have taken an important place (myocardium infarction, myocarditis, cardiomyopathy and other), resulting adapting-compensating remodeling and forming heart failure syndrome that leads mostly to lethal outcome [1].

Based on the estimations of experts, symptomatic heart failure ranges from 0.4 to 2% in population of Europe. Cases of heart failure are rapidly growing along with age. Middle age of patients is 74. At every age category there has been detected an immutable increase in death caused by heart failure. Total population of European Union countries is 900 million. Out of it 10 million is ill with heart failure and about the same number with myocardium

---

[*] E-mail address: giorgio532@gmail.com

dysfunction without clinical signs of failure. In case of inability of adequate correction of disease resulting formation of heart failure, prognosis is unfavorable despite the stage. Half of patients die within 4 years after diagnose. Due to acute heart failure the same number of patients dies within the first year [1].

Nowadays wide range of medicines and different non-medicine methods are been applied for treatment of heart failure, meanwhile the mentioned disease is becoming more frequent and wide-spread.

Therefore finding fundamentally new approaches in treatment of heart failure became the question of the day. Development of regenerative medicine turned out to be the topical question. Some earliest achievements in regenerative medicine have been detected in works of scientists in 60-ies.

Absence of reproductive ability in cardiomyositis has been well-known fact for a long time. Defects of cordial muscle are mainly replaced due to the proliferation of cells of strom (fibroblasts). In case of infarct of myocardium it is supposed to be irreversible and results remodeling of cordial muscle that leads to heart failure in several months or a year. And the only drastic method of treatment on the terminal stage is heart transplantation.

Besides, in recent years there have been piled up data revealing regenerative ability of myocardium. Reproductive ability of cordial muscle was examined by Kajtsura et al. (1998) on patients with transplanted heart and with dilatational cardiomyopathy by means of defining mitosis index of cardiomyopathy. Calculations carried out by the mentioned authors showed that if in the auricle of 45 years old healthy person there is a 5.8×109 nucleus in average, and mitosis index is 14 on 10^6, then 2×10^3 cardiomyositis nucleus is in the stage of mitosis. Duration of mitosis is about 1 hour, i.e. approximately 0.71×10^9 new nucleuses are formed in a left ventricle a year. Since the correlation of cardiomyositis with 1,2 and 4 nucleuses is 2/1/1, potential yearly increase in cardiomyositis can amount 0.4×10^9. According to the same author's data mitosis index of pathological heart (cardiomyopathy) is ten times more than normal myocardium. In author's opinion proliferation of cardiomyositis play significant role in compensatory-recovering process of pathological myocardium. In spite the revealed increase cannot compensate massive necrosis of cardiomyositis in case of myocardium infarct (since seam tissue more rapidly replenish defects), proliferation of cardiomyositis can create certain regenerative reserve of cordial muscle [2].

In recent years new directions and approaches have been developed to achieve cordial muscle regeneration. Development of biotechnology, molecular and cellular biology, has made a cell not only a main subject matter but also remedy for many diseases. It is supposed that the restoration of myocardium function can be achieved by increase in amounts of contractive elements and/or increase in functional reserve cardiomyositis of recipient with stimulation of cellular and intercellular regeneration processes. For this purpose implantation of cardiomyositis, smooth muscular cells, gene-modified cells (fibroblasts, myoblasts of skeleton), stem cells of marrow, embryonic stem cells and other is used. Term – "Cellular Cardiomyoplastic" unites all cellular technologies [3,4,5].

Transplantation of cells into cordial muscle is applied after myocardium infarct to replace necrosis cells or seam tissue with contractive cells in order to improve contractive function of a heart, regulate remodeling processes, and stimulate angiogenesis and other.

Stimulation of producing stem cells in marrow and transfer it to a systemic blood circulation can be alternative method of transplantation. Among the preparation groups used for this purpose we have to notice colonial stimulating factors: granulositis colonial stimulating factor and granulocytic-macrofagal colonial stimulating factor [12].

Orlik. D. et. Al. (2001) displayed that after experimental myocardium infarct in rats, hypodermic injection of granulocytical colonial stimulating factor results appearance of new cardiomyositis within the infarct area, improvement of hemodynamics indices and prolongation of lifetime of animals. In this connection mobilization of stem cells in blood and its significant increase in number takes on special interests, since non-hemopoezic stem cells have ability to replace no only damaged tissues of a heart but also tissues of other organs taking part in vicious circle.

A group of scientists (U.N. Belenkov, and co-author. 2003) in their clinical practice have treated 5 patients for heart failure of ischemic genesis by granulositis colonial stimulating factor. Noticeable improvement of clinical state and echoscope data was detected on one patient in whose blood were mobilized and transferred most of stem cells.

Hence we consider it important to define relation of stem cells and quantity indices of progenitors in peripheral blood with heart failure and seriousness of its state. In case of heart failure it is important to reveal mechanism of changes in composition of these cells.

The second question the paper concerns is connected to a diagnostics of mild cases of heart failure and asymptomatic heart failure by means of achievements of cellular biology.

In countries of European Union 10 million people suffer from myocardium dysfunction without clinical signs of failure. Researches of recent years have shown that prognosis of patients with asymptomatic dysfunction of myocardium and heart acute failure is unfavorable. Chronic heart failure almost in 40% of patients proceeds without symptoms in yearly stage. Besides, yearly death of patients from even I functional class of heart failure is 10-12% [1].

Therefore it is very important to reveal patients with mild cases of heart failure, also asymptomatic heart failure and mild cases of heart failure but patients disregard the diseases.

In different research institutes all over the world research works are being carried out on creating new screening tests for diagnostics of heart failure. In this regard data about high sensibility of natriuretic peptides, as screening tests for revealing heart failure, is noticeable. Researches of recent years have detected a role of n fraction of natriuretic hormone of cerebrum in diagnostics of heart congestion failure.

Hence creating new screening diagnostic tests for heart failure, based on achievements of cellular and molecular biology is of high interest and will enable us to reveal risk groups for further examination to evaluate myocardium dysfunction.

Tasks of Research

To achieve the aims defined above determined the following tasks:

- Studying percentage of cells CD34, CD34-133+, GP-A (CD235) in peripheral blood of patients with heart failure by flowing cytometer method and definition its relation with heart failure and the seriousness of the disease. (f.c. NYHA)
- Studying percentage of cells CD34, CD34-133+, GP-A (CD235) in peripheral blood of patients with chronic inactive hepatitis for the purpose of comparison.
- Studying percentage of cells CD34, CD34-133+, GP-A (CD235) in peripheral blood of healthy volunteers.

Aim of Research

Aim of research was to define a percentage of young forms of stem cells, progenitors and of erithroyd line in live nucleus cells of peripheral blood of patients with heart failure syndrome of different genesis and reveal its relation with heart failure and the seriousness of the disease.

Research Design and Methods

We have examined 110 patients with chronic heart failure of different genesis and 47 patients with chronic hepatitis (127 men and 30 women) aged from 19 to 65 (middle age ±3.7age). They have received outpatient and in-patient treatment at National Therapy Center within 2002-2005 years (they were selected by time, geography and other selection criteria). We offered patients participation in written form, with detailed description of examination and expected results, in accordance with Georgian law "About health protection" and recognized principles of medical ethics.

Positive Selection Criteria

- Consent of patient on participation in examination
- Patients with heart failure syndrome (I-IV f.c. NYHA) resulted from a local or diffusion damage of myocardium (ischemic heart disease – post-infarction cardio sclerosis, cardiomyopathy).
- Patients with chronic inactive chronic viral hepatitis (B, C) and for the moment without any complain testifying heart failure and clinical-diagnostic signs.
- Healthy volunteers (control group), feeling actually healthy for the moment of involvement in examination, have no signs of any pathology proceeded in organism and corresponded to comparison groups with demographic characteristic (age, sex, income).

Negative Selection Criteria

- Heart failure syndrome resulted from pathology of any organ (kidney failure, pulmonary heart and other).
- Clinical or laboratory data testifying acute inflammatory process in organism. (Concentration of fibrinogen, reactive protein C, leucocytes formula, precipitation reaction of erythrocytes).
- Pathology of hemogenic system
- Refusal of a patient to participate in examination.

Only 108 patients satisfied mentioned criteria (out of it 72 are with heart failure, 36 chronic inactive hepatitis) and 31 healthy volunteers. The rest of the patients have not taken part in the research because of different reasons. Particularly, 22 patients refused to take part, 21 patients appeared to have clinical and laboratory data testifying acute inflammatory process, pulmonary heart was detected in 4 patients and acute kidney failure was the reason of heart failure in 2 patients.

Patients were divided in three groups:

I group consisted of patients with chronic heart failure (n=72) I-IV f.c. (NYHA) (dilatational cardiomyopathy of viral etiology, ischemic heart disease), out of it 56 (77%) men and 16 (23%) women with age from 19 to 65 (middle age 45.2±9.3). Diagnosis of 35 (48.6%) patients in I group was dilatational cardiomyopathy (of viral etiology: Coxsakie B3 n=21, Adenovirus n=6, Epshtein-Bar virus n=7, Herpes virus n=1). Diagnosis of 37 (51,4%) was heart ischemic disease, post infraction cardio sclerosis. Echoscope and X-raying examinations of I group patients detected dilation of heart borders, ejection fraction from 48% to 20%. (average 35±4.1%). In all patients with post infarction cardio sclerosis cardiogram registered Q teeth. In 17 (23.6%) cases post infarction dysfunction was accompanied by pressure stenocardia I-III f.c. 5 (6.9%) patients had diabetes of second type. 25 (34,7%) patients had ventricle extra systole of II-IV gradation Lown. 2 (2.7%) patients had episodes of ventricle tachycardia, 4 (5,5%) patients had experienced balloon angioplasty. In patients of I group duration of heart failure was from 3 month to 6 years (in average 1.9±0.9 year).

II group consisted of patients with chronic viral hepatitis (B,C) (n=36). Out of it 29 (80.5%) men and 7 (19.44%) women aged from 25 to 51 (middle age 36±8.3). 17 (47.22%) patients were with chronic hepatitis C, 10 (27.7%) patients with B, 7 (19.44%) patients with B and C, and 2 (5.55%) patients with liver cirrhosis (of viral etiology, C)-according to Child-Pugh Class A (compensated) supplement #3. For the moment of examination II group patients did not have complaints on cardiovascular system, ultrasonic examination and X-raying did not reveal increasing of heart bounds, ejection fraction was not lower than 55% (average 57±1.7%). Disease duration in II group patients was from 1 to 7 year (in average 3±1.9).

III group consisted of healthy volunteers (n=31), feeling actually healthy at the moment of involvement in examination and didn't have any signs evidencing any pathology in organism. Among them 26 (8.87%) were men and 5 (16.13%) women aged from 22 to 59 (middle age 43±7.8). Ultrasonic examinations and X-raying did not show increase of heart bounds, ejection fraction was not lower than 55% (average 60±3.1%). Viral examinations of blood did not reveal HbsAg, anti-HBV, anti-HCV. ALT was not rise.

Against the background of non-medicament treatment I group patients (heart failure) were treated with –ACE inhibitors, diuretics, β-blockaders, aldosterone inhibitors, peripheral vasodilatations, digitalis, taking into account criteria provided by European Association of Cardiologists.

Against the background of non-medicament treatment II group patients were treated by interferon, hepatic protectors.

I group was divided into two subgroups. I a subgroup included patients with heart failure I-II f.c. n=32 (NYHA). Among them were 26 (81.25%) men and 6 (28.75%) women aged from 39 to 59 (middle age 49.3±7.8) and I b subgroup – patients with heart failure III-IV f.c.

(NYHA) n=40. Out of it 30 (75%) are men and 10 (25%) women aged from 19 to 58 (middle age 36.1±9.1). Characteristic of groups and subgroups, Table #1 and 2.

Table #1

Index	I group	II group	III group
n	72	36	31
Age	45.2 ± 9.3	36 ± 8.3	43 ± 7.8
Sex	56 men 16 women	29 men 7 women	26 men 5 women

Table #2

Index	I a subgroup	II b subgroup
Index	32	29
n	49.3 ± 7.8	36.1 ± 9.1
Age	26 men 6 women	30 men 10 women

At the hospital the following procedures have taken place: gathering anamnesis, evaluating complaints, electric cardiograph, echocardioscope, roentgenographic, blood clinical-diagnostic, bio-chemic examinations.

We diagnosed heart failure taking into account recommendations of European Association of Cardiologists. We defined functional class of heart failure of patients according to criteria provided by American Heart Association.

Apart from routine examinations we have conducted definition of young forms of erythroyd line, stem and progenitor cells in peripheral blood at laboratory of Immunology Department, Tbilisi State University.

We conducted immunophenotyping cells in peripheral blood by means of flowing cytometer by marking with monoclonal antibodies.

Such examinations have been conducted on some patients repeatedly after improving clinical state.

Examination was of double blind observation type. 3 persons have taken part in a trial. Laboratory assistant and operator processing data did not have information about research aim, expected results and participating patients. Third person estimated disease based on "Gold Standard" criteria (f.c. NYHA) of American Heart Association and was not aware about results of immunophenotyping peripheral blood.

Immunophenotyping Peripheral Blood

Carried out examination aimed at quantity studying stem cells, progenitors (CD34+, CD133+CD34-, CD117+cells) and young cells of erythroid line in live nuclear cells of peripheral blood [7,8,9,10,11,13, 15,16,19,22].

CD (CD34, CD133, CD253, CD177) methods of markers studying

Preparation of used solution:

1. Phosphate buffer: 1 pill of PBS (Phosphate Buffered Saline, Sigma) was dissolved in 200ml-distilled water.
2. 1% Albumin solution (PBS/BSA) 500mg of bull serum albumin-BSA (Bovine Serum Albumin) was dissolved in 50ml phosphate buffer.
3. 1% Para formaldehyde solution (PFA) 5g Para formaldehyde powder (Sigma) was dissolved in 500ml phosphate buffer.

Procedures:

1. 100mkl blood was transferred to test-tubes, then added 1-1 ml lysine-buffer warmed to 37^0C (0,9% NACI-FACS-Lysine Solution, diluted 1:10, Becton & Dickinson) and was incubated in thermostat, on to 37^0C, for 10 minutes.
2. After centrifuging 1500 turn/min, during 5 minutes sediment was washed ff twice with 200mkl 1% PBS/BSA solution 1500turn/sec and was added by 10 mkl monoclonal anti-bodies marked with fluorescent substance. Anti-bodies: FITC-anti CD34 Human, (Biotech), PE-anti CD133 Human (Biotech), PE-anti GP-A Human (Biotech) and for isotype control FITC-mouse lg G (Pharmingen).
3. Was incubated on 4^0C, 40 minutes, then washed twice by PBS/BSA solution and sediment was re-suspending in 1% Para formaldehyde solution.
4. Samples were analyzed by flowing cytometer of Beckon & Dickinson (USA).

Data were processed by means software SPSS. Distribution of indices was different to ordinary distribution and the difference between groups is defined by Manna-Witny criteria U (non parametric method). We defined the difference before and after treatment in patient groups by Wilcockson t criteria for dependent variables. Difference considered to be statistically trustful if $P<0.05$.

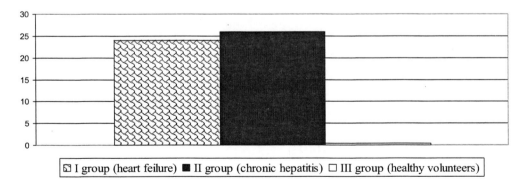

Diagram 1. Percentage consistence of CD34 cells in nucleus cells of peripheral blood.

Research Results

Percentage of CD34 Positive Cells In Live Nucleus Cells of Peripheral Blood

Results of conducted researches are presented in tables and diagrams (Table #10-11, diagram #1-2). There are presented indices average arithmetic±standard deviation.

Data processing revealed statistically trustful difference between I group (heart failure) and III group (healthy volunteers) in regard to percentage of CD34 cells (p=0.002). This index was high in group of patients with heart failure compare with a group of healthy volunteers (I group 24.15±6.15%, III group 0.24±0.18%).

Statistically trustful difference revealed also between groups II (chronic hepatitis) and III (healthy volunteers) in regard to percentage consistence of CD34 cells (p=0.001). Regarding to average meanings, higher percentage of CD34 positive cells was revealed in patient group with viral hepatitis than in a group of healthy volunteers (II group 26.55±7.68%, III group 0.24±0.18%).

Statistically trustful difference was not revealed between I group (heart failure) and I group (chronic hepatitis) regarding percentage consistence of CD34cells (p=0.54).

Statistically trustful difference was revealed between I a subgroup (I-II f.c. NYHA) and I b subgroup (III-IV f.c. NYHA) regarding percentage consistence of CD34cells (p<0.54). According to the average meanings lower percentage of CD34 positive cells was revealed in patients with slight form of heart failure compare to patients with acute heart failure (I a subgroup 15.75±9.13, I b subgroup 32.0±8.9%).

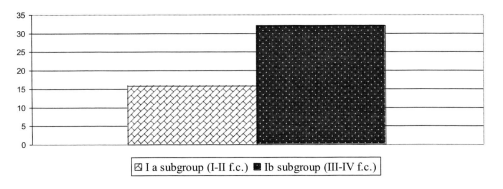

Diagram 2. Percentage consistence of CD34 cells in nucleus cells of peripheral blood.

Percentage of CD133+CD34- Cells in Live Nucleus Cells of Peripheral Blood

Results of conducted researches are presented in tables and diagrams (Table #3,4, diagram #3,4). There are presented indices average arithmetic±standard deviation.

Data processing did not reveal statistically trustful difference between I group (heart failure) and III group (healthy volunteers) in regard to percentage of CD133+CD34- cells (p=0.47). Following average meanings of this index was revealed in noticed groups (I group 1.83±0.9%, III group 0.84±0.46%).

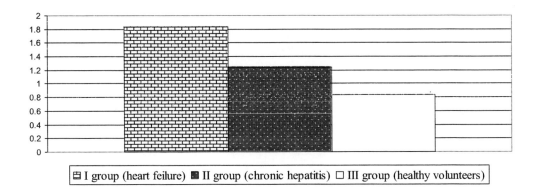

Diagram 3. Percentage consistence of CD34 cells in nucleus cells of peripheral blood.

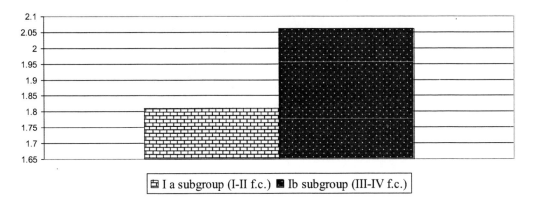

Diagram 4. Percentage consistence of CD34 cells in nucleus cells of peripheral blood.

Statistically trustful difference was not revealed also between groups II (chronic hepatitis) and III (healthy volunteers) in regard to percentage of CD133+CD34- cells (p=0.77). Following average meanings of this index was revealed in the noticed groups (II group 1.24±0.48%, III group 0.84±0.46%).

Table 3. Percentage consistence of CD34+ cells in live nucleus cells of peripheral blood. I group – patients with heart failure syndrome (I-IV f.c. NYHA), II group – patients with chronic hepatitis, III group – healthy volunteers

Index	I group	II group	III group
n	72	36	31
Age	45.2±9.3	36±8.3	43±7.8
Sex	56 men 16 women	29 men 7 women	26 men 5 women
CD 34	24.15±6.15%	26.55±7.68%	0.24±0.18%
CD133+CD34-	1.83±0.9%	1.24±0.48%	0.84±0.46
GP-A	23.44±7.11%	19.4±4.44%	11.4±8.43

Table 4. Percentage consistence of GP-A+ (CD235) + live nucleus cells in peripheral blood I a subgroup – patients with heart failure syndrome (I-II f.c. NYHA) I b subgroup – patients with heart failure syndrome (III-IV f.c. NYHA)

Index	I group	II group
n	32	29
Age	49.3±7.8	36.1±9.1
Sex	26 men	30 men
	6 women	10 women
CD34	15.75±9.13%	32.0±8.9%
CD133+CD34-	1.81±1.14%	2.06±1.85%
GP-A (CD235)	53.1±10.26%	15.08±5.44%

Statistically trustful difference was not also revealed between I group (heart failure) and II group (chronic hepatitis) regarding percentage of CD34cells (p=0.47).

Statistically trustful difference was not revealed between I a subgroup (I-II f.c. NYHA) and I b subgroup (III-IV f.c. NYHA) regarding percentage consistence of CD133+CD34-cells (p=0.27). According to the average meanings lower percentage of CD34 positive cells was revealed in patients with slight form of heart failure compare to patients with acute heart failure (I a subgroup 1.81±1.14%, I b subgroup 2.06±1.85%).

Percentage of GP-A (CD235) Cells in Live Nucleus Cells of Peripheral Blood

Results of conducted researches are presented in tables and diagrams (Table #14,15, diagram #5,6). There are presented indices average arithmetic±standard deviation.

Data processing did not reveal statistically trustful difference between I group (heart failure) and III group (healthy volunteers) in regard to percentage of GP-A cells (p=0.2). Following average meanings of this index was revealed in noticed groups (I group 23.44±7.11%, III group 11.4±8.43%).

Statistically trustful difference was not revealed also between groups II (chronic hepatitis) and III (healthy volunteers) in regard to percentage of GP-A cells (p=0.23). Following average meanings of this index was revealed in the noticed groups (II group 19.4±4.44%, III group 11.4±8.43%).

Statistically trustful difference was not also revealed between the I group (heart failure) and II group (chronic hepatitis) regarding percentage consistence of GP-A cells (p=0.64).

Statistically trustful difference was revealed between I a subgroup (I-II f.c. NYHA) and I b subgroup (III-IV f.c. NYHA) regarding percentage consistence of GP-A cells (p<0.01). According to the average meanings higher percentage of GP-A positive cells was revealed in patients with slight form of heart failure compare to patients with acute heart failure (I a subgroup 53.1±10.26%, I b subgroup 15.08±5.44%).

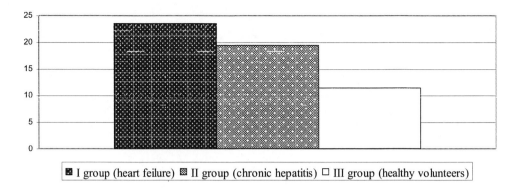

Diagram 5. Percentage consistence of GP-A(CD235) cells in nucleus cells of peripheral blood.

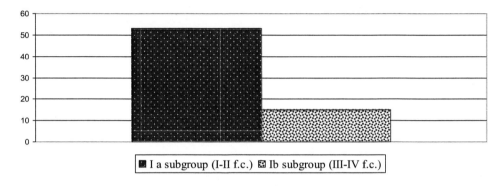

Diagram 6. Percentage consistence of GP-A(CD235) cells in nucleus cells of peripheral blood.

Statistically trustful difference (p<0.01) was revealed between I a subgroup (heart failure I-II f.c. NYHA) and III group (healthy volunteers). This index appeared to be higher in patients of I-II functional class NYHA, compare with healthy volunteers (I a subgroup 53.1±10.26%, III group 11.4±8.43%).

Statistically trustful difference (p<0.01) was also revealed between Ia subgroup (heart failure I-II f.c. NYHA) and II group (chronic hepatitis). This index appeared to be higher in patients of I-II functional class NYHA, compare with a group with chronic hepatitis (Ia subgroup 53.1±10.26%, II group 19.4±4.44%).

Percentage Consistence of CD34, CD34-CD133+ Cells in Live Nucleus Cells of Peripheral Blood Before and After Treatment

29 patients with heart failure syndrome (10 – III-IV f.c. and 11 – I-II f.c. NYHA) have been examined on surface antigens repeatedly. Results are presented in tables and diagrams (table 18, diagram #13,14), indices are average arithmetic±standard deviation.

As a result of processing data statistically trustful difference between indices was revealed before and after treatment regarding CD34 and CD34-CD133+ cells according to(p<0.05) and (p<0.05). These indices appeared to be higher before treatment, when their condition was evaluated with higher functional class of heart failure.

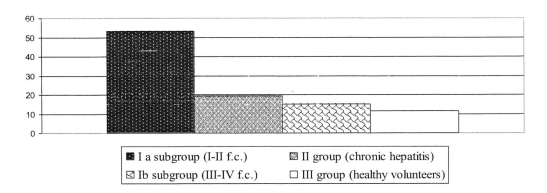

Diagram 7. Percentage consistence of GP-A(CD235) cells in nucleus cells of peripheral blood.

**Table 5. Percentage of CD34, CD133+CD34- cells in live nucleus cells
of peripheral blood before and after treatment n=29**

Index	Before treatment	After treatment
Average functional class	2.87	1.7
CD34+	36±6.54%	11.04±7.14%
CD34-CD133+	5.58±4.62%	1.12±0.81%

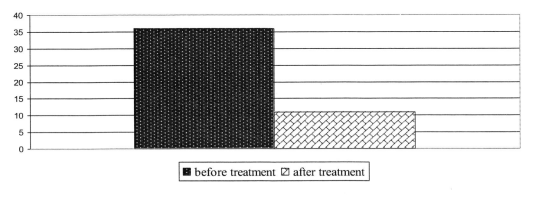

Diagram 8. Percentage consistence of CD34 cells in nucleus cells of peripheral blood.

Results Review

Results of conducted research have revealed that significant increase in amount of GP-A nucleus cells (erythroid precursors) take place in peripheral blood. Such an increase in young forms of erythroid line (GP-A) occurs only in case of relatively slight forms of heart failure – I-II f.c. and during III-IV f.c. heart failure amount of these cells considerably decreases in peripheral blood, that in our opinion, points to progress of vicious circle (….. researches of recent years have revealed that TNF-a has ability erythropoes inhibition, such decrease in amounts of cells during progressing heart failure possibly is connected to increase in concentration of TNF-a as negative regulator of erythroposing).

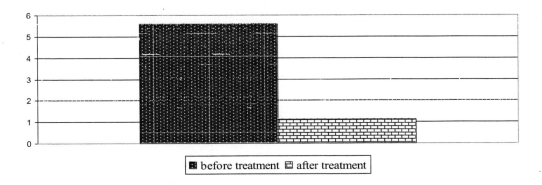

Diagram 9. Percentage consistence of CD133+CD34- cells in nucleus cells of peripheral blood.

We suppose GP-A can be used as a test for heart failure diagnostic. This opinion is based on revealed mechanism of changes of GP-A cells in relation with formation and process of heart failure, considerable increase of such cells during formation and slight forms of heart failure, also the fact that statistically trustful increase of GP-A cells have not been revealed in patients with chronic hepatitis. It is noticeable that in a research have taken part patients with heart failure of I functional class, having myocardium dysfunction but neglect it. In these patients high percentage of GP-A were detected in live nucleus cells of peripheral blood. Therefore we can conclude that mentioned tests could be effective for heart failure screening.

Table 6. Definition of sensibility and specificity of the test

Test results	Disease	
	exists	not exists
Positive	True-positive (TP) 21	False-positive (FP) 3
Negative	False-negative (FN) 9	True-negative TN) 66

After processing data of the research we have calculated operational characteristics of the mentioned test. Criteria provided by American Heart Association was set as gold standard for evaluating heart failure, particularly I-II functional class. Critical level of mentioned cells was 50%. Taking this into account we have calculated operational characteristics of diagnostic test – sensibility and specificity. According to the research results sensibility of the test for revealing patients with heart failure of I-II f.c. is 70%, specificity – 95%.

Table #6 represents prognostic values of positive and negative tests (using theory of Bayes) for results of sensibility and specificity (sensibility 70% and specificity 95%) taking into account initial probability (prevalence) of disease.

Results of conducted research showed that during heart failure syndrome there is an increase in amounts of hemopoezic stem cells and progenitors (CD34) in peripheral blood compare with control group (healthy volunteers). Statistically trustful rise of the mentioned cells have been detected also in patients with chronic hepatitis. Consequently, we can suppose that increase of hemopoezic stem cells and progenitors occur during differ pathological processes that have to be confirmed in other researches.

In patients group before and after treatment statistically trustful decrease of CD34 cells was noticed after effective treatment, while their condition was evaluated as low functional class of heart failure.

As for CD133+CD34- stem cells and progenitors, presently considered to be non-hemopoezic system cells (different examinations have proved pluri-potentiality of these cells), in conducted examination we have revealed increase of average value in patients with heart failure and with chronic hepatitis compare with control group but the mentioned changes are not statistically trustful. Besides, in patients group before and after treatment statistically trustful decrease of the mentioned cells was noticed after effective treatment, while their condition was evaluated as low functional class of heart failure.

Since in carried out researches in patients group before and after treatment decrease of stem cells and progenitors after effective treatment have been detected almost in all cases, we consider it possible to use mentioned amount changes of CD34 and CD133+CD34- for monitoring heart failure treatment.

Conducted research showed that in peripheral blood of healthy patients amount of CD133+CD34- (nonhemopoetic cells) is higher than of CD34 (hemopoezic) cells. Researches of different scientists have proven that in marrow on 35-75% of CD34, antigen CD113 is expressed, and CD133+CD34- cells are in low quantity. We suppose that higher level of CD133+CD34- cells than of CD34 cells, is caused by demand of organism for CD133+CD34- cells as "building materials". In healthy persons presently unknown mechanisms are ensuring transfer of such cells in peripheral blood circulation. During pathological processes their amount in peripheral blood increases. AS to CD34 cells, their number in peripheral blood of healthy persons is minimal, but increases several times during pathological processes.

Results of conducted studies showed that statistically trustful increase take place in amounts of stem hemopoezic cells and progenitors (CD34) in peripheral blood during heart failure syndrome compare with control group.

Results of conducted studies showed that there is an increase in amounts of stem nonhemopoezic cells and progenitors (CD133+CD34-) in peripheral blood during heart failure syndrome compare with control group (according to average meanings) but this change is not statistically trustful.

In patient group before and after treatment statistically trustful changes in amounts of CD34, CD34-CD133+ cells take place simultaneously with seriousness of state of heart failure.

It has been revealed that in peripheral blood of healthy control group amount of CD133+CD34 cells (nonhemopoezic stem cells and progenitors) is higher than of CD34 cells (hemopoezic stem cells and progenitors) CD34/CD133<1. In case of disease (damage of organs) this proportion changes to contrarily CD34/CD133>1.

Conclusions

- Conducted studies revealed that in case of heart failure syndrome statistically trustful increase in amount of hemopoezic stem cells (CD34) take place in peripheral blood compare with control group.
- Results of conducted studies showed that there is an increase in amounts of stem nonhemopoezic cells and progenitors (CD133+CD34-) in peripheral blood during heart failure syndrome compare with control group (according to average meanings) but this change is not statistically trustful.
- It has been revealed that in peripheral blood of healthy control group amount of CD133+CD34 cells (nonhemopoezic stem cells and progenitors) is higher than of CD34 cells (hemopoezic stem cells and progenitors) CD34/CD133<1. In case of disease (damage of organs) this proportion is changed to contrarily CD34/CD133>1.
- In patient group before and after treatment statistically trustful changes in amounts of CD34, CD34-CD133+ cells take place simultaneously with seriousness of state of heart failure.
- The study has detected statistically trustful increase of young forms of erythroid line (GP-A nucleus cells) in patients with heart failure I-II f.c. (NYHA) compare with control group. And in patients (III-IV f.c. NYHA) increase of such cells is not statistically trustful compare with control group.

References

[1]   Remme, W. J. and Swedberg, K. Comprehensive guidelines for the diagnosis and treatment of chronic heart failure. Task force for the diagnosis and treatment of chronic heart failure of the European Society of Cardiology. *European journal of heart failure* (ISSN: 1388-9842) January 1, 2002 - Volume *4*, Issue 1

[2]   Kajstura, J., Leri, A., Finato, N., et al. Myocyte proliferation in endstage cardiac failure in humans. *Proc Natl Acad Sci USA,* 1998, 95, 8801 - 8805.

[3]   Morayma Reyes, Arkadiusz Dudek, Balkrishna Jahagirdar, Lisa Koodie, Paul, H. Marker and Catherine, M. Verfaillie Origin of endothelial progenitors in human postnatal bone marrow. *J. Clin Invest,* February 2002, Volume 109, Number 3, 337-346

[4]   Yin, A. H., Miraglia, S., Zanjani, E. D., et al. AC133, a novel marker for human hematopoietic stem and progenitor cells. *Blood* 1997, 90, 5002-12.

[5]   Miraglia, S., Godfrey, W., Yin, A. H., et al. A novel five-transmembrane hematopoietic stem cell antigen: Isolation, characterization and molecular cloning. *Blood,* 1997, 90, 5013-21.

[6]   Bühring, H. J., Seiffert, M., Bock, T. et al. Expression of novel surface antigens on early hematopoietic cells. *Ann NY Acad Sci,* 1999, 872, 25-39.

[7]   DeWynter, E. A., Buck, D., Hart, C., et al. CD34+AC133+ cells isolated from cord blood are highly enriched in long-term culture-initiating cells, NOD/SCID-repopulating cells and dendritic cell progenitors. *Stem Cells* 1998, 16, 387-96.

[8]   Gehling, U. M., Ergün, S., Schumacher, U., et al. *In vitro* differentiation of endothelial cells from AC133-positive progenitor cells. *Blood,* 2000, *95*, 3106-12.

[9]   Peichev, M., Naiyer, A. J., Pereira, D., et al. Expression of VEGFR-2 and AC133 by
      circulating human CD34+ cells identifies a population of functional endothelial
      precursors. *Blood*, 2000, 95, 952-8.

[10] Corbeil, D., Roper, K., Hellwig, A., et al. The human AC133 hematopoietic stem cell
      antigen is also expressed in epithelial cells and targeted to plasma membrane
      protrusions. *J Biol Chem*, 2000, 275, 5512-20.

[11] Nobuko Uchida*, , David W. Buck*, Dongping He*, Michael J. Reitsma*, Marilyn
      Masek*, Thinh V. Phan*, Ann S. Tsukamoto*, Fred H. Gage , and Irving L. Weissman
      Direct isolation of human central nervous system stem cells (2000) *Proc. Natl. Acad.
      Sci.* USA 97, 14720-14725

[12] Kaufman, Dan S., Eric, T. Hanson and Rachel, L. Lewis, Robert Auerbach, and James
      A. Thomson. Hematopoietic colony-forming cells derived from human embryonic stem
      cells. *Proc. Natl. Acad. Sci.* USA. 2001, 98,10716-10721

[13] Pluripotency of mesenchymal stem cells derived from adult marrow Yuehua Jiang,
      Balkrishna N. Jahagirdar, R. Lee Reinhardt, Robert E. Schwartz, C. Dirk Keene, Xilma
      R. Ortiz-Gonzalez, Morayma Reyes, Todd Lenvik, Troy Lund, Mark Blackstad, Jingbo
      Du, Sara Aldrich, Aaron Lisberg, Walter C. Low, David A. Largaespada, *Catherine M.
      Verfaillie Nature*, Vol 418 July 4 2002, 41-49

[14] Kuci, S., Wessels, J. T., Buhring, H. J., Schilbach, K., Schumm, M., Seitz, G, Loffler,
      J., Bader, P., Schlegel, P. G., Niethammer, D., Handgretinger, R. (2003) Identification
      of a novel class of human adherent CD34- stem cells that give rise to SCID-
      repopulating cells. *Blood*,101, 869-876.

[15] Mario Peichev, Afzal J. Naiyer, Daniel Pereira, Zhenping Zhu, William J. Lane,
      Mathew Williams, Mehmet C. Oz, Daniel J. Hicklin, Larry Witte, Malcolm A. S.
      Moore, and Shahin Rafii Expression of VEGFR-2 and AC133 by circulating human
      CD34+ cells identifies a population of functional endothelial precursors *Blood*, Vol. 95
      No. 3 (February 1), 2000: pp. 952-958

[16] Gehling, U. M. et al.In vitro differentiation of endothelial cells from AC133-positive
      progenitor cells. *Blood* , **95**, 3106-3112 (2000).

[17] Stamm, C., et al.; "Autologous bone-marrow stem-cell transplantation for myocardial
      regeneration"; *The Lancet*, **361**, 45-46, 4 January 2003

[18] Chasis, J. and Mohandas, N. 1992. Red blood cell glycophorins. *Blood*, **80**, 1869-1879

[19] Chishti, A Y., Palek. J., FiSher, D. 1996. Reduced invasion and growth of Plasmodium
      falciparum into elliptocytic red blood cells with a combined deficiency of protein 4.1,
      glycophorin C, and p55. *Blood*, **87**, 3462-3469

[20] Chien, S. Red cell deformability and its relevance to blood flow. *Annu Rev Physiol* **49**,
      177-192, 1987

[21] Bruce, L. J., Ghosh, M., King, D. 2002. Absence of CD47 in protein 4.2-deficient
      hereditary spherocytosis in man: an interaction between the Rh complex and the band 3
      complex. *Blood*, **100**, 1878-1885

[22] Hassoun, H. and Hanada, T.. 1998. Complete deficiency of glycophorin A in red blood cells
      from mice with targeted inactivation of the band 3 (AE1) gene. *Blood*, **91**, 2146-2151.

[23] Mohandas, N., Phillips, W. M., Bessis, M. Red blood cell deformability and hemolytic
      anemias. *Semin Hematol*, **16**, 95, 1979

[24] Loris zamia, et al. TNF-related apoptosis-indusing ligand (TRAIL) as a negative regulator
      of normal human erythropoiesis. *Blood*. Vol. 95 No (June 15) 2000, pp. 3716-3724.

In: Focus on Science and Technology...
Editor: Sergo Gotsiridze, pp. 195-204

ISBN: 978-1-61209-970-5
© 2011 Nova Science Publishers, Inc.

Chapter 19

EFFECT OF PRE-FERMENTATION ENZYME MACERATION OF GEORGIAN RED WINE MADE FROM GRAPE VARIETY, "SAPERAVI" ON QUANTITY AND EXTRACTION OF RESVERATROL, ANTHOCYANINS, AND TOTAL PHENOLS

M. Khomasuridze[1], G. Datukishvili[1], N. Shakulashvili[2], G. Dakishvili[3], E. Kalatozishvili[4] and L. Mujiri[4]

[1] Department of Chemical and Biological Technology,
Georgian Technical University, Tbilisi, Georgia
[2] LTD Wine Laboratory, Georgia
[3] Schuchmann Wines Georgia, Ltd., Telavi, Georgia
[4] Georgian Institute of Winemaking, Georgia

Abstract

Investigations were arranged to evaluate the effect of pre-fermentation enzyme maceration on the quantity of phenol compounds, to research the influence of the used enzyme preparation on red wine quality, to determine resveratrol and to enhance its content in Georgian wine, derived from grape variety "Saperavi".

Folin - Ciocalteu spectrophotometric method was applied for determination of total phenols. Resveratrol and anthocyanins were determined by HPLC; Volatile and titriable acids, SO_2, sugar and alcohol were measured by conventional methods.

The obtained results reveal, that pre-fermentation enzyme maceration with the above mentioned enzyme preparation increased the content of anthocyanins (by 221mg/l) and phenol compound (by 790 mg/l); Also were increased of alcohol content (by 1, 2 vol. %) and the amount of dry extract (by 3, 2 g/l). Both cis- and trans- forms of resveratrol are presented in Georgian wine prepared from grape variety "Saperavi". It was found, that the enzymatic treatment caused the additional extraction of this compound in wine. There were no big differences between titriable and volatile acids in control and experimental samples. Organoleptic evaluation confirmed that pre-fermentation enzyme maceration with LAFASE ® HE GRAND CRU did not give additional astringency and bitterness.

LAFASE ® HE GRAND CRU can contribute to improve Georgian wine quality.

Keywords: Resveratrol, Enzyme preparation, Maceration, Red wine, Saperavi, Total Phenols, Anthocyanins.

Introduction

The enzyme preparations imported in Georgia from different European countries are used for native grape varieties, but their influence on the Georgian wine components has not been studied. Determination of phenol and polyphenol compounds in wine is common approach and recently was performed by many researchers almost in all Georgian wine sorts, but determination of resveratrol in Georgian wines, made from different grape varieties, is rather new approach. The scientific confirmation of presence of significant quantity of resveratrol in Georgian wines made from different grape sorts, will be helpful for Georgian wine producers to promote their output.

"Saperavi" is a red grape variety unique to Georgia and presented specifically in the Kakheti region. This grape sort is the most important for Georgian wine culture and produce deep red wines which are suitable for extended periods of aging, up to 50 years. It is one of the distinguished representatives of the red grape varieties of the world and is cultivated in specific micro zones of the Kakheti region. It gives particularly splendid flavor to all red wines (Saperavi, Mukuzani, Napareuli, Akhasheni and Kindzmarauli). Saperavi grapes are known for their dark pink flesh and very dark skins, from which they get their name (meaning, literally "paint" or "dye").

Maceration is the winemaking process where the phenolic and polyphenolic materials of the grape -- tannins, coloring agents and flavor compounds --- are extracted from the grape skins, seeds and stems into the wine. The process of maceration begins, to varying extent, as soon as the grapes skins are broken and exposed to some degree of heat. Temperature is the guiding force with higher temperatures encouraging more breakdown and extraction of phenols from the skins and other grape materials. Maceration continues during the fermentation period and can last well past the point when the yeast has converted all sugars into alcohol [1].

The preparations of pectic enzymes are used for a more efficient extraction of desirable red grape pigments and other phenol compounds which are bound in plant cells and can be faster released by the action of pectolytic enzymes. The enzyme preparations in current commercial cans contain diverse amount of cellalytic, β-glycosidic, proteolytic and other species of enzymes apart from the main pectolityc compounds. For the production of these enzymes mainly a fungus line of Aspergillius is used. The enzymes are released into the production medium and are further purified and concentrated. Pectins are present in tissues of all higher plants and differ in composition. The plant tissues are formed by cells which are separated by wall cells, contrary to the animal cells pectin is present as intercellular putty and together with hemicellulose forms a part of the wall. Pectin belongs to the group of polysaccharides.The main polysaccharide chains are short and long, straight or branched. The pectolytic enzymes are able to split those chains and saccharidic bonds between the chains. Release of grape pigments and aroma compounds can be quicker in such a way. The pectin, due to its structure, acts further as a stabilizer of cloudy stuffs and retards to speed of settling and filtration [2].

Red wine pomace and black currant juice residues provide a new source of antioxidants that may provide nutritional benefits. The extraction of antioxidants could be enhanced by enzyme treatment and may be useful in upgrading of wine and fruit juice press residues for production of nutritional supplements or novel food ingredients [3]. Phenolic antioxidants are usually grouped into flavonoids and non-flavonoids, according to their structure. With regard to the tannic character, phenolic antioxidants are further subdivided to tannic phenols and non-tannic phenols. Collectively, these compounds contribute to the high antioxidant capacity of wine. Phenolic compounds contribute to the sensory characteristics of wines and particularly of red ones. Anthocyanins and their derivates are predominant pigments in red wine. Flavanols and hydroxycinnamates are bitter than proanyhocyanidins, which are polymers of flavan-3ol units and are also cold-condensed-tannin, contribute bitterness and astringency [4]. Phenolic compounds are present mainly in the skins and seed in red berries. Anthocyanins and flavanols are found in the skins. Hydroxycinnamates are found also in the skin. On the other hand flavan-3-ol monomers and majority of gallic acids in red wine are likely organized in the seeds. Wine making practices influence the extraction of phenolic compounds [5].

In addition these compounds have recently received much attention because of their potential contribution to human health due to their antioxidant, antimicrobial, antiviral, and anticancerogenic characteristics [6].

Resveratrol (3,5,4'-trihydroxystilbene) has attracted special attention in recent years, because of its diverse pharmacological properties such as cancer chemoprevention [5] and protection against cardiovascular and neurodegenerative disease [7]. Wine contains significant amounts of resveratrol, which constitutes the single most important contributor to the ''French paradox'' phenomenon — the low incidence of cardiovascular disease in France despite a high-fat diet [8].

Resveratrol (3,5,4'-trihydroxystilbene) is a polyphenolic phytoalexin. It is a stilbenoid, a derivate of stilbene, and is produced in plants with the help of the enzyme stilbene synthase. It exists as two geometric isomers: cis- (Z) and trans- (E), with the trans-isomer shown in the top image. The trans- form can undergo isomerisation to the cis- form when exposed to ultraviolet irradiation [9]. Trans-resveratrol in the powder form was found to be stable under "accelerated stability" conditions of 75% humidity and 40 degrees C in the presence of air [10]. Resveratrol content also stayed stable in the skins of grapes and pomace taken after fermentation and stored for a long period [11].

In grapes, resveratrol is found primarily in the skin, in muscadine grapes — also in the seeds. The amount found in grape skins also varies with the grape cultivar, its geographic origin, and exposure to fungal infection. Fermentation time, i. e. time of wine contact with grape skins is an important determinant of its resveratrol content [12]. The amounts of resveratrol found in several food varies greatly. Red wine contains between 0.2 and 5.8 mg/L depending on the grape variety [13]. The trans-resveratrol concentration in forty Tuscan wines tasted, ranged from 0.3 to 2.1 mg/L for 32 red wines, and had a maximum of 0.1 mg/L for tested 8 white wines. Both, the cis- and trans-isomers of resveratrol were detected in all tested samples. Cis-Resveratrol levels were comparable to those of the trans-isomer. They ranged from 0.5 mg/L to 1.9 mg/L in red wines and had a maximum of 0.2 mg/L in white wines [14].

Methods and Materials

Wine analyses for titriable and volatile acids, dry extract, sugar, alcohol, and SO_2 was conducted according to standard methods.

We applied the Folin-Ciocalteu spectrophotometric method for determination of total phenols [15] and HPLC for Anthocyanins. The HPLC analysis was carried out in the following conditions: Injection Volume - 50 μl.; Flow rate - 0.8 ml/min; Temperature - 40°C; Run time: 45 minutes; Post time: 5 minutes; Detection - 518 nm. These Analyses were performed according to Compendium of International Methods of wine and must Analyses (International Organization of Wine and Vine) [16].

The content of resveratrol was determined in control and trial samples by High Pressure Liquid Chromatography in "Wine Laboratory Ltd." (Tbilisi, Georgia). Chromatographic conditions were: Equipment - HPLC (Knauer); Detection mode - UV; Detection wavelength - 280 nm; Injection volume - 20 μl; Elution order - isocratic; Oven temperature - 40°C; Flow rate - 0.8 ml/min; Mobile phase - 25 mM sodium dihydrophospate adjusted with HCl to pH-3.1 / ACN - 70 / 30 - v/v; Analysis time - 20 minutes.

Grape variety: Grape was soured from the vineyards which belonged to "Telliany Valley PLC" (Georgia, Kaheti Region, Telavi district) and harvested in 2008.

Enzyme preparation: Commercial pectic enzyme preparation and yeast strain were used in the experiments. They were obtained from Laffort. LAFASE ® HE GRAND CRU is purified, pectolytic enzyme preparation for the production of structured wines for cellaring that is rich in coloring matter and with supple tannins. It is specific for traditional macerations (with or without pre-fermentative maceration), and increases the extraction of more stable phenolic compounds (more polymerized anthocyanins and tannins) and ageing potential.

Yeast: **Zymaflore F10** is the strain for superior red wines which are defined by their elegance, combining structure, volume on the palate and intense color. A new technique, directed breeding (non-GMO cross breeding), applied to the F10 strain, has provided increased resistance to membrane sterol deficiency, thus ensuring high fermentation security. It is particularly recommended for producing premium to icon wines, notably Cabernet Sauvignon or Merlot.

Vinification: The control and experimental samples were prepared in winery, "Telliany Valley PLC". Grapes were picked in technical maturity in September 2008, delivered in winery Telliany Valley PLC" (Georgia, Kaheti region, Telavi district), processed by standardized vinification procedures - destemmed and crushed. The grape must was sulphidized with 30mg/l SO_2 prior to alcoholic fermentation and divided in two equal portions.

Preparation of the control sample: The must was placed in fermentation reservoir and yeast strain immediately inoculated. The dry yeast was dissolved according to manufacturer's recommendation.

Preparation of the experimental sample: The must was placed in pre-fermentation reservoir and corresponding quantities of enzyme preparation was added. Pre–fermentation maceration was conducted during 16 hours at 20 °C temperature and then the yeast strain was inoculated.

The temperature was controlled 6 times in 24 hours and the must was plunged and overpowered 3 times per day during alcoholic fermentation. The temperature of alcoholic fermentation varied between 20 °C and 25 °C. The content of sugar and alcohol were checked twice in every 48 hour. The parameter of completed active alcoholic fermentation was assumed the stage when sugar content in wine materials was 0.25%. At completion of fermentation the wine were pressed and 50 mg/L SO_2 was added. The samples were racked twice every three weeks and than after three months, adding each time 30 mg/L SO_2. All samples were clarified before the measurements at the end of March. Samples were sulphidized again before the clarification with 30 mg/l SO_2. All samples were treated with . All samples were treated with gelatin **Colperl** (producer - Institute Oenologique De Champagne), in order to precipitate the unstable substances and to prepare wine for sensory evaluation and chemical analyses. The content free SO_2 was checked and increased up to 40 mg/l At bottling wine was sterile filtered in membrane filter and stored at 12 °C in a cellar.

Results

Analytical analyses and organoleptic evaluation were carried out after 3 month from bottling. The results received from sugar, volatile and titriable acids, dry extract and alcohol determination are shown on table I.

Table I. Chemical analyses of experimental and control sample

Samples	Alcohol (Vol. %)	Volatile acids (G/L)	Titriable acids (G/L)	Dry extract (G/L)	Sugars (Vol. %)
Experimental sample	13.8	0.45	5.74	31.8	0.14
Control sample	12.6	0.33	4.93	28.6	0.91

There are no significant differences between sugars, volatile and titriable acids content in experimental and control samples. In control sample volatile acids content is a little lower (by 0.12 g/l) then in experimental. It might be related to the additional contact of must with air. During the enzyme treatment, at first must was placed in pre-fermentation reservoir and than pumped in fermentation tank. Besides this it was plunged and overpowered 3 times more that control sample. On the other hand these technological steps contributed to activation the intensity of alcoholic fermentation. More sugar stayed in control sample than in experimental wine. In addition the enzyme treatment retarded the initial increase in alcohol content by 1.2 %. Compare with control sample in experimental sample titriable acids content is higher by 0,84g/l and dry extracts by 4,2g/l.

Besides, appreciable changes are observed in the content in total phenols and anthocyanins. The nine major anthocyanins were detected in both samples. Group 1 - "Nonacylated anthocyanidin-3-glucosides"; delphinidol-3-glucoside; cyanidol-3-glucoside; petunidol-3-glucoside; peonidol-3-glucoside; malvidol-3-glucoside. Group 2 - "Acetylated anthocyanidin-3-glucosides": peonidol-3-acetylglucoside; malvidol-3-acetylglucoside. Group 3 - "Coumarylated anthocyanidin-3-glucosides": peonidol-3coumarylglucoside; malvidol-3-coumarylglucoside.

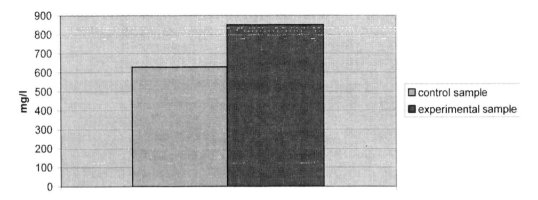

Figure 1. The sum of total major anthocyanins in control and experimental wine samples.

The total sum of these compounds in control sample was 629 mg/l and in experimental one - 850 mg/l. dissimilarity is apparent on figure 1. The amount of anthocyanins is higher in experimental sample by 221mg /l.

The data gathered from spectrophotometric analyses - the quantity of total phenols is evident on Figure 2. The content of total phenols in experimental is sample was elevated by 790 mg/l.

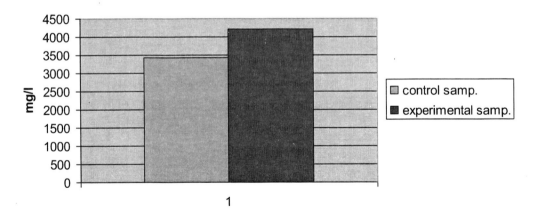

Figure 2. The amount of total phenols in control and experimental samples.

The detection of resveratrol in the red wine made from Georgian grape variety "Speravi" is quite new approach and very important for our wine makers. The proved existence of this antioxidant will support Georgian red wine to find the proper place in world market and may be used for its advertising. In order to ensure customers and winemakers that resveratrol really exists in Saperavi wine we have decided to include the pictures of chromatogram in this article – Figure 3 and Figure 4 The data is also given in table II.

The achieved data confirms that resveratrol mainly exists in grape skin in Georgian Grape variety Sapaeravi, and additional maceration process promotes its solution in wine.

The increase quantity of dry extract, total phenols, anthocyanins and resveratrol in experimental sample must be caused by the ability of pectolytic enzymes to split the main polysaccharide chain and sachharide bonds between the chains, releasing grape pigments and other compounds [17].

The promotion of intensive color of red wine by macerating enzyme treatment of the must may be due to the breakdown of protective polysaccharide-protein colloids by hemicellulase and protease activities in enzyme preparation [18].

**Table II. The content of trans- and cis-resveratrol in control
and experimental samples**

Sample	trans- resveratrol	cis-resveratrol
control	0.23	0.4
experimental	1.84	1.66

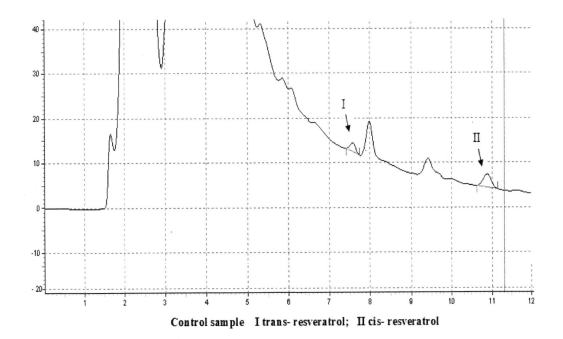

Control sample I trans- resveratrol; II cis- resveratrol

Figure 3. Chromatogram of trans- and cis- resveratrol in control sample.

Along with the analytical research we carried out organoleptic evaluation. Beside us, four winemakers from "Teliani Valley PLC" were included in testing commission. The maximum score was 5. The summarized results are declared in table III. Both wines were suitable to high quality young dry wines. They were strong bodied and had deep purple color, without special astringency and bitterness.

Experimental sample had more intensive fruit aroma and flavor than control one. In this experiment we had not researched the aromatic compounds. The maceration process in technology of elaborating flavor wines obviously influenced by temperature and the length of contact between the must and solid parts of grape [19]. Maceration process, temperatures and duration influenced definitively the quantitative aroma extraction [20].

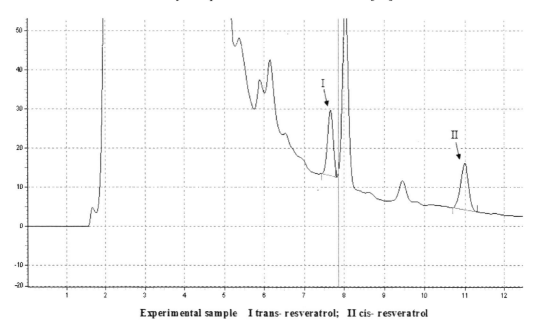

Experimental sample   I trans- resveratrol;   II cis- resveratrol

Figure 4. Chromatogram of trans- and cis- resveratrol in experimental sample.

Table III. The results of organoleptic evaluation

Wine examples	color	aroma	taste	total score
control sample	4.1	3.9	3.0	3.66
experimental sample	4.5	4.3	3.5	4.1

Conclusion

Obtained results from conducted analyses and organoleptic evaluation confirmed that: Enzyme preparation LAFASE ® HE GRAND CRU is suitable for production of wine from Georgian Grape variety "Saperavi". Pre-fermentation enzyme maceration promotes the increase of dry extract, total phenols, anthocyanins and resveratrol in red wine. The enzyme preparation treatment enhances wine aroma and does not emphatically changes the wine standard parameters, varietal aroma and character. The considerable amount of cis and trans resveratol exists in red wine derived from Georgian grape variety "Saperavi". Enzyme preparation LAFASE ® HE GRAND CRU causes the additional extraction of resveratrol and raises its concentration in wine.

In conclusion, Pre-fermentation enzyme maceration ameliorates the physical and chemical parameters of red wine and may be recommended for Georgian wine producers.

References

[1]   Robinson, J. (ed) *"The Oxford Companion to Wine"* Third Edition. Oxford University Press; New York; 2006, 414-415, 688-689.

[2]   Capounova, D. and Drdak, M. "Comparison of some commercial pectic enzyme preparations applicable in wine Technology". *Czech Journal of Food Sciences;* 2002 vol. 20, no 4, 131-134.

[3]   Meyer, A. S. and Landbo, A. K. "Enzyme-Assisted Extraction of Antioxidative Phenols from Black Currant Juice Press Residues"; *Journal of Agriculture and Food Chemistry,* 2001, 49 (7), 3169–3177.

[4]   Weiingerl, V., Stric, M., Kocar, D. "Comparison of Methods for Determination of Polyphenols in Wine by HPLC-UV/VIS, LC/MSlMS and Spectrophotometry" *Acta Chimica Slovenica;* 2009, vol. 56, no3, 698-703.

[5]   Vidal, S., Francis, L., Guyot, S., Marnet, N., Kwiatkowski, M., Gawel, R., Cheynier, V. and Waters, E. J. "The Mouth Feel Properties of grape and apple proanthocyandins in wine like medium". *Journal or the Science Food Agriculture* 2003, 83, 564-573.

[6]   Cheynier, V. "Polyphenols in food are more complex than often thought"; *.American Journal of Clinical Nutrition;* 2005,Vol. 81, 67-72.

[7]   R. Jackson. "Wine Health and Food". In "Wine Science" ed. Teyor, S. Academic Press, San Diego; 2000, 591-607

[8]   Farina, A., Ferranti, C. and Marra, C. "An improved synthesis of resveratrol". *Nat. Prod. Res.* 2006, 20 (3): 247-52...

[9]   Milner, J. A., McDonald, S. and Anderson, D. "Molecular Targets for Nutrients Involved in Cancer Prevention", *Nutrition and Cancer* 2002 41(1-2):1-16.

[10]   Stewart, J. R., Artime, M.C. and O'Brian, C. A. "Resveratrol a candidate nutritional substance for prostate cancer prevention." *J Nutr.* 2003,133 (7Suppl): 2440-2443.

[11]   Lamuela-Raventos, R. M. "Direct HPLC Analysis of cis- and trans-Resveratrol and Pieced Isomers in Spanish Red Vitis vinifera Wines" (Scholar search). *J. Agric. Food Chem.* 1995, 43, 281–283.

[12]   Prokop, J., Abrman, P., Seligson, A. L. and Sovak, M. "Resveratrol and its glycon pieced are stable polyphenols". *J Med Food* 2006, 9 (1), 11–4.

[13]   Bertelli, A. A., Gozzini, A., Stradi, R., Stella, S. and Bertelli, A. "Stability of resveratrol over time and in the various stages of grape transformation". *Drugs Exp Clin Res* 1998, 24 (4), 207-11.

[14]   Mozzon, M., Frega, N., and Pallotta, U. "Resveratrol content in some Tuscan wines". *Italian journal of food science,* 1996, vol. 8, no2, 145-152

[15]   Compendium of International Methods of wine and must Analyses (International Organization Of wine and Vine). *Folin-Ciocalteu Index.* Volume 1; Section 2; (ed. 2003).

[16]   Compendium of International Methods of wine and must Analyses (International Organization Of wine and Vine). HPLC-Determination Of Nine Major *Anthocyanins In Red And Rose Wines.* Volume 2; Section 3.15; (ed 2003)

[17]   Rexova-Benkova, F. Pectic enzymes. *J. adv. Carbohydr.* Chem.Bioch. 1976, *33,* 323-385.

[18]   Heatherbell, D., Dicey, M., Goldsworthy, S. and Vanhanen, L. Effect of pre-

fermentation cold maceration on the composition, color and flavor of Pinot noir wine. Proceeding of New Zealand Grape and Wine Symposium, New Zealand Society for Viticulture and Oenology, P>O> Box 90-276, Auckland Mail Centre, Auckland 1, New Zealand1997. *Steans,* G.F. (ed), 30-42.

[19] Stoica, F. and Gheogrhita, M.,, The Influence of the maceration process on the terpens and polyphenols complexes of the flavor wines. Bulletin UASVM, *Agriculture* 2008 65 (2), 33-39.

[20] Stoica, F. and Mountean, C. Researches concerning the enzymatic preparation influence on the flavors extraction from Tamaioasa romaneasca by Dregasini during the maceration process, *4th Croatian and 1st International Symposium on Agriculture,* Opartija, Croatia, 2006. 753-762

In: Focus on Science and Technology...
Editor: Sergo Gotsiridze, pp. 205-212

ISBN: 978-1-61209-970-5
© 2011 Nova Science Publishers, Inc.

Chapter 20

DEPOSIT PROGNOSIS BY NEURON NETWORKS PRACTICED FOR DIAMONDS

Avtandil A. Kvitashvili[1], Rafael G. Tkhinvaleli[1], Erekle G. Gambashidze[1] and Lev M. Natapov

[1]Department of Applied Pattern Recognition Systems, Institute of Cybernetics of the Georgian Academy of Sciences, Tbilisi 1086, Georgia
[2]National Key Centre for Geochemical Evolution and Metallogeny of Continents Mcquarie University, Sydney, NSW, 2109, Australia

Abstract

An accumulated experience in geology shows ineffectiveness of the existing approaches to make a prognosis of the deposits when the available crust characteristics do not have explicit relationships to a given deposit. That takes generally place in cases of more valuable deposits as gold, diamond and so forth. The present paper proposes a use of an Artificial Neuron Network for prognosis of diamond deposits that was successfully carried out about 20 years ago while the worked out method makes it possible its application to any other ores or natural resources.

Keywords: neuron network, kimberlite, deposit features, stochastic approximation.

Introduction

To perform a trustworthy prognosis of any deposit as oil, gas, various ores or minerals is very important, and, it is hard to overestimate its meaning. Geological information about the deposit classes to be considered as static data since a hypothetic understanding of the formation processes of the Earth are going on for hundreds of millions of years. On the one hand, it might make somehow misleading imagination about a simplicity of analysis of the geological information, although, on the other hand, because of the lack of real understanding of the cause-and-effect relationships between the mentioned processes, and, subsequent incorrectness of proposed hypotheses we can not often determine the straight correlations among various ingredients even by means of the numerous available implicit and distant observed properties.

Perhaps a one of the reasons of the disadvantage in successful processing of the available geological information (may be excluding oil and gas cases) is caused by, either the existing pattern recognition methods are not used in geological survey, or they are ineffectively applied, particularly for deposit prognosis. The present paper proposes a principle of using the adaptive artificial neuron network for a deposit prognosis, that today are successfully exploited in such a very different and difficult prognosis problem as it is the forecast of Dow Jones Industrial or S&P500 /1/.

Note that all kind of underground resources we can classify as the easy-prognostic and heavy-prognostic deposits. In the first case we have a set of properties or features correlated with the deposit, when it is easy to make not only prognosis of this deposit, as well as its amount or capacity. To this class of resources we can assign oil, gas, iron ores and so forth. As a rule such kind of deposits are occupying large underground layers and spaces that also influences their market cost. In the case of heavy-prognostic deposits we deal with relatively small or very small amounts of the searched entities having often too weak correlations with its surrounding substances' geophysical and geochemical properties.

Current geological searching practice is equipped by very powerful surveillance equipment comprising magneto-metric, gravimetric, radar and other distant devices installed on the planes or satellites. These data often contains important information about different pursued objectives. Last decades a distribution of plants' cover or visual analysis of the explored territory together with a geophysical, geochemical, distant and other information might represent useful data for correct prognosis of a deposit.

Therefore, as we see the observed features may be expressed by quantitative as well as qualitative data, the number of which might be different – more or less depending on a kind of deposit. There are different ways of transformation qualitative features into numerical ones, and, for convenience, as a rule, it is necessary further normalization of all these quantities to place them into interval between 0 and [1]. We should note also that depending on the kind of searched deposits the territory maps have to be digitalized into pixels, sizes of which are determined by the resolution properties of the surveillance devices and available data relevant to the pixels.

Now, if we can suppose that according to the density of distribution for deposits expressed into an amount of searched entity per single volume of a relevant territory ground, then it becomes obvious that the highest density is peculiar for such deposits as coal, oil, gas, and so on. For relatively inexpensive non-ferrous metals, and some minerals this density is much lower, and, finally, from gold, platinum and similar expensive deposits to such precious stones like ruby, emerald and diamond this density falls down in a highest extent. That actually is as well connected with a substantial decreasing and vanishing necessary correlated features in the data available.

Therefore the present paper is dedicated to a development of the prognosis for the most heavy-prognostic mineral as it is diamond, that reflects the results of 10-year intensive work in 80s carried out by the Department of Applied Pattern Recognition Systems of the Institute of Cybernetics of the Georgian Academy of Sciences and the former Corporation COSMO-AERO-GEOLOGY of the Ministry of Geology of USSR (does not exist almost 20 years)[2].

Searching works relating to the diamond geology mostly are based on very poor and often intuitive approaches of the different groups of geologists worked out along their own experience. However, we are not sure that the situation with the prognosis of other rare and expensive deposits is much better, or, in other words, they have more or less successful scientifically based prognosis methodologies. It seems because of absence of an application of the current non-parametric statistical methods such successfully used to solve business or various biological or other scientific problems as it was mentioned above. In addition by the end of 70s we proposed to use the artificial neuron network that might be also adjusted to the problem of diamond deposit prognosis, and, consequently we have realized the prognosis model and examined it successfully.

Artificial Neuron Networks' Approach

Many years of investigations of the artificial neuron networks were carried out at the Department of Applied Pattern Recognition Systems relating to their application for various problems as visual pattern recognition, classification of the random time series like the diagnostics of the psycho-physiological state of an operator by its bioelectrical activity, and so forth. We tried to use such a network for prognosis of diamond deposits that usually are associated with so-called kimberlite pipes – ancient underground formations which among others contain the diamond crystals.

It is well known that the neuron network is one of the best performer of non-parametric statistics procedures accumulating, rationally distributing, and extracting information about cause-and-effect relationships among numerous variables and the searched target, object or pattern. While those relationships are hidden for us in the neuron network as a result of long iterative procedures of adjusting weight coefficients we can obtain a final state of the neuron network corresponding to the best prognosis or forecast model. Besides, such a model makes it possible to determine one among the features is actually important (or correlated) to the deposit.

According to the accumulated practice of simulating neuron networks, and, depending on a problem and available input information to be processed it is necessary to use different number of layers of neurons, and, each layer itself comprises various number of neurons. Not penetrating in detail, simply saying, here we are trying to apply a nonlinear transform operator or threshold (sometimes it has a sigmoid form) to the sum of various groups of values of input features, or, to the sum of outputs of previous layer neurons (in case of two or more layers). In other words, it is a permanent averaging of sums and their subsequent weighting, that is probably one of the main and most important properties of a living brain.

Since for solution of any kind of problems it is essential that initial information (as a rule, including a noise of various character) must contain some useful (or correlated) information about the deposit. Unfortunately the diamond assigned to the third group of above mentioned deposits don't have a clearly expressed correlation to any other features except the deep underground beaks and the satellite minerals as purop (garnet) and picroilmenite. Relating to the deep breaks, all kimberlite pipes, as a rule, are placed on such breaks. However, despite the highest correlation with those breaks, because of a high density of the deep breaks' net over a surveyed territory the mentioned correlation actually almost loses any sense. Considering the second kind of features represented by satellite minerals they are scattered on the surface, and, therefore, since they have been moving many millions of years under various influences it makes difficulties to locate their origins. As we see the mentioned information represents poor bases for prognosis of kimberlites. Although the surveyor geologists experienced in kimberlites are considering various other features presumably correlated with diamond deposits. For example, such implied information carriers are magnetic and gravimetric anomaly curves, as well many other different kinds of maps often representing artificially transformed colorful photos and so on.

Subsequently, initially we got about forty different types of maps that we scrutinize and at the first stage we reduce their number essentially. We elaborate two new conditional maps for two above mentioned satellite minerals. They were representing combined features as a result of grouping several maps (relief map with rivers, mountain chains, etc.). Although separately each of those maps a priori did not have any correlation particularly with kimberlites, however their combination might carry useful information about diamonds. From this point no matter what kind of deposit prognosis problem we are considering. We have carefully studied available data and its character. That allows finding out a way of combining the separate maps or features, a priori quite evidently not having any connection with a searched objective. Nevertheless their combination as a new feature might turn out to be substantially helpful for a prognosis.

Learning Sequences and Design of an Adaptable Network

In Figure 1 a general scheme of neuron network for prognosis of any kind of deposits is shown, that was also used for prognosis of diamond deposits. Let us first consider the initial presentation of data that is designated as a vector Y_j. As we mentioned before there are cases when we have various cartographic, categorical or any other kind of information about the correlated features, for example, about accompanying substances. From the first glance they seem do not have a direct relation to an original location of the searched deposit because they are scattered and found too far from it. However, we might combine some other maps in different ways based on some logic of experience which are approximately indicating an implied trajectories of replacement of those satellites, and, therefore, they, to a certain extent, may determine a presumable place of the deposit's location. Those maps could be used for obtaining the new conditional features - Y_q^j, where $q = 1,2,...k$, and, are representing a result of transformed vectors $X_1^j, X_2^j,...$ and so on. Here k is a possible number of conditional new secondary features obtained as a result combinations of various cartographical, geophysical or other available information, and, might be presented as a map of new conditional features containing very useful information. The second group of input features X_r^j are representing numerical data, which might be simply summed up and averaged by comparing with a threshold value θ. All those are determined while modeling the learning process by gradual sorting out process according to the feedbacks from the Supervision and Control of the Network Learning and Minimization unit. In other words, the modeling shows are there or not such possibly grouped inputs that afterwards are also representing the new secondary features as variables $Y_{k+1}^j, Y_{k+2}^j,...,$ and so on. In general case of a large amount of initial variables the conditional or non-conditional (direct) variables respectively are $X_1^j, X_2^j,...,$ and

$X^j_{k+l+1}, X_{k+l+2}, ..., $ or $X^j_{k+1}, X^j_{k+2}, ..., $ and $X^j_{k+l+m+1}, X^j_{k+l+m+2}, ...,$ and, the combining or non-combining variables respectively are $X^J_1, X^j_2, ..., $ and $X^j_{k+1}, X^j_{k+2}, ..., $ or $X^j_{k+l+1}, X^j_{k+l+2}, ...,$ and, $X^j_{k+l+m+1}, X_{k+l+m+2}, ...$

Here all presumable X variables are determined taking into account that the number of them the larger the better, because, as we mentioned above, a priori we do not often know possible relationships between them and searched deposit. However, later, during the network learning and minimization process the network structure and its parameters are determined as a result of influence of the three feedbacks (block arrows in Figure 1). They imply the weight changes (the first feedback), selection of the combined and non-combined variables and finding their better transforms appropriately into the secondary variables (the second feedback), and, elimination of superfluous components keeping error within satisfactory limits (the third feedback).

Note that the index $j = 1$ or 2 which corresponds respectively to the set of data for presence and absence the searched deposit at a given pixel of territory. The weights are changing according the modified stochastic approximation method that provides the least square error in Bayesian sense. In other words, the prognosis function is the optimal linear function in the secondary feature space, coefficients of which are determined as a result of the following iterative procedure

$$\Delta \mathbf{W}_{n+1} = - \alpha^*_n [\mathbf{W}^T_n \mathbf{Y}^j_n - \rho_1 \beta_1 - \rho_2 \beta_2] \mathbf{Y}^T_n$$

where: $\Delta \mathbf{W}_{n+1}$ = the increment of weight vector $\mathbf{W}$ at n + 1 step of learning,

α^*_n = the series of positive numbers satisfying the following conditions:

$$\lim_{k \to \infty} \alpha_k = 0, \quad \sum_{k=1}^{\infty} \alpha_k = \infty, \quad \sum_{k=1}^{\infty} \alpha_k < \infty$$

T = transposition,

$$\rho_1 = \begin{cases} 1, j = 1, \\ 0, j = 2, \end{cases} \text{ and } \rho_2 = \begin{cases} 1, j = 2, \\ 0, j = 1, \end{cases}$$

$\beta_1 = -\beta_2 = 1$ = the probability distribution mean for the deposits, and, -1 for the points of their absence.

The learning process implies sequential and multiple presentation of learning samples or input vectors comprising all numerical data about a given processed pixel. The learning sequence of vectors consists of the data characterizing pixels belonging to presence of the deposit as well as to its absence. Except, it is necessary to have an additional set of the samples of both classes not used in learning but using for examination of the network's capability of correct assignment of the diamond containing and not containing pixels.

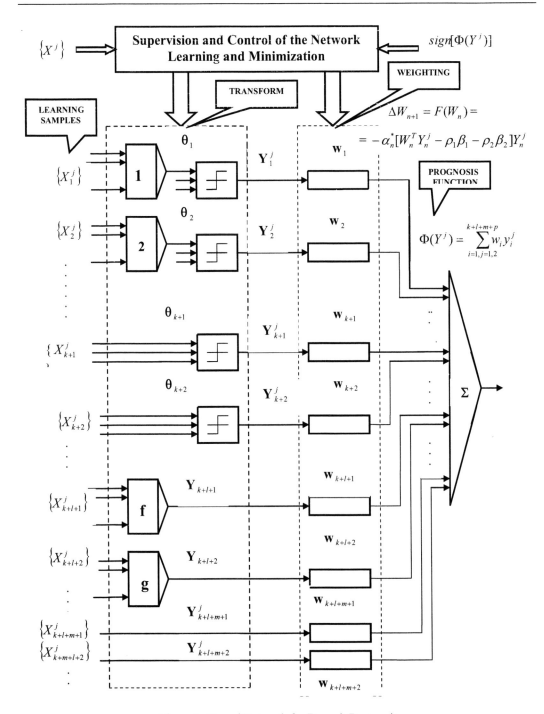

Figure 1. Neural Network for Deposit Prognosis.

The network's adaptation is completed when we achieve an accepted error of prognosis for the minimum structural components of the neuron network. This network realizes the linear function of secondary Y_i^j variables, coefficients of which are representing the optimal vector components. Afterwards, according to substitution of all the rest pixels' vectors into the prognosis function and computation of its values we get numbers for entire area showing the probability of presence of diamond deposit at this pixel.

The above considered neural network is slightly different from well known homogeneous multilayer networks. Its structure is most likely oriented to the geological surveillance problem in general, and, that was successfully practiced for prognosis of diamond deposits in Yakutia (Siberia) in 80s /3/. We carried out a long consideration of the set of available geophysical, geochemical, remote sensing data, magnetic anomaly, gravimetric, deep breaks, scattered satellites and many other maps. Taking into account various experts' suggestions we concluded that the diamonds have much more uncertain relationships with the observable variables than any other deposits. Despite the all kimberlite pipes are situated along deep breaks the concentration of which all over Yakutia is very high and this kind of information turned out to be useless. Similar situation seems to be concerning the widely scattered satellite minerals, though they helped us to construct very valuable new conditional maps. It was reflected in the prognosis function's expression where their secondary or transformed variables got higher values of coefficients than other ones.

Diamond Deposit Prognosis Using Stochastic Approximation

Subsequently, we processed over 120,000 sq. km of surveyed area that was divided into 2×2 sq. km pixels. Initially we together with experts had selected 21 initial maps that we transformed into numerical data and processed. The learning sequence included 176 pixels and for verification we used 200 examination pixels. Both pixel sets were equally containing and not containing diamond deposits. The simulation of the network learning and minimization finally determined 10 input maps that were transformed into two conditional combined secondary variables. They imply possible trajectories of moves for purop and picroilmenite. The next we had the combined secondary features including different maps relating to concentration and main direction of deep breaks, the feature of non-condensing consolidated crust, the gravimetric anomaly and so called ring-like structures.

In Figure 2 a piece of the prognosis map is shown. Here triangles are showing the industrial kimberlites and black spots other kimberlits. Results of prognosis show colored pixels that according to values of prognosis function corresponding to the probabilities for location of kimberlite pipes. The numbers 1,2,3,4 and 5 in circles are indicating values of probabilities beginning respectively from the maximum probability to their lower values. The probability of correct prognosis for all known 366 pixels, that included the samples used, and, as well not used in the network learning process, was equal 96%, while it was equal to 94.25% for the pixels of unknown properties to us as well for the neuron network, or, the samples not used in learning.

Finally, we should notice that according to experts point about relationships of high values of gravimetric indicators and diamond deposits turned out to be quite opposite – in the expression for prognosis function we got the minus sign of its appropriate weighted variable showed that the higher gravimetric index the less probability of kimberlite presence. This unexpected result may have various hypothetical explanations.

This paper authors are dedicating to memory of a distinguished person, our friend and collaborator Sasha Ufland.

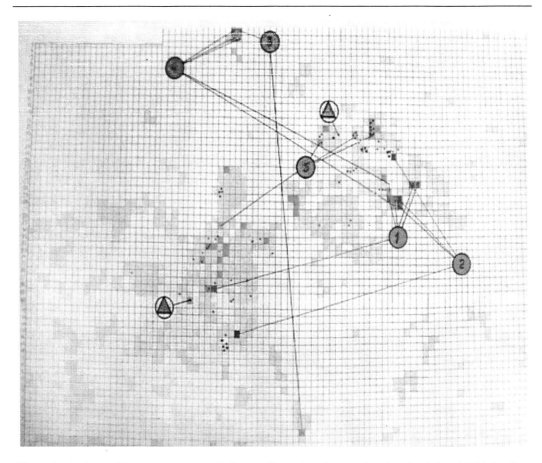

Figure 2. A piece of the prognosis map is shown. Here triangles are showing the industrial kimberlites and black spots other kimberlits. Results of prognosis show colored pixels that according to values of prognosis function corresponding to the probabilities for location of kimberlite pipes. The numbers 1,2,3,4 and 5 in circles are indicating values of probabilities beginning respectively from the maximum probability to their lower values. The probability of correct prognosis for all known 366 pixels, that included the samples used, and, as well not used in the network learning process, was equal 96%, while it was equal to 94.25% for the pixels of unknown properties to us as well for the neuron network, or, the samples not used in learning.

References

[1]   Widrow, B. Neural Networks Applications in Industry, Business and Science, *Communications of the ACM*, **37**, no. 3, March 1994, pp. 93-105.

[2]   Kvitashvili, A. A., Natapov, L. M., Tkhinvaleli, R. G., and Gambashidze, E.G. *Kimberlite Fields Prognosis by Nonparametric Statistics,* Reports of All-soviet Meeting on Diamonds, Moscow, 1980, p.9.

[3]   Kvitashvili, A. A. *A Universal Approach to the Rational Design of Problem-Oriented Pattern Recognition Systems Based on Neural Networks*, Proc. 3-rd Int. Conf. on Application of Fuzzy Systems and Soft Computing, Wiesbaden, Oct. 5-7, 1998, pp. 168-173.

In: Focus on Science and Technology…
Editor: Sergo Gotsiridze, pp. 213-226

Chapter 21

INVESTMENT ENVIRONMENT IN GEORGIA AND THE PROSPECTS OF ITS DEVELOPMENT

T. Shengelia and R. Mindeli
Tbilisi State, University Str. 2, Tbilisi 0143, Georgia

From the contemplation of economic development, Georgia is now on the transitional level. So, for the improvement of the economic and social state of the country, the investment flow has the great importance. The main aim of the country nowadays is integration in the world economic sphere, which can be reached by harmonization of business environment to the world main demands in this field. If we look through the history we will see that the only way to overcome unemployment and poverty is holding the institutional reforms, growing investment activity. From this point of view the possibilities of the local funds are scant. That is why the main priority for the country is to attract the foreign investments, which is quite difficult as the strategic products, like oil and gas, interesting for the foreign investors, are not presented in great volume in Georgia. This is the motivating reason for Georgia to use open economic policy, which is based on the experience of the countries, which developed quickly since 60th of XX century. We are talking about "Asian tigers" – Thailand, Singapore, and Hong Kong, which used the open economic policy and development of attractive business environment, in order to reach success. But these actions are to be based on the important institutional changes. In Georgia such institutional changes must be directed at harmonization with the economic sphere of the world. Besides it is clear that any successful convergence does not happen only as a result of market economy, but it takes place as a result of free market economy. That is why it is very important to take into the consideration the institutional reforms, being executed in the country, as they may become the foundation of the fast economic development of Georgia.

The economic reforms of the recent years gained Georgia the image of one of the fast developing countries all over the world. The factors of success are the following:

- The economic reforms are systematic, causing the country to engage the leading place in this field;

- The strategically geopolitical location of the country gives it the opportunity to use this important and rather useful factor for acceleration of its economic development.
- The macroeconomic environment got stabilized.
- The comfortable trading environment was formed.
- Effective tax reform was executed.
- The procedure of licensing and permission issuing was simplified.
- The aggressive privatization policy was maintained.
- The environment free of corruption was formed.
- The mechanism of law became more active in business.

In regards to the information of the World Bank issued on 2008, according to the economic development Georgia transferred from 118th place to 11th. By the evaluation of the experts of the same organization, no other country in the world executed so many reforms so sequentially and fast in 2005-2008, as it was made in Georgia. In the united researches of World Bank and EBRD "Business environment and industry" it is indicated that in 2004-2008 years Georgia executed the greatest corruption decreasing among other countries on the developing level of economy. According to the economic freedom, Georgia took the 21st place among 161 countries and 20th place among 41 European countries. As it may be mentioned at the scheme #1, the indexes of business, trade, fiscal, independence from the government, monetary, investment, financial and freedom of the working force varies between 99 and 60. In the whole, in seven from all ten indexes, Georgia remains the high rates.

Georgia is located at the crossroads between Europe and Asia. Therefore the country with its transport highways is the bridge, connecting important regions of the world. These are: Europe – 495 millions, Countries of Black Sea coast – 243 millions, Turkey – 73 millions, Caucasus region – 16 million populations. The shortest road between these countries of Central Europe and Eastern Asia for oil, gas and other cargos is stretched through it. The bulk pipeline for oil and gas, well organized railway system, Black sea ports and International airports play an important part in connecting Western and Eastern countries. In the recent years the bulk pipeline system became more developed in the country. An English company "British Petroleum" and its Georgian partners invested 5 billion USD in the rehabilitation-construction of 3 oil and gas pipelines. In 2005 the construction of Baku-Tbilisi-Ceyhan oil pipeline was finished. 4 billion USD were spent on this construction and this pipeline can pump over 10 million barrels of oil a day. This is one of the greatest pipelines in the world and its profit for the country in 2006 was 7.8 million USD. In 2009 this index will rise to 25 million and in future to 50 millions. The South Caucasus gas pipeline, opened in 2006 maintains gas transit from Shakh-Deniz gas deposit through Georgia, Azerbaijan and Turkey to the Caspian Sea. This pipeline is one of the strategic safety terms for Georgia, which gives us the opportunity to get gas supply independently from the Russian gas. The share of Georgia in such gas supply will grow up to 800 million m^3. The Georgian railway, which connects Europe and Asia, has the great importance for the whole Euro-Asian transport corridor. In 2009 only Georgian railway carried 3.9 million passengers and 22.6 million cargoes. It is also the knot, connecting Armenia, Azerbaijan and Russia. On the basis of the resolution of Georgian, Azerbaijan and Turkish government, the construction of Baku-Kars-

Tbilisi railway was stated and it will be implemented in 2010-2012. Its freight turnover will consist 15 thousand tons annually. Poti and Batumi ports play an important part in TRACECAS (Transport corridor – Europe, Caucasus and Asia) project. In 2009 Georgian government supported 459 million GEL for road rehabilitation. These funds were used in the following way – 1474 km. of international roads, 326km. state roads, 15439 km. local roads construction. In February 2007 a new International Air Terminal was opened in Tbilisi, which is one of the most modern in the world. Air transportation in Georgia is served by 4 national and 15 foreign companies.

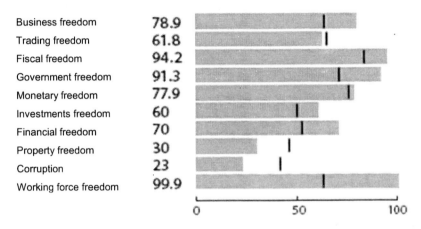

Scheme 1. Ten economic freedom indexes of Georgia.

The greatest managers of the world sea ports compete in order to receive the management on Poti port, as it is the free economic zone. The city of Poti as well as its neighboring 400 hectares, is to become reachable for the foreign business, which has not still assimilated the advantages of Georgia as opposite to other countries, like its geopolitical location and simplified trading terms.

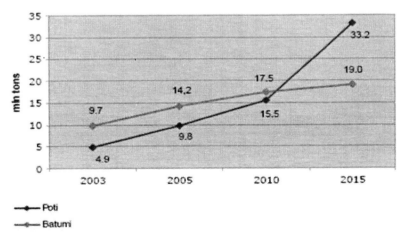

Growth of Batumi and Poti ports capacity.
Source: Data of the Ministry of Finance of Georgia. Tbilisi. 2009

Scheme 2. Capacity of Batumi and Poti ports.

Important steps from the point of view of liberalization of trade politics were made in the country. This was expressed on following:

- The low import tariffs were stated and they were canceled for 90% of products;
- Only three tariffs from 16 actives remained;
- The import-export quotes were canceled;
- VAT for local and imported products got equal;
- The same for the excise;
- The number of export and import licenses decreased from 14 to 8.

The suitable market terms were formed in Georgia, and this made it more attractive for the world market. This fact is stated by: the first - the fact that Georgia, together with the members of World Trade Organization worked out the single tariffs system, which is rather low in comparison to other countries in the world. And the second – the advantage trading regime was stated with USA, Canada, Switzerland and Japan.

In 2005 Georgia issued a new Tax Code, which strongly decreased tax rates. An important institutional change was executed in Tax system. (See scheme #3):

- In 2007 the number of taxes was decreased to $7 - 5$ of them are state and 2 are local. The state taxes are income tax, excise, VAT, customs, profit tax. The local taxes are stated by the local governmental boards.
- In 2008 the social tax counted 20%, paid by business and income tax counted 12% were combined and formed one income tax counted 25%, which became 20% in 2008 and 15% since 2009.
- The corporate profit tax from 20% became 15% since 1st of January 2008, and dividends were exempted from it.
- VAT counted 18% is obligatory for those businessmen whose annual turnover increases 100 thousand GEL.
- Export international transfers, tourism and other services expose to null VAT.

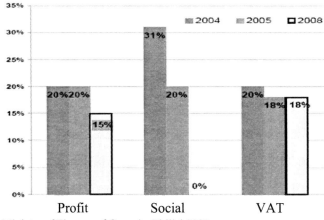

Source: Data of the Ministry of Finance of Georgia. Tbilisi 2009.

Scheme 3. The dynamics of tax change in Georgia.

In order to motivate business development in the country, the terms of license and permission issuing became more simplified, the number of licenses decreased by 98%, 756 licenses were void; the new regime of licenses issuing was established. The licenses are to be issued for the following fields: safety and health defense, safety of apartments, state and social interests' defense. The time limit for license issuing became 30 days, permission issuing only 20 days.

The most dynamically developing field in Georgia nowadays is bank field. In 1994-1997 the process of state bank privatization was over, and afterwards the private banks were found. For nowadays this field is 100% independent. In 2009 the number of deposits in bank sphere increased by 35%. The whole turnover was counted 3-5%. Georgia is among the countries, which executed the liberalization of accounting capitals operations and currency regimes. The terms for currency operations for the citizens of Georgia and foreign countries are the same. The citizens of Georgia have the opportunity to open bank accounts in the foreign banks. The citizens of Georgia and foreign citizens have the equal terms while investing in Georgian banks. The following great investments took place in Georgian banks:

- In November 2006 the Bank of Georgia became the second CIS bank at London stock exchange. It successfully presented its share in form of global deposit income.
- In March 2006 "Societe General" group received the control packet of shares of Bank Republic.
- In October 2007 Kazakh bank "TuranAlen" purchased the great share at "Silkroad Bank".
- In January 2008 Russian VTB Bank purchased 51% shares of United Georgian Bank.
- On the territory of Georgia the branches of leading Greek, German, Turkish banks were established.

Before the Rose Revolution the government of Georgia was widely involved in corruption. Afterwards the main achievement of the new government was overcoming the systematic corruption, the share of shade economy in the GDP was limited (see scheme #4). The processes being executed in the private and social fields in this regards meet the international demands. It is stated by the following terms:

- In the report of 2006 – "Anticorruption transformation", the World Bank greets the achievements of Georgia in anticorruption activities.
- The World Bank and EBRD in their report state that among the private firms the corruption decreased from 60% to 15%. According to the data of 2005-2008, the number of firms paying bribes by any reasons decreased from 44 to 11%.
- According to the data of the Century Change Millennium Program in 2008 the indexes of corruption state that the country fully satisfies the demands of corruption control standards. Georgia, in the group of similar countries has higher grades.
- USA International Republican Institute social research stated that 95% of questioned population confirms that they have not given bribes in 2009.

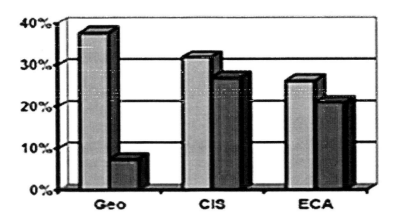

Scheme 4. Frequency of bribes.

There are practically no limits for the foreign citizens to purchase Georgian firms. According to the Tax Code the process of establishing business by the foreign citizen is easy, cheap and effective. Its duration does not exceed three days. The number of registered business in 2007 grew by 30.6% in comparison to 2006. The simplification of company registration had positive influence on growth of private business.

Despite all these achievements still the state of people's life is rather hard, which is expressed in low incomes, changeable inflation and unemployment, and it is rather a heavy load for the population. But also it must be mentioned that the institutional reforms and their results do not match in time. The reforms are aimed on long term perspectives. Therefore the policy of the government in the field of institutional reforms may become the basis of economic progress of Georgia, and foreign investments play the great part in this process as well.

Georgia, together with its resources and location always had the potential to become attractive spot for foreign investment project execution. Following the fact that Georgia began moving on the way of market economic development, in order to stimulate investment activity, the laws about "support of government for the foreign investors" and the law "about the foreign investments" was revised. Afterwards the gross share of the foreign investments in Georgia grew a lot.

Since 2008 the global financial crisis covered the whole world, and this caused the reduction of investments all over the world, including Georgia. This process was also aggravated by the war between Russia and Georgia in August 2008, followed by annexation of Georgian territories by Russia.

For the developing countries, like Georgia, attracting the investments has the great importance. In case if the country has no possibility to improve its economy by source of its own resources, the international investments play the great part in economic growth of the country.

In Georgia the investment projects are regulated by the law about "Governmental support to the investments", which has the aim to support investments, improvement of industrial activity using invested funds, and besides additional legal terms for that. This law regulates all investments maintained on the territory of Georgia. This law presents additional support for the foreign investments in order to make it more attractive.

In 2008 the volume of investments reached 1.563 million USD. Despite the data of statistic department, the part of experts does not trust this information. The indexes of investments in Georgia increase the same indexes in Armenia twice, but the same indexes are four time lower than in Azerbaijan. In his interviews the president of Georgia announces that the greatest part of investments come from USA and European countries, and the prime-minister states that such investments are the result of well organized and correct policy of the Georgian government. He announced that according to the investment program for 2010 year the volume of investments will reach 1,5 – 2 millions, which is equal to 21% of GDP. We can also see that the mentioned volume of the investments really was reached but it did not support the stated index of GDP.

The five biggest investors of Georgia in 2008 are the following:

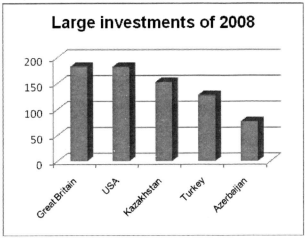

Source: Data of the Department of statistics of Georgia. Tbilisi 2009.

Scheme 5. The biggest investors in Georgia for 2008. (in thousands of USD).

In that period many skeptics were doubtful in regards to the Kazakh investments, as they thought that behind such investments the Russian interests were hiding. The part of Kazakh investments in Georgia were served in governmental organizations. Besides in the Kazakh Company "Kazmungas" Russian share holders are greatly presented. That is why according to the statement of experts such investments are not always secure. "Behind Kazakh investments Russian funds are widely presented, that is why we always have to be very careful with such investments, as we have already let Russian fund in Georgia a lot, and we have to know how profitable it will turn out to be for the country." – announced the expert in this field Ramaz Sakvarelidze.

In comparison to 2008, in 2009 the volume of investments grew from 1.563 billion up to 2.015 billion. And the mentioned investments were made between the following countries:

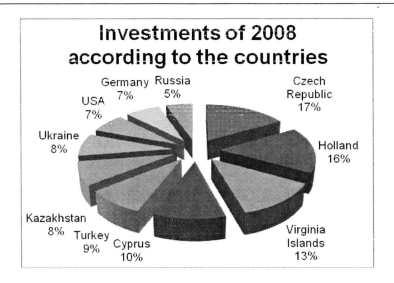

Source: Data of statistic department of Georgia. Tbilisi. 2009

Scheme 6. The structure of investments among the countries.

It is to be indicated specially that even during the military actions, the auctions for useful licenses were still being held, and this means that investors did not fear to make investments into Georgia economy, as they knew that this military conflict would not be long. It is well known that the government of Georgia is extremely interested in attracting investors to the country. Besides it is also indicated that Georgia has suitable environment for business, and this is caused by the successful reforms held by the government. In regards to this, the interest of the foreign investors towards Georgia is growing. The political risks in our country really grew since the number of Russian investments increased. The risks also increased during the August conflict, as there were different uncontrolled territories in Georgia, but despite all this, the mentioned conflict did not influence the number of foreign investments in our country.

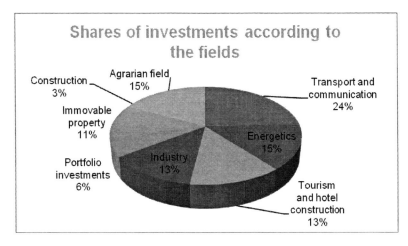

Source: data of Statistic department of Georgia. Tbilisi 2009.

Scheme 7. Investments shares according the specific fields.

The investments structure in Georgia is also very interesting, as it shows which field is more attractive and interesting for the investors. (scheme #7). As you can see on this scheme, the leading place in the investment field structure is held by transport and communications, with 24%, the second place is occupied by agrarian field – 15%, the increased investments are attracted to tourism – 13%, as this field has a great priority for the country, but because of the global financial crisis the investments in construction are decreased.

In 2008 the direct investments in Georgian economy counted 1.365 billiard USD, which was the greatest sum for the country, the whole GDP of which was 10.2 billion USD. The Georgian government was expecting even more investments to be done in 2008, but because of the Russian war, the number of investments decreased. Despite the government expected to increase this index till the end of the year. Nowadays we can freely state that this optimism of the government was justified.

According to the data of the statistics department, the index of direct foreign investments was the following:

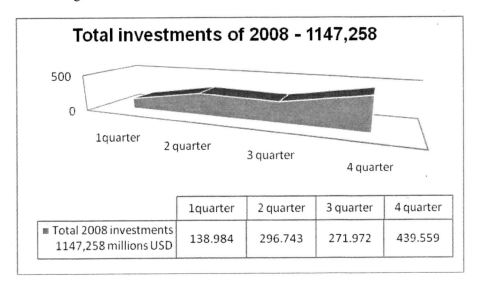

Total investments of 2008 - 1147,258

	1quarter	2 quarter	3 quarter	4 quarter
▪ Total 2008 investments 1147,258 millions USD	138.984	296.743	271.972	439.559

Source: data of Statistic department of Georgia. Tbilisi 2009.

Scheme 8. The structure of investments for the quarters of 2008.

As we can see in the given scheme, the investments are of the increasing term, but the direct investor, which is entering the country need an additional guaranty of liquidity of his investment and this can be reached by source of additional environment establishing. The establishing of attractive environment for the investors means creation of liquidity market of securities, as the serious investor will never invest his funds, before he analyses the ways of leaving this business. Also it is also important to prepare information – investment proposals for venture foundations, and after research of such proposals, it will be possible to choose the most suitable variant. Without foundation of effective risk financing system, it will not be possible to create perfect investment market in the country, as exactly venture foundations must support the risks of the minor firms, in order to create attractive and successful environment for bigger investors.

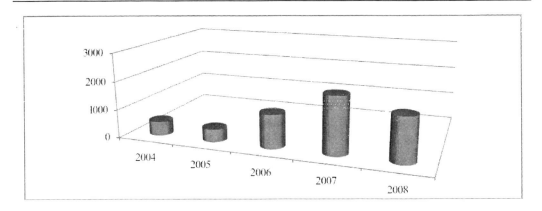

Source: data of Statistic department of Georgia. Tbilisi 2009.

Scheme 9. Direct foreign investments in Georgia. Millions of USD.

On the scheme #9 in is clearly shown that from 2004 till 2008 in Georgia there is a tendency for the direct foreign investments. In 2004 499.1 million USD entered Georgia as a direct foreign investments. In 2005 – 449.8 million USD, in 2006 – 1190.4 million USD, in 2007 – 2014.8 million USD, and in 2008 – 1564.0 million USD. As it is indicated in the given scheme the volume of the direct foreign investments from 2004 till August 2008 was increasing, but after August 2008 it began decreasing. According to the U.N.O experts, in order to attract DFIs each country is obliged to create the proper environment for future investments. This can be reached by decreasing administrative and bureaucratic barriers, which will case great attraction for the foreign investors.

In accordance to the business, general economic state and controlled regime, influence not only the foreign firms, but also local investments a lot. But in comparison to the local firms, foreign investors are more sensitive, as they have the opportunity to compare the environment for the investments in different countries in order to choose the best. That is why it is very important for the receiving country to crash all possible bureaucracy barriers and stimulate the investment activity.

Still many foreign firms do not know much about investment environment of Georgia and possible partners for them here. Such terms slow down the flow of investments to Georgia. In order to supply the potential investors with the suitable information, it is necessary to create modern, updating system of information, representing full and trustful data about the investment environment in Georgia. The presence of such system and regulation of DFI is absolutely obligatory, as without it the country is even not researched as a potential spot for future investments by donors. In order to interest the investors and attract them it is necessary to establish special agency of DFI regulation policy, which will execute the following actions:

- Work out the strategy of DFI attraction in coordinated cooperation with the government.
- The strategy for DFI attraction must be adapted to the world market and economic changes of the receiving country.
- The strategy of DFI must be founded on the possibilities of the country, depended on its attraction, and possibility of success.

The participation of Georgia in such international projects, like TRACEKA, gives the country the opportunity to become integrated in World Economic Union System. It is expected that as a result of integration in this project the GDP volume will grow up to 43 million tons in 2010, which will make it easier for it to integrate the World market. The great part in attracting of foreign capitals will be played by "Great Oil", which is connected to the production of energetic resources in Caspian Sea. The oil products are the main priority for Georgian economics development, which will give the country the opportunity to attract great foreign investments in this field and reach the increasing of economy of the country.

The great authority distribution is taking place in the world arena nowadays, and it is rather useful for Georgia. United anti terrorist complain increases the cooperation of Georgia with Western countries, and this helps us to escape the politic isolation and stimulate the foreign investments entering the country. The foreign debt of Georgia, as the factor of negotiations with creditors, is the chance of receiving the foreign investments, as the pay back of the foreign debt is possible only by source of increasing of economic development level.

The attraction of investors is also possible by activating the "Northern-Southern" scheme. The famous economists state that Russia is dropped behind country if you take into the consideration its economic and democratic state, but if you consider its geopolitical location, the "Northern-Southern" scheme may not be left out of the account. Georgia, in this case, is the connecting chain of North – Russia, and South – Islamic world. Therefore the two way production flow from the North to the South has its perspective, but its real activating is possible if the following problems are solved: occupation of Georgian territories (Abkhazia and Southern Ossetia) by Russia, economic embargo, political claims and etc. In regards to the attractiveness of national economic investment, it is also very important to decrease the terms, preventing the flow of DFI into the country, state them and try to reject the obstacles causing them.

Many developing countries execute the reforms in economic and political fields in such a way that the foreign investors do not have the opportunity to get acquainted to them. The foreign investors may feel confused in regards to the economic state of the country being in the transitional state. It is well known that Trans National Corporations, first of all choose the region for the new investment, and further on state the country in the given region, receiving the mentioned investment funds. So for the activating the investment environment for DFI, it is necessary to reach effective communicational strategy, the integrated part of which is – public relations. In the field of DFI public relations in Georgia must have the aim to create the positive image of the best investment zone. In order to create such an image, the information interesting for the potential investors must be widely spread. It includes the information about the economic, political, legal state of the country, legislation about DFI and its regulating regimes and the business stimulating processes in the whole. In order to execute the stated aims, it is necessary to create suitable structural units and attract qualified staff for its successful management, having experience in foreign business, knowing well the culture and traditions of the partner countries. Such function may be maintained by the DFI agency, which will have the opportunity to evaluate the competitiveness of the country in DFI field and coordinate the international projects.

On the primary level of market relations forming in Georgia, foreign capital played the leading role neither from the point of view of privatization, nor in direct financing of the projects. This was caused by such problems of the country, which took place in that period. An unexpected liberalization of the economy and spontaneous privatization caused decreasing of the country competitiveness. The government presented in the country since 2003 and its conscious actions caused the increasing of trust of the foreign investors. The effective economic politics of receiving country decreases the possibility of destabilization, but economic instability, as it is well known, negatively influences the activity of local and foreign investors.

The low and predicted inflation is an important terms for the macroeconomic stability. The world experience shows us that the financial stability is reached in case if the annual growth of prices does not exceed 20%, and monthly – 2%. In case, if this index is high, the investment activity decreases, and the same must be said in accordance to the economic growth and state of living of population. The experience of Georgia clearly showed us that long term inflation decreasing is not enough term for investment activity. It must be accompanied by the economic and political stability and development of fund market. In the field of foreign investments, policy is integrated part of industrial strategy of the country, stating the rout for developing of the particular field in the country. In the recent years the correlation in the field of the foreign investments between policy and macroeconomic policy is straightened. According to the explanations of "UNCTAD" the modern foreign investment attracting strategy includes two circles: first – the policy in the field of investments and trade liberalization and the second – macroeconomic policy in the field of monetary-credit, currency and tax. In this field of state policy, the united position and close coordination of governmental activity is obligatory.

The stability of the currency rate is very important in order to overcome inflation and stimulate the economic growth. Certainly, the policy of hard currency supports the modernization of the economy of the country, improvement of foreign debt service. The growth of currency rate in fact decreases the outflow of capital, the growth of investments, including foreign ones.

The establishment of national currency – GEL stimulates the import of machines and equipment, especially from the twin foreign countries, which is an important factor for the modernization of industry inside the country. Though, in case of leading such a policy, it is obligatory to keep the balance between establishment of the currency and maintained institutional, structural and social reformations. First of all the improvement of tax system is implied, decreasing the tax load, serious improvement of administrative taxes, conversion of industries, development of competitive market and etc.

It is abolutely clear that if the high rate of GEL will not be supported by the productivity of labour increasing, further on keeping such rate will cause slowing-down the economic development of the state.

The lightening of the tax load will surve development of the industrial strength, but the interest rate reached as a result of using this method, at the given moment will not cause the flow in of the foreign capital because of the simple reason – capital in Georgia is less mobile, as despite privatisation of industries, all strong investment scourse – stock exchange, practically does not function in the country.

The world experiense showed us that we would need important and long term growth of local investments in order to involve foreign investors into the active development of the country. The foreign investor will not invest funds in the country, having paralised inner investment environment. We can indicate the development of inner investment activity, as one of the terms for ecomonic development of the country, as well as the catalist for the foreign investments.

The growth of investment activity is strongly connected to the market of securities and its development, which may be stated, together with the banking system, to be a mechanism of transforming investments into the financial recources. In the terms of investments resources, together with the direct crediting the particual projects, it is possible to issue securities as a hopefull guarantee. Agreement, reached with the foreign investor, growth of investment activity, these are hardly connected to the security market development, which may be stated, together with the banking system, to be a mechanism of transforming investments into the financial recources. In the terms of investments resources, together with the direct crediting the particual projects, it is possible to issue securities as a hopefull guarantee. The standard method for agreement to be reached with the foreign investor, is presenting the strong bank guarantees for him. The standard method for agreement to be reached with the foreign investor, is presenting the strong bank guarantees for him. The creation of suitable investment environment and long term resource investment mechanism is hardly depended on the activities, based on optimal portions of governmental regulations over market and investment activities. The state must finance the fields, which are scaled, long term and of low income greatly, as they can not be assimilated by the private business. The terms for the normal business activity and stabile general-economic investment climat must be presented. In this order, it is suitable to use an experience of the leading countries in "Aimed" financing. There must be established a net of special foundations, which will participate in financing of diffetent social and business development porgrams.

The diferentiating specification of investment process state strategy is investment of funds in "administrative capital"; that is why the government of Georgia has the great responsibility for the creation of the main factors, like primary and secondary education system, market infrastructure of the country and the field of scientific researches.

The main principle of the state policy in regards to the private investments is that it must not get involved into their activity, maximally fit the legal base to the investment active processes and support the safety of invested funds. It is obligatory for the government to execute a flexible policy; the given field deffers a lot and it is not depended only on indexes of economic development. The foreign policy of the country plays here a great role, as well as a political consent, society mentality, as the foreign investments are directed to the countries, where there is already established governmental, business and social strictly formed position directed on cooperation with foreign producers; where the investment climat is not only created, but already probated.

References

[1]   Foreign Direct Investment. Washington; IFC, *Foreign Investment Advisory Servise*, 2008.128p.

[2]   Pedro Belli. Jock R. Anderson. Economic Analisis of Investment Operation: Analitical tools a practical applicationns. Wasington: *The World Bank.* 2008. 198p.

[3]   Melasvili, M. *Transpacific Foreign Direct Investment and the Investment Development Part.* Tb. 2009. 219p.

[4]   Sengelia, T. *Treade and Foreign Direct Investment in Georgia.* Tb. 2009.101p.

[5]   *Statistical collection – Investment in Georgia.* Tb., 2009.42.p.

[6]   International Finance Corporation: 2009, Annual Report_ Wasington; World Bank
       2008. Vol 1; *Investing in sustainable private sector*. 120p.
[7]   Unated Nations, World Investment Report, 2008. New York 2009

In: Focus on Science and Technology...
Editor: Sergo Gotsiridze, pp. 227-231

ISBN: 978-1-61209-970-5
© 2011 Nova Science Publishers, Inc.

Chapter 22

GEORGIA'S FOREIGN TRADE DEVELOPMENT TRENDS

T. Shengelia[1] and Xatuna Berishvili[2]

[1]Economic science, Tbilisi State University
[2]Economic science, Sokhumi State University

Globalization has resulted in increasing levels of trade and investment throughout the world. Reductions in the levels of domestic production, an increase in the number of trade blocs and agreements, and improved technology have facilitated this increase in trade and investment. Strengthening the world's trade system is a determinant factor for democratic processes throughout the world. When a trade system is free and open, the economy has a better chance to increase, which conditions high standards of living. Globalization has significantly boosted *economic growth* in East Asian economies such as Hong Kong (China), the Republic of Korea, and Singapore. But not all *developing countries* are equally engaged in globalization or in a position to benefit from it. In fact, except for most countries in East Asia and some in Latin America, developing countries have been rather slow to integrate with the world economy. An actively trading country benefits from the new technologies that "spill over" to it from its trading partners, such as through the knowledge embedded in imported production equipment. These technological spill over's are particularly important for developing countries, because they give them a chance to catch up more quickly with the *developed countries* in terms of *productivity.* Former centrally planned economies, which missed out on many of the benefits of global trade because of their politically imposed isolation from market economies, today aspire to tap into these benefits by reintegrating with the global trading system. But active participation in international trade also entails risks, particularly those associated with the strong competition in international markets. The countries, where trade liberation happened, and who have studied competition in global markets, economy growth and development are becoming more advanced, compared with those countries where local markets are protected.[1]

In the early 90s, after the political and economic disintegration of the Soviet Union, Georgia had economy difficulties, but with the help of the international financial organization the country made deep liberalization and developed foreign trade. In 2000, Georgia became a member of the World Trade Organization (WTO). Subsequently export -import quotas were abolished. Since 2003 Georgia's foreign trade has been growing rapidly as a result of aggressive policy reforms to make it easier and less expensive to trade across borders. The new Tax Code was accepted at the end of 2004 and was put into action by 1st January 2005, abrogating the tax code of 2007. According to the new code, the number of taxes has sharply decreased from 21 to seven. Tax rates have also been decreased, which made Georgia the country with the lowest tax rates in the region.[2] Liberalization policy was followed by a massive privatization process: The legislative base of the process was prepared and enacted before the privatization process started in 1992. Up to 18,000 enterprises and other entities were already privatized by the beginning of 2009. Although after the Russian aggression in Georgia in August 2008, economy activity was decreased in the country. For the last few years, there has also been an upward trend in the number of Georgia's trading partners, during the last few years; the Russian Federation unilaterally imposed economic restrictions on Georgia through restriction of trade, transport, post and other spheres.

To estimate Georgia's foreign trade relation, consider the country's trade turnover, export-import, the important trade partners, goods of export and import and etc. Foreign trade turnover of Georgia was USD 3483,2 million in 2009 out of which export was equal to USD 729,1 million, and import – USD 2753,6 million.(see table 1) Despite that foreign trade turnover has decreased, export is growing but not significantly , which apprise that trade deficit will diminish.

A significant part of Georgia's foreign trade turnover is related to the Commonwealth of Independent States (CIS). Georgia has a free - trade regime with members of CIS, who's other members are: Azerbaijan, Armenia, Belarus, Kazakhstan, Kyrgyzstan, Moldova, Russia, Tajikistan, Turkmenistan, Uzbekistan, and Ukraine. The CIS countries have traditionally been Georgia's largest trade partners and although their share of trade with Georgia has declined and was equal to USD 106,0 million (see.table 2), they accounted for 25,3% of export and 39,6% of import in 2009.

Table 1. Georgia's foreign trade turnover 2006-2009 (USD millions)

	2006	**2007**	**2008**	**2009**
The whole turnover	4674	6456,9	5334,15	3483,2
Growth compared with previous year	39%	38,1%	-17,38%	-34,7%
export	993	1240,2	067,4	729,1
Growth compared with previous year	14,1%	24,9%	-13,93%	-31,7%
import	3681	5216,7	4262,53	2753,6
Growth compared with previous year	48%	41,7%	-18,29%	-35,4%
net	-2688	-3976,5	-3195,13	-2024,5

Georgia terminated its membership of the CIS organization[3] while maintaining its right to remain a member of the CIS free trade-area arrangements, and the bilateral agreements. In 1994, Georgia signed an FTA with other CIS member states that envisaged the trading of goods without customs tariffs. It was ratified by all the CIS member countries except Russia. However, the FTA remained ineffectual and, as a result, preferential trading relations among CIS countries have been established and determined in bilateral free-trade agreements. In

2005 and 2006, Russia banned imports of Georgian agricultural products, mineral water, and wine, for which Russia was Georgia's largest market. In September 2006, Russia cut all direct transport links with Georgia. Gazprom, the Russian gas monopoly, doubled the price of natural gas supplied to Georgia but new supplies of natural gas from Azerbaijan and increased hydroelectric generating capacity are making Georgia much less dependent on Russian energy sources. In 2009, the denoted diminishing of trade turnover within CIS countries was conditioned with scathing trade relations with Russia.

It is very important for Georgia to enhance its trade relations with the European Union (EU). The EU market is one of the world's biggest markets, its share is almost one fifth of the world trade. The share of the EU is growing but not significantly. In 2009 the import from EU countries was firstly overrated compared with import from CIS. It should be emphasised that 27, 8% of foreign trade turnover came from trading with European Union in 2009. Relations between the EU and Georgia started in 1992, when Georgia regained its sovereignty after the break-up of the Soviet Union, and bilateral relations have further intensified since 2004. The legal foundation for EU-Georgia relations is the Partnership and Cooperation Agreement (PCA), negotiated at the same time as Georgia's WTO accession, and entering into force in 1999. The PCA regulates political dialogue, trade, investment, economic, legislative, and cultural cooperation. It envisages progressive regulatory approximation of Georgia's legislation and practises to the most important EU trade-related regulatory acquis. It is known that the main priority for the Georgian republic is to join with the EU, which will assist in the formation of the European post-industry society.

Table 2. Georgia - CIS trade turnover 2006-2009 (USD millions)

	2006	2007	2008	2009
Turnover	1799	2323,3	1756,6	1061,0
Growth compared with previous year	28%	29,1%	24,39%	39,6%
export	395	469,4	362,2	270,6
Growth compared with previous year	-5%	18,8%	-22,8%	-25,3%
import	1404	1853,9	1394,0	790,4
Growth compared with previous year	41%	32%	24,8%	43,3%
net	-1009	-1384,5	-1031,8	-519,8

In 2009, Georgia's most important trading partner was Turkey. The trade turnover with Turkey was equal to USD 164,3 million[4]. A free-trade agreement between Georgia and the Republic of Turkey was signed on 21 November 2007 and entered into force on 1 November 2008. According to this agreement, customs tariffs on industrial products have been fully eliminated but a number of agricultural products are excluded by both parties and some are subject to tariff quotas by Turkey. The FTA is set to facilitate the development of trade and economic relations between the two countries, encourage entrepreneurs to gain access to new markets, and support implementation of investment projects. In addition our country's largest trading partners are: Ukraine, Azerbaijan, Germany, Russia, the USA, Bulgaria, China, Italy, Romania etc. Georgia's export is less diversified. The basic export commodities remain to be raw materials, and with significant delays (the same wine and citrus or even black scrap metal, nut, vegetable etc). Imports are becoming more active. Due to the domination of import and the suppression of local production, export resources are decreasing further. The main part of import production is not produced in Georgia, though there are also a lot of productions in import structure, which could be produced in the country with good quality. Reducing the trade deficit is one of the key priorities of the Georgian government .A small scale of export is directly linked with low level of production in the country, since this entails not only a small scale of export production, but about a high level of imports. Permanent increase of imports demonstrates that locally manufactured goods cannot satisfy local markets either, which, in its turn, causes unemployment. Therefore, the focus on low indicator of exports and on growing imports and orienting economic emphasis on the dynamics of export-import is important for overcoming the principal social-economic challenge of the country - poverty. The local firms oriented to the export cannot develop, which condition the poverty and problems that are linked with unemployment. Increasing the import caused by zero tariff is positive on the one hand, as it can facilitate the trading among the countries, Import is becoming cheaper for local population, through, on the other hand the boosts decreasing of local producing ability, in the same time. The strengthening national currencies is negatively reflected on export that stimulate the import and on the contrary, makes export expensive. For Georgia it is strategically important to diminish distinction among export-import, encourage export and it is possible with monetary policy. Along with other important factors (limited support of Georgian firms, their weak positions compared with foreign firms, high expenditures to preserve their position and active searching of new markets) all these problems are causing the weak development of export. The country's priority should be heightening qualities of food production, development of export potential and competitiveness of local firms. Therefore, a sharp change of export-import configuration in terms of trade deficit growth is a long-term problem of the Georgian economy, which may and can result in negative outcomes. Therefore, the necessity for undertaking a comprehensive analysis of the balance of trade and respective preventive actions is evident. Georgia like other developing countries must achieve competitive advantage not only in traditional, but also in high technological spheres, in this way the country can achieve long-term competitiveness, which will increase incomes, salaries and standards of living. Liberalization of world economic creates the ability to profit with advantages of market economy, the activity of WTO and trade policy of national economies conditioned the social and economic development throughout the world. As has already been mentioned in 2000 Georgia became a member of the WTO, from 2001 to 2007, Georgia benefitted from trade-related assistance funded by international institutions and bilateral donors, to the tune of US$119.1 million[5]. Assistance significantly increased in 2006 when Georgia received almost US$47 million, mainly in the form of grants and concessional loans, with major contributions from the United States through its business climate reform project. . In addition, Georgia attaches great importance to targeted assistance to help strengthen human resource capacities in the private sector and diversify its economy, focusing on high-value products, optimization of energy resources, and development of telecommunications, transport, and infrastructure.

In June 2004, Georgia joined the European Neighbourhood Policy (ENP). The aim of this policy is to increase the country's stability, security and welfare. In 2005, the EU provided Georgia with an expanded opportunity to make use of tariff privileges (under GSP plus), enabling Georgia to export about 7,200 items to the EU market with zero tariff rates. Previously Georgia was allowed to export only 3,300 products without any customs duty and 6,900 products under certain preferences. Georgia qualifies for the enhanced preferences for good governance and sustainable development and, in order to continue benefiting from the system after January 2009; Georgia ratified the two outstanding conventions listed in GSP regulations, namely the UN Convention on Anti-Corruption, and the Cartagena Protocol on Biosafety. As a result Georgia was granted the Special Incentive Arrangements under GSP Plus for 2009-2011.

On the modern stage, according to the low level of foreign trade, in order to increase the economic development, it will be dependent on the intensification of foreign trade, which requires and demands diplomatic intensification on its side with regard to foreign economic relations.

References

[1] B. Ohlin. "Interregional and International Trade". CAMB. 1997 P 13.
[2] D.Narmania, E Khokrishvili, Georgian Tax System (part 1), Next Economists, Special Edition, 2006, PP.4-5
[3] Georgia terminated its membership of the 1991 Agreement on the formation of the CIS, the 1993 CIS Charter, and the 1993 Agreement on the creation of economic relations.
[4] www.statistics.ge
[5] Joint WTO/OECD TRTA/CB database, *Beneficiary Country: Georgia.* Viewed at: http://tcbdb.wto. org/benef_country.aspx

In: Focus on Science and Technology...
Editor: Sergo Gotsiridze, pp. 233-238

ISBN 978-1-61209-970-5
© 2011 Nova Science Publishers, Inc.

Chapter 23

MEAN CONVERGENCE OF CESARO (C, α)-MEAN FOURIER-LAPLACE SERIES

S.B. Topuria [*] *and V.V. Khocholava* [†]
Department of Mathematics, Georgian Technical University,
77 Kostava St. 0175 Tbilisi 75, Georgia

Abstract

It is stated for what α (summability exponent) and for what classes of functions the mean convergence of Cesaro (C, α)-mean Fourier-Laplace series takes place on a sphere.

2010 Mathematics Subject Classification: Primary 42Cxx, 42Axx; Secondary 42C05, 42A38, 42C10

Keywords and phrases: Convergence of Cesaro Mean, Fourier-Laplace series, Legendre and Jacobi polynomials

1. Notation, Definitions and Auxiliary Statements

We use the following notations:

R^k is the k-dimensional Euclidean space $(k = 1, 2, \ldots)$;

$x = (x_1, x_2, \ldots, x_k), y = (y_1, y_2, \ldots, y_k)$ are the points (vectors) of the space R^k;

$(\rho, \theta_1, \theta_2, \ldots, \theta_{k-2}, \varphi)$ are the spherical coordinates of the point $x = (x_1, x_2, \ldots, x_k)$;

$(x, y) = \sum\limits_{i=1}^{k} x_i y_i$ is a scalar product of vectors from R^k;

$|x| = \sqrt{(x, x)}$ is length of the vector x;

$S^{k-1} = \{x : x \in R^k; \ |x| = 1\}$ is the unit sphere, and $|S^{k-1}|$ is its area;

$D^{k-1}(x; h) = \{y : y \in S^{k-1}, (x, y) > \cos h, \ 0 < h \le \pi\}$;

$L_p(S^{k-1}), 1 \le p < \infty$ is the space of functions defined on S^{k-1} and integrable in the p-th degree with a standard norm.

---

[*]E-mail address: topur@list.ru

[†]E-mail address: lado54@mail.ru

Definition 1.1 Let $f \in L(S^{k-1})$, $k \geq 3$. Its Fourier-Laplace series is called the series

$$S(f;x) = \sum_{n=0}^{\infty} Y_n^{\lambda}(f,x), \tag{1.1}$$

where $Y_n^{\lambda}(f,x)$ is the hyperspherical harmonics of the function f of order n defined by

$$Y_n^{\lambda}(f,x) = \frac{\Gamma(\lambda)(n+\lambda)}{2\pi^{\lambda+1}} \int_{S^{k-1}} P_n^{\lambda}[(x,y)]f(y)dS^{k-1}(y), \quad n = 0, 1, 2, \ldots,$$

where $\lambda = \frac{k-2}{2}$ is a critical exponent, $\Gamma(\lambda)$ is Euler's integral of second kind, $P_n^{\lambda}(t)$ are the Gegenbauer (ultraspherical) polynomials of order n; $dS^{k-1}(x)$ is an element on the surface of S^{k-1}.

Definition 1.2 By $\sigma_n^{\lambda,\alpha}(f;x)$ we denote Cesaro means (C,α), $\alpha > -1$ of the series (1.1), i.e.,

$$\sigma_n^{\lambda,\alpha}(f;x) = \frac{1}{A_n^{\alpha}} \sum_{v=0}^{n} A_{n-v}^{\alpha-1} S_v^{\lambda}(f;x) = \frac{\Gamma(\lambda)}{2\pi^{\lambda+1}} \int_{S^{k-1}} \Phi_n^{\lambda,\alpha}(\cos\gamma) f(y) dS^{k-1}(y),$$

where

$$A_n^{\alpha} = \frac{(\alpha+1)(\alpha+2)\cdots(\alpha+n)}{n!};$$

$$\Phi_n^{\lambda,\alpha}(\cos\gamma) = \frac{1}{A_n^{\alpha}} \sum_{v=0}^{n} (v+\lambda) A_{n-v}^{\alpha} P_v^{\lambda}(\cos\gamma);$$

$\cos\gamma = (x,y)$, $x(\theta_1, \theta_2, \ldots, \theta_{k-2}, \varphi) \in S^{k-1}$, $y(\theta_1', \theta_2', \ldots, \theta_{k-2}'\varphi') \in S^{k-1}$; $S_n^{\lambda}(f;x)$ is a partial sum of the series (1.1), i.e.,

$$S_n^{\lambda}(f;x) = \sum_{v=0}^{n} Y_v^{\lambda}(f;x) = \frac{\Gamma(\lambda)}{2\pi^{\lambda+1}} \int_{S^{k-1}} \Phi_n^{\lambda}(\cos\gamma) f(y) dS^{k-1}(y),$$

$$\Phi_n^{\lambda}(\cos\gamma) = \sum_{v=0}^{n} (v+\lambda) P_v^{\lambda}(\cos\gamma).$$

Note that if $\theta_1 = 0$, that is x coincides with a pole, then $\cos\gamma = \cos\theta_1'$, and hence $\gamma = \theta_1'$. Consequently, transforming the system of coordinates in such a way that the pole coincides with x, we obtain $\theta_1' = \gamma$.

Of special importance in the problems of convergence and summability of Fourier series are the so-called Lebesgue constants which for the Cesaro (C,α), $\alpha > -1$ means of the series (1.1) have the form

$$L_n^{\lambda,\alpha}(x) = \frac{\Gamma(\lambda)}{2\pi^{\lambda+1}} \int_{S^{k-1}} |\Phi_n^{\lambda,\alpha}(\cos\gamma)| dS^{k-1}(y) =$$

$$= C_k \int_0^{\pi} |\Phi_n^{\lambda,\alpha}(\cos\gamma)| \sin^{2\lambda}\gamma d\gamma,$$

where C_k is defined by [1]

$$C_k = \frac{\Gamma(\lambda)}{\Gamma(\lambda + i/2)\sqrt{\pi}}.$$

The dependence of the Lebesgue constant for the Cesaro (C, α)-means on the space dimension and on the summability exponent α is expressed as follows (see [1], [2], [3]).

Theorem 1.1. *For the Lebesgue constant of the Fourier-Laplace series there are the estimates:*

$$C_1 n^{\lambda - \alpha} < L_n^{\lambda,\alpha} < C_2 n^{\lambda - \alpha}, \quad for \quad \alpha < \lambda;$$
$$C_1 \ln n < L_n^{\lambda,\alpha} < C_2 \ln n, \quad for \quad \alpha = \lambda;$$
$$0 < L_n^{\lambda,\alpha} < C, \quad for \quad \alpha > \lambda.$$

The following lemmas are valid ([4], pp. 249-253).

Lemma 1.1. *Let* $1 < p \le 2$. *If*

$$\sigma_*^{\lambda,\alpha}(f; x) = \sup_{n \ge 0} |\sigma_n^{\lambda,\alpha}(f; x)|,$$

then

$$\|\sigma_*^{\lambda,\alpha}(f; x)\|_{L_p(S^{k-1})} \le C_{p,\alpha} \|f(x)\|_{L_p(s^{k-1})} \tag{1.2}$$

for

$$\alpha > (k-2)\left(\frac{1}{p} - \frac{1}{2}\right). \tag{1.3}$$

Remark 1.1 If $2 \le p < \infty$, then the conclusion of Lemma 1.1 is valid for

$$\alpha > (k-2)\left(\frac{1}{q} - \frac{1}{2}\right), \quad \frac{1}{p} + \frac{1}{q} = 1.$$

Lemma 1.2 *If* $f \in L \ln^+ L$, *then*

$$\int_{S^{k-1}} |\sigma_n^{\lambda,\lambda}(f; x)| dS^{k-1}(x) < C_1 \int_{S^{k-1}} |f(x)| \ln^+ |f(x)| dS^{k-1}(x) + C_2.$$

2. Main Results

Relying on Lemmas 1.1-1.2 and Theorem 1.1, we prove the following theorems.

---

[1] In what follows, C will, generally speaking, denote various positive constants.

Theorem 2.1.

  Let $1 < p \leq 2$. *Then*

$$\lim_{n \to \infty} \|\sigma_n^{\lambda,\alpha}(f;x) - f(x)\|_{L_p(S^{k-1})} = 0, \tag{2.4}$$

if α *satisfies the condition* (1.3).

Remark 2.1 If $2 \leq p < \infty$, then the conclusion of Theorem 2.1 is valid for

$$\alpha > (k-2)\left(\frac{1}{q} - \frac{1}{2}\right), \quad \frac{1}{p} + \frac{1}{q} = 1.$$

Remark 2.2. It follows from Theorem 2.1 that the means $\sigma_n^{\lambda,\lambda}(f;x)$ converge in the metric of $L_p(S^{k-1})$ to the function $f \in L_p(S^{k-1})$ if $1 < p < \infty$ (an analogue of the M. Riesz theorem).

Remark 2.3. For $p = 2$, the equality (2.4) is fulfilled for $\forall \alpha \geq 0$ (if $\alpha = 0$, then this is the Bonami-Clerc theorem [5]). The works due to Bonami [6], Clerc [7], and Bonami, Clerc [8] are also devoted to these problems.

Theorem 2.2. *If* $f \in L \ln^+ L$, *then*

$$\lim_{n \to \infty} \|\sigma_n^{\lambda,\lambda}(f;x) - f(x)\|_{L(S^{k-1})} = 0.$$

Remark 2.4. In [15], Pollard shows that any function can be expanded in a Fourier series in Legendre polynomials in the space L_p if $\frac{4}{3} < p < 4$, and gives examples of the functions for which the series diverges if $1 \leq p < \frac{4}{3}$ or $p > 4$.

  In Newman et al. [9], it is shown that for $p = \frac{4}{3}$ and $p = 4$ the series of functions $f \in L_p$ can diverge in L_p.

  V. P. Motorny [10] and P. K. Suetin [11] found the sufficient conditions which have to be satisfied by the function f in order for the Fourier-Legendre series to be convergent in L_p for $p \in \left[1, \frac{4}{3}\right] \cup [4, \infty]$ and $C[-1, 1]$. The expansion of functions in a series in Jacobi polynomials in the space L_p is investigated with sufficient completeness in Pollard [12-14], Wing [16], Muckenhoupt [17], Askey et al. [18], Badkov [19] and others. In particular, in Askey [20] it is proved that Cesaro (C, α)-means $\sigma_n^\alpha(f;x)$ of the Legendre series converge in the metric of $L_p[-1, 1]$ when

$$\frac{2}{\frac{3}{2} + \alpha} < p < \frac{2}{\frac{1}{2} - \alpha}.$$

References

[1] Gronwall T.H., Uber die Summierbarkeit der Reihe von Laplace und Legendre. *Math. Ann.* **75**(1914), 321-375.

[2] Kogbetliantz E., Recherches sur la summabilite des series ultraspheriques par la methode des moyennes arithmetiques. *J. Math. Pures Appl.* **9**, No 3(1924), 107-187.

[3] Khocholava V.V., On Lebesgue constants of Fourier-Laplace series. *Trudy Gruz. Politechn. Inst. No* **5**(237), 1981, 24-34 (in Russian).

[4] Topuria S.B., *The Fourier-Laplace series on a sphere.* Tbilisi, Izdat. Tbiliss. Gos. Univ (in Russian).

[5] Bonami A., Clerc J.L., Sommes de Cesaro et multiplicateurs des developpements en harmoniques en spheriques. *Trans. Amer. Math. Soc.,* **183**(1973), Sept.223-263.

[6] Bonami A., Multiplicateurs des series ultraspheriques. *C.R. Acad. Sci.,* **273**, No 3(1971).

[7] Clerc J.L., Les sommes partielles de la decomposition en harmoniques spheriques ne convergent pas dans L_p $(p \neq 2)$. *C.R. Acad. Sci.,* **274**, No 1(1972), A59-A61.

[8] Bonami A., Clerc J.L., Sommes de Cesaro et multeplicateurs des developpements en harmoniques. *C.R. Acad. Sci.,* **274**, No 24(1972), A1707-A1709.

[9] Newman J., Rudin W., Mean convergence of orthogonal series. *Proc. Amer. Math. Soc.,* **3**(1952), 219-222.

[10] Motorni V.P., On the convergence in the mean of Fourier series in Legendre polynomials. *Izd. AN SSSR, ser.* Math., **37**, No 1(1973), 135-147 (in Russian).

[11] Suetin P.K., On the representation of continuous and differentiable functions by Fourier series in Legendre polynomials. *Dokl. AN SSSR,* **158**, No 6(1964), 1275-1277 (in Russian).

[12] Pollard H., The mean convergence of orthogonal series of polynomials. *Proc. Nat. Acad. Sei. USA,* **32**(1946), 8-10.

[13] Pollard H., The mean convergence of orthogonal series. 1. *Trans. Amer. Math. Soc.,* **62**(1047), 387-403.

[14] Pollard H., The mean convergence of orthogonal series. II. *Trans Amer. Math. Soc.,* **63**(1948), 355-367.

[15] Pollard. H., The mean convergence of orthogonal series. III. *Duke Math. J.,* **16**, No 1(1949), 189-191.

[16] Wing G.M., The mean convergence of orthogonal series. *Amer. Math. J.,* **72**(1950), 792-807.

[17] Muckenhoupt B., Mean convergence of Jacobi series. *Proc. Amer. Math. Soc.,* **23**, No 2(1969), 306-310.

[18] Askey R., Hirshman I.I., Mean summability for ultraspherical polynomials. *Math. Scand.,* **12**, No 2(1963), 167-177.

[19] Badkov V.M., Convergence in the mean and almost everywhere of Fourier series in polynomials, orthogonal on a segment. *Math. Sbornik,* **95**, No 2(1974), 229-262 (in Russian).

[20] Askey R., Norm inequalities for some orthogonal series. *Bull. Amer. Math. Soc.,* **72**, No 5(1966), 808-823.

In: Focus on Science and Technology...
Editor: Sergo Gotsiridze, pp. 239-247

ISBN 978-1-61209-970-5
© 2011 Nova Science Publishers, Inc.

Chapter 24

ON UNIFORM MEASURES IN BANACH SPACES

Gogi Pantsulaia[*†]
Department of Mathematics, Georgian Technical University,
Kostava Street-77, 0175 Tbilisi-75, Georgia
I.Vekua Institute of Applied Mathematics, Tbilisi State University,
University Street - 2, 0143 Tbilisi-43, Georgia

Abstract

We prove that a non-existence of a measurable cardinal implies a non-existence of such a translation-invariant Borel measure μ in an arbitrary infinite-dimensional Banach space for which the closed unit ball has μ measure 1. This answers negatively to the question of D. Fremlin [3]. For an arbitrary infinite parameter set α, we construct a uniform measure in the Banach space ℓ^α and show that this measure has no a uniqueness property.

2010 Mathematics Subject Classification: Primary 28xx, 03xx; Secondary 28C10, 03Exx

Keywords and phrases: Banach space, translation-invariant Borel measure, uniform measure, measurable cardinal.

1. Introduction

D. Fremlin [3] posed the following questions.

Question 1.1. *Which Banach spaces have the property that there exists a translation-invariant Borel measure μ such that the closed unit ball has measure 1?*

Question 1.2. *Let ℓ^∞ be a Banach space of all real-valued bounded sequences equipped with norm $||\cdot||_\infty$ defined by* .

$$||(x_k)_{k\in N}||_\infty = \sup_{k\in N} |x_k|.$$

*The designated project has been fulfilled by financial support of the Georgia National Science Foundation (# GNSF / ST $08/3 - 391$, # GNSF / ST $09_144 - 3 - 105$).
†E-mail address: gogi_pantsulaia@hotmail.com

Let $\mathbb{B}_{||\cdot||_\infty}(\ell^\infty)$ be a Borel σ-algebra of subsets of ℓ^∞ generated by the norm $||\cdot||_\infty$. Further, let B_0 be a closed unite ball defined by

$$B_0 = \{(x_k)_{k\in\mathbb{N}} : (x_k)_{k\in\mathbb{N}} \in \ell^\infty \ \& \ ||(x_k)_{k\in\mathbb{N}}||_\infty \leq 1\}.$$

Does there exists a translation-invariant measure μ defined on the measure space

$$\left(\ell^\infty, \mathbb{B}_{||\cdot||_\infty}(\ell^\infty)\right)$$

such that $\mu(B_0) = 1$?

The affirmative answer to the Question 1.2 has been obtained in Solovay model(see, [7]).

In the present paper, Questions 1.1–1.2 will be objects of our consideration in the theory ZFC.

2. Auxiliary Propositions

Let us recall a definition of the standard product of non-negative real numbers $(\beta_j)_{j\in\alpha} \in [0, +\infty]^\alpha$.

Definition 2.1 A standard product of the family of numbers $(\beta_j)_{j\in\alpha}$ is denoted by **(S)** $\prod_{j\in\alpha} \beta_j$ and defined as follows:

(S) $\prod_{j\in\alpha} \beta_j = 0$ if $\sum_{i\in\alpha^-} \ln(\beta_j) = -\infty$, where $\alpha^- = \{j : ln(\beta_j) < 0\}$ [1], and
(S) $\prod_{j\in\alpha} \beta_j = e^{\sum_{j\in\alpha} \ln(\beta_j)}$ if $\sum_{j\in\alpha^-} \ln(\beta_j) \neq -\infty$.

Let α be an arbitrary infinite parameter set and let $(\alpha_i)_{i\in I}$ be its any partition such that α_i is a non-empty finite subset of the α for every $i \in I$. Let μ_j be a σ-finite continuous Borel measure defined on a Polish space (E_j, ρ_j) for $j \in \alpha$.

We denote by $\mathcal{R}_{(\alpha_i)_{i\in I}}$ the family of all measurable rectangles $R \subseteq \prod_{j\in\alpha} E_j$ of the form $\prod_{i\in I} R_i$ with the property $0 \leq$ **(S)** $\prod_{i\in I} \tau_i(R_i) < \infty$ such that at most $\aleph_0$ of them are noncompact.

We suppose that there exists $R_0 = \prod_{i\in I} R_i^{(0)} \in \mathcal{R}_{(\alpha_i)_{i\in I}}$ such that

$$0 < \text{(S)} \prod_{i\in I} \tau_i(R_i^{(0)}) < \infty.$$

We say that a Borel measure $\nu_{(\alpha_i)_{i\in I}}$ defined on $\mathcal{B}(\prod_{j\in\alpha} E_j)$ is a standard $(\alpha_i)_{i\in I}$-product of the family of σ-finite continuous Borel measures $(\mu_j)_{j\in\alpha}$ if for every

$$R = \prod_{i\in I} R_i \in \mathcal{R}_{(\alpha_i)_{i\in I}}$$

we have

$$\nu_{(\alpha_i)_{i\in I}}(R) = \text{(S)} \prod_{i\in I} \tau_i(R_i),$$

---

[1]We set $ln(0) = -\infty$

where $\tau_i = \prod_{j \in \alpha_i} \mu_j$ for $i \in I$.

Lemma 2.1 ([10], Theorem 1, p.6) *Let μ_j be a σ-finite continuous Borel measure defined on a Polish space (E_j, ρ_j) for $j \in \alpha$, where α is an arbitrary infinite parameter set. Let $(\alpha_i)_{i \in I}$ be any partition of the α such that α_i is a non-empty finite subset of the α for every $i \in I$. Let us suppose that there exists $R_0 = \prod_{i \in I} R_i^{(0)} \in \mathcal{R}_{(\alpha_i)_{i \in I}}$ such that*

$$0 < (\mathbf{S}) \prod_{i \in I} \tau_i(R_i^{(0)}) < \infty.$$

Then there exists a standard $(\alpha_i)_{i \in I}$-product of the family $(\mu_j)_{j \in \alpha}$.

In the sequel we denote a standard $(\alpha_i)_{i \in I}$-product of the family $(\mu_j)_{j \in \alpha}$ by

$$(\mathbf{S}, (\alpha_i)_{i \in I}) \prod_{j \in \alpha} \mu_j.$$

Lemma 2.2([10], Theorem 3, p. 8) *Under assumption of Lemma 2.1, if each measure μ_j is G_j-left-invariant, where G_j denotes a group of Borel transformations of the E_j for $j \in \alpha$, then the measure $(\mathbf{S}, (\alpha_i)_{i \in I}) \prod_{j \in \alpha} \mu_j$ is a $\prod_{j \in \alpha} G_j$-left-invariant.*

Remark 2.1 For $j \in \alpha$, we set $E_j = \mathbb{R}$ and $\mu_j = m$, where m denotes a linear Lebesgue measure on $\mathbb{R}$.

Let $(\alpha_i)_{i \in I}$ be any partition of α such that α_i is a non-empty finite for every $i \in I$.

It is clear that $\prod_{j \in \alpha} [a_j, b_j] \in \mathcal{R}_{(\alpha_i)_{i \in I}}$ if $0 \le (\mathbf{S}) \prod_{i \in I} m^{\alpha_i}(\prod_{j \in \alpha_i} [a_j, b_j]) < \infty$, where m^{α_i} is a Lebesgue measure on $\mathbb{R}^{\alpha_i}$.

Then the measure $(\mathbf{S}, (\alpha_i)_{i \in I}) \prod_{j \in \alpha} \mu_j$ has the following property

$$((\mathbf{S}, (\alpha_i)_{i \in I}) \prod_{j \in \alpha} \mu_j)(\prod_{j \in \alpha} [a_j, b_j]) = (\mathbf{S}) \prod_{j \in \alpha} (b_i - a_i).$$

The measure $(\mathbf{S}, (\alpha_i)_{i \in I}) \prod_{j \in \alpha} \mu_i$ is called a standard "$(\alpha_i)_{i \in I}$-Lebesgue measure" on $\mathbb{R}^\alpha$.

When $\text{card}(\alpha_i) = 1$ for every $i \in I$, then $(\mathbf{S}, (\alpha_i)_{i \in I}) \prod_{j \in \alpha} \mu_i$ is called a standard Lebesgue measure on $\mathbb{R}^\alpha$ and is denoted by m^α.

Let f be any permutation of α. A mapping $A_f : \mathbb{R}^\alpha \to \mathbb{R}^\alpha$, defined by $A_f((x_i)_{i \in \alpha}) = (x_{f(i)})_{i \in \alpha}$ for $(x_i)_{i \in \alpha} \in \mathbb{R}^\alpha$, is called a canonical permutation of $\mathbb{R}^\alpha$.

Note that in our situation $\mathcal{R}_{(\alpha_i)_{i \in I}}$ is the family of all measurable rectangles $R \subseteq \mathcal{B}(\mathbb{R}^\alpha)$ of the form $\prod_{i \in \alpha} Y_i$ with the property $0 \le (\mathbf{S}) \prod_{i \in \alpha} m(Y_i) < \infty$ such that at most $\aleph_0$ of them are noncompact (i.e., $\text{card} \{i : i \in I \ \& \ Y_i \text{ is not compact in } \prod_{i \in \alpha_i} E_i\} \le \aleph_0$.). It is obvious to see that a measure m^α is invariant under a group $\mathcal{P}(\mathbb{R}^\alpha)$ generated by shifts and canonical permutations of $\mathbb{R}^\alpha$ and

$$m^\alpha(\prod_{i \in \alpha} Y_i) = (\mathbf{S}) \prod_{i \in \alpha} m(Y_i).$$

3. On Questions 1.1-1.2

The following result gives a negative answer to the Question 1.1 for infinite-dimensional Radon Banach spaces.

Theorem 3.1 Let $\mathbb{B}$ be an infinite-dimensional Radon ^{2} Banach space. Then there does not exist a translation-invariant Borel measure μ in $\mathbb{B}$ such that the closed unit ball has μ-measure 1.

Proof. Assume the contrary and let μ be such a translation-invariant Borel measure in $\mathbb{B}$ which gets a numerical value 1 on the unite closed ball $\mathbf{B}(0,1)$. Because the space is infinite dimensional, by using Riesz well known construction (see, [5], Theorem 12.16, p. 483) one can construct an infinite sequence of disjoint open balls $(B(x_k, \frac{1}{4}))_{k \in \mathbb{N}}$ of radius $\frac{1}{4}$ which are contained in the ball $\mathbf{B}(0,1)$. It is clear that $\mu(B(0, \frac{1}{4})) = 0$ which follows from the following condition

$$+\infty \times \mu(B(0, \frac{1}{4})) = \sum_{k \in \mathbb{N}} \mu(B(x_k, \frac{1}{4})) \leq \mu(\mathbf{B}(0,1)) = 1.$$

Since the restriction $\mu_{\mathbf{B}(0,1)}$ of the μ to the $\mathbf{B}(0,1)$, defined by

$$(\forall X)(X \in \mathcal{B}(B) \rightarrow \mu_{\mathbf{B}(0,1)}(X) = \mu(B(0,1) \cap X)),$$

is a Borel probability measure on Radon metric space $\mathbb{B}$ we claim that there exists a compact set $F \subseteq \mathbf{B}(0,1)$ of non-zero finite $\mu_{\mathbf{B}(0,1)}$-measure. Hence,

$$0 < \mu_{\mathbf{B}(0,1)}(F) < 1 \Leftrightarrow 0 < \mu(F) < 1.$$

Now let consider a covering $(B(0, \frac{1}{4}) + x)_{x \in F}$ of the F. Since F is compact there exists a finite family $(x_k)_{1 \leq k \leq n}$ of elements of the F such that $F \subseteq \cup_{1 \leq k \leq n} B(x_k, \frac{1}{4})$. By using translation-invariance of the μ we get

$$0 < \mu(F) \leq \mu(\cup_{1 \leq k \leq n} B(x_k, \frac{1}{4})) \leq \sum_{1 \leq k \leq n} \mu(B(x_k, \frac{1}{4})) = n\mu(B(0, \frac{1}{4}) = 0.$$

The latter relation is a contradiction, and thus, Theorem 3.1 is proved. $\square$

We give some examples of infinite-dimensional Banach spaces for which Question 1.1 is solvable negatively in the theory ZFC.

Example 3.1 Let $\mathbb{B}$ be a separable infinite-dimensional Banach space. Since a topological weight of the B is equal to $\aleph_0$, by famous result of Ulam (see, for example, [4]) we claim that the $\mathbb{B}$ is Radon. By Theorem 3.1 we claim that there does not exist a translation-invariant Borel measure μ in $\mathbb{B}$ such that the closed unit ball has μ-measure 1.

Example 3.2 Let I be a set of cardinality $\aleph_1$. Let denote by $(L_2(I), || \cdot ||)$ an infinite-dimensional Banach space of all real-valued square-summable sequences on I, where a norm $|| \cdot ||$ is defined by

$$||(x_i)_{i \in I}|| = \left(\sum_{i \in I} x_i^2 \right)^{\frac{1}{2}}$$

---

^{2}A complete metric space $\mathbb{E}$ is called Radon if every Borel probability measure is tight, i.e. for every $\epsilon > 0$ there is a compact set C_ϵ such that $\mu(C_\epsilon) > \mu(E) - \epsilon$.

for $(x_i)_{i \in I} \in L_2(I)$. It is obvious that a topological weight of the $L_2(I)$ is equal to $\aleph_1$. By famous result of Ulam (see, [4]) we claim that the topological weight of $L_2(I)$ being equal to the $\aleph_1$ is not a real-valued measurable cardinal. By Ulam's another well known result asserted that *all complete metric space with non-measurable topological weight is Radon*(cf. [4]) we deduce that $L_2(I)$ is Radon. By Theorem 3.1 we claim that there does not exist a translation-invariant Borel measure μ in $L_2(I)$ such that the closed unit ball has μ-measure 1.

As a simple consequences of Theorem 3.1 we have the following assertions.

Corollary 3.1 *Let the continuum is not a real-valued measurable cardinal (or (CH)). Then the answer to Question 1.2 is no.*
Proof. Since topological weight of the Banach space ℓ^∞ is equal to the continuum, we claim that under our assumption the condition of Theorem 3.1 holds for the ℓ^∞ (it follows from the condition that topological weight of the ℓ^∞ is not a real-valued measurable cardinal and from the Ulam's well known result). Hence, there does not exist a translation-invariant Borel measure μ in ℓ^∞ such that the closed unit ball has μ-measure 1.

Under (CH), by Ulam's well known result *asserted that the $\aleph_1$ is not a real-valued measurable cardinal* one can get the validity of the same result for the ℓ^∞.   □

Corollary 3.3 *Let B be a Banach space whose topological weight is not a real-valued measurable cardinal. Then there does not exist a translation-invariant Borel measure μ in $\mathbb{B}$ such that the closed unit ball has μ-measure 1.*

Corollary 3.4 *Let there does not exist a real-valued measurable cardinal. Then the answer to the Question 1.1 is no for every infinite-dimensional Banach space.*

Remark 3.1 Since the answer to Question 1.2 is *yes* in Solovay model, and is *no* in the theory $ZFC + CH$, we claim that the Question 1.2 is not solvable within the theory ZF. We do not know whether an analogous result is valid for the Question 1.1.

4. On Uniform Measures on ℓ^α

On a Polish space (X, d) a measure u is uniformly distributed if the u measure of a ball depends only of the radius and not of the center. There need not be any Group structure, i.e. the group of isometrics might not be transitive even if a uniformly distributed measure exists. But the uniformly distributed measure are analogous to Haar measure. More precisely, such a measure is unique up to a positive scalar factor.

One can extend a notion of uniformly distributed measure in non-separable Banach space B as follows:

Definition 4.1 Let $\mathcal{S}$ be a translation-invariant σ-algebra of subsets of B such that

$$\mathcal{F}(B) \subseteq \mathcal{S},$$

where $\mathcal{F}(B)$ denotes a minimal σ-algebra of subsets of B generated by all open balls.

We say that a measure u defined on $\mathcal{S}$ is uniformly distributed if the u measure of a ball depends only of the radius and not of the center.

It is clear that an arbitrary translation-invariant measure on B with $\mathcal{F}(B) \subseteq \text{dom}(\mu)$ is uniformly distributed.

It is natural to consider the following

Problem 4.1 Let μ_1 and μ_2 are two uniform measures on Banach space B such that

$$\mathcal{F}(B) \subseteq (dom)(\mu_1) = (dom)(\mu_2)$$

and the values of measures μ_1 and μ_2 coincide on an arbitrary ball.

Can we assert that the measures μ_1 and μ_2 coincide?

The purpose of the present section is to give a negative answer to Problem 4.1 for the ℓ^∞ when domains of measures μ_1 and μ_2 coincide with $\mathcal{B}(\ell^\infty) \cap \mathcal{B}(\mathcal{R}^\infty)$.

We need the following auxiliary lemmas.

Lemma 4.1 ([8], Theorem 15.3.2, p. 209) *There exists a semi-finite inner regular translation-invariant Borel measure G_M in $\mathbf{R}^\infty$ which satisfies the following conditions:*

1) $G_M([0,1]^\infty) = 1$;

2) there exists a rectangle $\prod_{k=1}^\infty (a_k, b_k)$ with $0 < \prod_{k=1}^\infty (b_k - a_k) < \infty$ such that

$$G_M(\prod_{k=1}^\infty (a_k, b_k)) = 0;$$

3) the measure G_M is absolutely continuous with respect to Baker's measure G_B [2] (see, [8],Remark 15.3.2, p. 211).

Lemma 4.2 *Every rectangle $\prod_{k=1}^\infty [-a_k, +a_k]$ with $0 < \prod_{k=1}^\infty (2a_k) < \infty$ belongs to the intersection $\mathcal{B}(\ell^\infty) \cap \mathcal{B}(\mathcal{R}^\mathcal{N})$.*

Proof. The validity of the relation $\prod_{k=1}^\infty [-a_k, a_k] \in \mathcal{B}(\mathbf{R}^\infty)$ is obvious.

Let us show that $\prod_{k=1}^\infty [-a_k, a_k] \in \mathcal{B}(\ell^\infty)$.

Since $0 < \prod_{k=1}^\infty (2a_k) < \infty$ we claim that the sequence of numbers $(a_k)_{k \in \mathbf{N}}$ is bounded by any positive number $C > 0$, i.e., $0 \leq a_k \leq C$. It means that $\prod_{k=1}^\infty [-a_k, +a_k]$ is a bounded subset of ℓ^∞.

Now let $x \in \ell^\infty \setminus \prod_{k=1}^\infty [-a_k, a_k]$. Since $\prod_{k=1}^\infty [-a_k, +a_k]$ is a bounded subset of ℓ^∞ there exists k_0 such that $x_{k_0} \notin [-a_{k_0}, a_{k_0}]$. We set $\delta = \frac{|x_{k_0}| - |a_{k_0}|}{2}$. We put $U_x = \{y : \|y - x\|_\infty < \delta\}$. Let show that $U_x \cap \prod_{k=1}^\infty [-a_k, a_k] = \emptyset$. Indeed, if we assume the contrary then for any $z = (z_k)_{k \in \mathbf{N}}$ we get $z_k \in [-a_{k_0}, a_{k_0}]$ for every $k \in \mathbf{N}$. In particular, for k_0 we get $z_{k_0} \in [-a_{k_0}, a_{k_0}]$. It implies that $\|z - x\|_\infty \geq |z_{k_0} - x_{k_0}| > \delta$, which contradicts to the condition that $z \in U_x$. Thus, we have proved that $U_x \cap \prod_{k=1}^\infty [-a_k, a_k] = \emptyset$ for every $x \in \ell^\infty \setminus \prod_{k=1}^\infty [-a_k, a_k]$. Now it is obvious that $\prod_{k=1}^\infty [-a_k, a_k] = \ell^\infty \setminus \cup_{x \in (\ell^\infty \setminus \prod_{k=1}^\infty [-a_k, a_k])} U_x$. The latter relation means that the set $\prod_{k=1}^\infty [-a_k, a_k]$ is complement of the open set and Lemma 4.2 is proved. $\square$

The main result of the present section is formulated as follows.

Theorem 4.1 *There exist two translation-invariant uniform measures μ_1 and μ_2 defined on the σ-algebra $\mathcal{B}(\ell^\infty) \cap \mathcal{B}(\mathcal{R}^\infty)$ such that*

(i) for every closed ball $B(x,r)$ with center at x and radius r we have $\mu_1(B(x,r)) = \mu_2(B(x,r))$;

(ii) the measure μ_1 is absolutely continuous with respect to μ_2, and μ_1 and μ_2 are not equivalent.

Proof. For $X \in \mathcal{B}(\ell^\infty) \cap \mathcal{B}(\mathcal{R}^\mathcal{N})$, we put $\mu_1(X) = G_M(A^{-1}(X))$ and $\mu_1(X) = G_B(A^{-1}(X))$, where $A : \mathbf{R}^\infty \to \mathbf{R}^\infty$ is defined by $A((x_k)_{k\in\mathbf{N}}) = (2x_k - 1)_{k\in\mathbf{N}}$ for $(x_k)_{k\in\mathbf{N}} \in \mathbf{R}^\infty$.

We have

$$\mu_1(B(x,r)) = G_M(A^{-1}(B(x,r))) = G_M(A^{-1}(\prod_{k\in\mathbf{N}}[x_k - r, x_k + r])) =$$

$$G_M((\prod_{k\in\mathbf{N}}[\frac{x_k - r + 1}{2}, \frac{x_k + r + 1}{2}])) =$$

$$G_M((\prod_{k\in\mathbf{N}}[\frac{x_k - r + 1}{2}, \frac{x_k + r + 1}{2}]) - (\frac{x_k - r + 1}{2})_{k\in\mathbf{N}}]) =$$

$$G_M([0, r]^\infty) = \lim_{n\to\infty} r^n.$$

Analogously, we get

$$\mu_1(B(x,r)) = G_B(A^{-1}(B(x,r))) = \lim_{n\to\infty} r^n.$$

Now it is obvious to show the validity of Item (i).

The proof of Item (ii) employs Lemmas 4.1–4.2. $\quad\square$

Theorem 4.2 *For an arbitrary non-empty parameter set α, we denote by the ℓ^α a Banach space of all real-valued bounded functions defined on the α and equipped with a norm $||\cdot||_\alpha$ defined by*

$$||(x_i)_{i\in\alpha}||_\alpha = \sup_{i\in\alpha} |x_i|$$

for $(x_i)_{i\in\alpha} \in \ell^\alpha$.

Let $\mathbb{F}(\ell^\alpha)$ be a minimal σ-algebra of subsets of the ℓ^α generated by the family of all open balls in ℓ^α.

Further, let B_0 be a closed unite ball defined by

$$B_0 = \{(x_i)_{i\in\alpha} : (x_i)_{i\in\alpha} \in \ell^\alpha \ \& \ ||(x_i)_{i\in\alpha}||_\alpha \leq 1\}.$$

Then there exists a translation-invariant measure μ defined on the measure space $(\ell^\infty, \mathbb{F}(\ell^\alpha))$ such that $\mu(B_0) = 1$.

Proof. Let $A : \mathbb{R}^\alpha \to \mathbb{R}^\alpha$ be defined by $A((x_i)_{i\in\alpha}) = (2x_i - 1)_{i\in\alpha}$. We set

$$(\forall X)(X \in \mathcal{B}(\ell^\alpha) \cap \mathcal{B}(\mathbb{R}^\alpha) \to \mu(X) = m^\alpha(A^{-1}(X))).$$

It is obvious that

$$\mathcal{F}(\ell^\alpha) \subseteq \mathcal{B}(\ell^\alpha) \cap \mathcal{B}(\mathbb{R}^\alpha)$$

and the restriction λ of the measure μ to the class $\mathcal{F}(\ell^\alpha)$ is a translation-invariant measure for which $\lambda(B_0) = 1$. $\quad\square$

Theorem 4.3 *For an arbitrary infinite parameter set α, there exist two uniform measures λ_1 and λ_2 defined on the Banach space ℓ^α such that*

(i) $(dom)(\lambda_1) = (dom)(\lambda_2) = \mathcal{F}(\ell^\alpha)$;

(ii) the values of measures λ_1 and λ_2 coincide on an arbitrary ball such that $\lambda_1(B_0) = \lambda_2(B_0) = 1$;

(iii) the measures λ_1 and λ_2 are not equivalent.

Proof. Let $\alpha_0 = (i_k)_{k \in \mathbb{N}}$ be a countable family of the α. Let G_B be a Baker measure on $\mathbb{R}^{\alpha_0}$(see, [2]). We set

$$(\forall X)(X \in \mathcal{B}(\ell^\alpha) \cap \mathcal{B}(\mathbb{R}^\alpha) \to$$

$$\mu_1(X) = m^\alpha(A^{-1}(X)) \,\&\, \mu_2(X) = (G_B \times m^{\alpha \backslash \alpha_0})(A^{-1}(X))),$$

where $A : \mathbb{R}^\alpha \to \mathbb{R}^\alpha$ is defined by $A((x_i)_{i \in \alpha}) = (2x_i - 1)_{i \in \alpha}$.

Let us denote by λ_1 and λ_2 restrictions of measures μ_1 and μ_2 to the class $\mathcal{F}(\ell^\alpha)$, respectively. Then it is obvious that λ_1 and λ_2 are translation-invariant measures defined on the σ-algebra $\mathcal{F}(\ell^\alpha)$ for which $\lambda_1(B_0) = \lambda_2(B_0) = 1$. Thus, the validity of items (i)-(ii) is proved.

For a set

$$X = A\Big(\prod_{k=1}^{\infty} [0, e^{\frac{(-1)^k}{k}}] \times [0,1]^{\alpha \backslash \alpha_0} \Big),$$

we have $X \in \mathcal{F}(\ell^\alpha)$, $\mu_1(X) = 0$ and $\mu_2(X) = 1$. The latter relation implies the validity of the item (iii). □

References

[1] Baker, R., "Lebesgue measure" on R^∞. *Proc. Amer. Math. Soc.*, **113(4)** (1991), 1023–1029. MR1062827 (92c:46051),Zbl 0741.28009

[2] Baker, R., "Lebesgue measure" on $\mathbb{R}^\infty$. II. *Proc. Amer. Math. Soc.*,**132(9)**, (2004), 2577–2591 (electronic). MR2054783 (2005d:28012), Zbl 1064.28015

[3] Fremlin D.H., [p98*] *Measure theory,* in preparation (draft available by anonymous ftp.at ftp. essex. ac. uk /pub / measure theory).

[4] Oxtoby J.C., *Measure and Category*, A survey of the analogies between topological and measure spaces. Second edition. Graduate Texts in Mathematics, 2. Springer-Verlag, New York-Berlin (1980), x+106 pp.

[5] Bruckner A.M., Bruckner, J. B., Thomson B.S., *Real Analysis* (2007), http: www.classicalrealanalysis.com.

[6] Halmos P.R., *Measure theory*, Princeton, Van Nostrand (1950).

[7] Pantsulaia G.R., Relations between shy sets and sets of ν_p-measure zero in Solovay's Model, *Bull. Polish Acad.Sci.*, vol.52, No.1, (2004).

[8] Pantsulaia G.R., *Invariant and Quasiinvariant Measures in Infinite-Dimensional Topological Vector Spaces, Nova Science Publishers, Inc* (2007), xiv+231.

[9] Pantsulaia G.R., On ordinary and Standard Lebesgue Measures on R^∞, *Bull. Polish Acad.Sci.*, **73**(3) (2009), 209-222.

[10] Pantsulaia G.R., On a standard product of an arbitrary family of σ-finite Borel measures with domain in Polish spaces, *Theory Stoch. Process*,1-10 (accepted).

[11] C.A. Rogers, *Hausdorff Measures*, Cambridge Univ. Press, 1970.

In: Focus on Science and Technology...
Editor: Sergo Gotsiridze, pp. 249-260

ISBN 978-1-61209-970-5
© 2011 Nova Science Publishers, Inc.

Chapter 25

RIEMANN INTEGRABILITY AND UNIFORM DISTRIBUTION IN INFINITE-DIMENSIONAL RECTANGLES

Gogi Pantsulaia[*]
Department of Mathematics, Georgian Technical University,
Kostava Street-77, 0175 Tbilisi-75, Georgia
I.Vekua Institute of Applied Mathematics, Tbilisi State University,
University Street - 2, 0143 Tbilisi-43, Georgia

Abstract

We consider a concept of the uniformly distribution for increasing sequences of finite subsets in an infinite-dimensional rectangle, and by using the technique of the infinite-dimensional Lebesgue measure, introduce a notion of Reimann integrability for functions defined on the entire rectangle. Further, we prove an infinite-dimensional version of Weyl famous result.

2010 Mathematics Subject Classification: Primary 28xx, 03xx; Secondary 28C10, 03Exx

Keywords and phrases: Uniformly distributed sequence, Riemann integral, infinite dimensional rectangles, Infinite-dimensional Lebesgue measure

1. Introduction

Following [6], a sequence $s_1, s_2, s_3, \cdots$ of real numbers from an interval $[a, b]$ is said to be equidistributed, or uniformly distributed in that interval if the proportion of terms falling in a subinterval $[c, d]$ is proportional to the length of that interval. Such sequences are studied in Diophantine approximation theory and have applications to Monte Carlo integration (see, for example, [9], [11], [6]).

Let $\mathcal{R}$ be the class of all infinite dimensional rectangles $R \in \mathcal{B}(\mathbf{R}^\infty)$ of the form

[*]E-mail address: gogi_pantsulaia@hotmail.com. The designated project has been fulfilled by financial support of the Georgia National Science Foundation (Grants: # GNSF / ST $08/3 - 391$, # GNSF / ST $09 _ 144 - 3 - 105)$).

$$. \ R = \prod_{i=1}^{\infty} [a_i, b_i), \quad -\infty < a_i < b_i < +\infty$$

with $0 < \prod_{i=1}^{\infty}(b_i - a_i) < +\infty$, where

$$\prod_{i=1}^{\infty}(b_i - a_i) := \lim_{n \to \infty} \prod_{i=1}^{n}(b_i - a_i).$$

Let λ be an infinite dimensional-Lebesgue measure defined on the infinite-dimensional topological vector space of all real valued sequences $\mathbf{R}^{\infty}$ equipped with Tychonoff topology (see, for example, [1], [2], [7]).

In terms of the measure λ, we introduce a notion of Reimann integrability for functions defined on above mentioned rectangles.

We consider a new concept of the uniformly distribution in such rectangles and prove an infinite-dimensional version of Weyl famous result [12].

The paper is organized as follows.

In Section 2 we give Weyl some classical results. In Section 3 we give proof of main results.

2. Uniform Distribution of Sequences in an Interval $[a, b]$

Definition 2.1 A bounded sequence $s_1, s_2, s_3, \cdots$ of real numbers is said to be equidistributed or uniformly distributed on an interval $[a, b]$ if for any subinterval $[c, d]$ of the $[a, b]$ we have

$$\lim_{n \to \infty} \frac{\#(\{s_1, s_2, s_3, \cdots, s_n\} \cap [c, d])}{n} = \frac{d - c}{b - a},$$

where $\#$ denotes a counter measure.

Definition 2.2 (Weyl [12])   The sequence $s_1, s_2, s_3, \cdots$ is said to be equidistributed modulo 1 or uniformly distributed modulo 1 if the sequence $(s_n - [s_n])_{n \in N}$ of the fractional parts of the $(s_n)_{n \in N}$'s, is equidistributed (equivalently, uniformly distributed) in the interval $[0, 1]$.

Example 2.1 ([6], Exercise 1.12, p. 16) The sequence of all multiples of an irrational α

$$0, \alpha, 2\alpha, 3\alpha \cdots$$

is uniformly distributed modulo 1

Example 2.2 ([6], Exercise 1.13, p. 16) The sequence

$$\frac{0}{1}, \frac{0}{2}, \frac{1}{2}, \frac{0}{3}, \frac{1}{3}, \frac{2}{3}, \cdots, \frac{0}{k}, \cdots, \frac{k-1}{k}, \cdots$$

is uniformly distributed modulo 1.

Example 2.3 The sequence of all multiples of an irrational α by successive prime numbers

$$2\alpha, 3\alpha, 5\alpha, 7\alpha, 11\alpha, \cdots$$

is equidistributed modulo 1. This is a famous theorem of analytic number theory, proved by I. M. Vinogradov in 1935 (see, [15]).

Remark 2.1 If $(s_k)_{k \in N}$ is uniformly distributed modulo 1, then $((s_k - [s_k])(b-a) + a)_{k \in N}$ is uniformly distributed in an interval $[a, b)$.

Lemma 2.1 (Weyl [12]) *The following three conditions are equivalent:*

(i) $(a_n)_{n \in N}$ is equidistributed modulo 1;

(ii) For every Riemann integrable function f on $[0, 1]$

$$\lim_{n \to \infty} \frac{1}{n} \sum_{j=1}^{n} f(x_j) = \int_{[0,1]} f(x)dx;$$

(iii) For every nonzero integer k

$$\lim_{n \to \infty} \frac{1}{n} \sum_{j=1}^{n} e^{2\pi i k a_j} = 0.$$

The third condition is known as Weyl's criterion. Together with the formula for the sum of a finite geometric series, the equivalence of the first and third conditions furnishes an immediate proof of the equidistribution theorem.

Now let X be a compact Polish space and μ be a probability Borel measure on X. Let $\mathcal{R}(X)$ be a space of all continuous functions on the X.

Definition 2.3 A sequence $s_1, s_2, s_3, \cdots$ of elements of the X is said to be μ-equidistributed or μ-uniformly distributed on the X if for every $f \in \mathcal{R}(X)$ we have

$$\lim_{N \to \infty} \frac{1}{N} \sum_{n=1}^{N} f(s_n) = \int_X fd\mu.$$

Lemma 2.2 ([6], Lemma 2.1, p. 199) *Let $f \in \mathcal{R}(X)$ and $\mu_\infty = \mu^\infty$. Then, for μ_∞-almost every sequences $(s_n)_{n \in N} \in X^\infty$, we have*

$$\lim_{N \to \infty} \frac{1}{N} \sum_{n=1}^{N} f(s_n) = \int_X fd\mu.$$

Lemma 2.3 ([6], pp. 199-201) *Let S be a set of all μ-equidistributed sequences in X^∞. Then*

(i) $\mu_\infty(S) = 1$;

(ii) S is a set of first category;

(iii) S is everywhere dense in Tychonoff topology.

Definition 2.4 Let $(E, \mathcal{F}, \nu)$ be a probability measure space. Let $T : E \to E$ be a measurable transformation.

The ν is called T-invariant if $\nu(T^{-1}(A)) = \nu(A)$ for every $A \in \mathcal{F}$.

The transformation T is called ergodic if, for $A \in \mathcal{F}$, an equality $T^{-1}(A) = A$ implies that $\nu(A) = 0$ or $\nu(A) = 1$.

Example 2.4 Let us define $T : X^\infty \to X^\infty$ by

$$T(x_0, x_1, x_2, \cdots) = (x_1, x_2, x_3, \cdots)$$

for $(x_0, x_1, x_2, \cdots) \in X^\infty$. Then the T is called a shift operator. It is obvious to see that the shift operator is ergodic with respect to the measure μ^∞.

Lemma 2.4 (An individual ergodic lemma, [6], p. 200) *Let $(E, \mathcal{F}, \nu)$ be a probability measure space and $T : E \to E$ be an ergodic transformation of the E with respect to the ν. Then, for every ν-integrable function f on E, we have*

$$\lim_{N \to \infty} \frac{1}{N} \sum_{n=1}^{N} f(T^n(y)) = \int_E f d\mu$$

for ν-almost all $y \in E$.

Example 2.5 For every Borel set $E \subseteq [0, 1]^\infty$, we have

$$\lim_{N \to \infty} \frac{1}{N} \sum_{n=1}^{N} I_E(T^n(y)) = \lambda(E),$$

for λ-almost all $y \in [0, 1]^\infty$, where T is the shift operator of the $[0, 1]^\infty$.

3. Uniform Distribution of an Increasing Sequence of Finite Sets in Infinite-dimensional Rectangles

Let $s_1, s_2, s_3, \cdots$ be a uniformly distributed in an interval $[a, b]$. Setting $Y_n = \{s_1, s_2, s_3, \cdots, s_n\}$ for $n \in N$, the $(Y_n)_{n \in N}$ will be such an increasing sequence of finite subsets of the $[a, b]$ that, for any subinterval $[c, d]$ of the $[a, b]$, the following equality

$$\lim_{n \to \infty} \frac{\#(Y_n \cap [c, d])}{\#(Y_n)} = \frac{d - c}{b - a}$$

will be valid.

This remark raises the following

Definition 3.1 An increasing sequence $(Y_n)_{n \in N}$ of finite subsets of the $[a, b]$ is said to be equidistributed or uniformly distributed in an interval $[a, b]$ if, for any subinterval $[c, d]$ of the $[a, b]$, we have

$$\lim_{n \to \infty} \frac{\#(Y_n \cap [c, d])}{\#(Y_n)} = \frac{d - c}{b - a}.$$

Definition 3.2 Let $\prod_{k \in N} [a_k, b_k[\in \mathcal{R}$. A set U is called an elementary rectangle in the $\prod_{k \in N} [a_k, b_k[$ if it admits the following representation

$$U = \prod_{k=1}^{m} [c_k, d_k[\times \prod_{k \in N \setminus \{1, \cdots, m\}} [a_k, b_k[,$$

where $a_k \leq c_k < d_k \leq b_k$ for $1 \leq k \leq m$.

It is obvious that

$$\lambda(U) = \prod_{k=1}^{m}(d_k - c_k) \times \prod_{k=m+1}^{\infty}(b_k - a_k),$$

for every elementary rectangle U.

Definition 3.3 An increasing sequence $(Y_n)_{n \in N}$ of finite subsets of the infinite-dimensional rectangle $\prod_{k \in N}[a_k, b_k[\in \mathcal{R}$ is said to be uniformly distributed in the $\prod_{k \in N}[a_k, b_k[$ if for every elementary rectangle U in the $\prod_{k \in N}[a_k, b_k[$ we have

$$\lim_{n \to \infty} \frac{\#(Y_n \cap U)}{\#(Y_n)} = \frac{\lambda(U)}{\lambda(\prod_{k \in N}[a_k, b_k[)}.$$

Theorem 3.1 *Let $\prod_{k \in N}[a_k, b_k[\in \mathcal{R}$. Let $(x_n^{(k)})_{n \in N}$ be uniformly distributed in the interval $[a_k, b_k[$ for $k \in N$. We set*

$$Y_n = \prod_{k=1}^{n}(\cup_{j=1}^{n} x_j^{(k)}) \times \prod_{k \in N \setminus \{0,1,\cdots,n\}} \{a_k\}.$$

Then $(Y_n)_{n \in N}$ is uniformly distributed in the rectangle $\prod_{k \in N}[a_k, b_k[$.

Proof. Let

$$U = \prod_{k=1}^{m}[c_k, d_k[\times \prod_{k \in N \setminus \{1,\cdots,m\}} [a_k, b_k[$$

be an elementary rectangle in the $\prod_{k \in N}[a_k, b_k[$.

Since $(x_n^{(k)})_{n \in N}$ is uniformly distributed in the interval $[a_k, b_k[$ for $k \in N$, we have

$$\lim_{n \to \infty} \frac{\#(\{x_1^{(k)}, x_2^{(k)}, \cdots, x_n^{(k)}\} \cap [c_k, d_k[)}{n} = \frac{d_k - c_k}{b_k - a_k}.$$

Hence, we have

$$\lim_{n \to \infty} \frac{\#(Y_n \cap U)}{\#(Y_n)} = \lim_{n \to \infty} \prod_{k=1}^{m} \frac{\#(\{x_1^{(k)}, x_2^{(k)}, \cdots, x_n^{(k)}\} \cap [c_k, d_k[)}{n} =$$

$$\prod_{k=1}^{m} \lim_{n \to \infty} \frac{\#(\{x_1^{(k)}, x_2^{(k)}, \cdots, x_n^{(k)}\} \cap [c_k, d_k[)}{n} =$$

$$\prod_{k=1}^{m} \frac{d_k - c_k}{b_k - a_k} = \frac{\lambda(U)}{\lambda(\prod_{k \in N}[a_k, b_k[)}.$$

$\square$

Remark 3.1 In context with Theorem 3.1, it is natural to ask *whether there exists an increasing sequence of finite subsets* $(Y_n)_{n\in N}$ *such that*

$$\lim_{n\to\infty} \frac{\#(Y_n \cap U)}{\#(Y_n)} = \frac{\lambda(U)}{\lambda(\prod_{k\in N}[a_k, b_k[)}$$

for every infinite-dimensional rectangle $U = \prod_{k\in N} X_k \subset \prod_{k\in N}[a_k, b_k[$, *where, for each* $k \in N$, X_k *is a finite sum of pairwise disjoint intervals closed from the left and open from the right.*

Let us show that the answer to this question is no.

Indeed, assume the contrary and let $(Y_n)_{n\in N}$ be such an increasing sequence of finite subsets in $\prod_{k\in N}[a_k, b_k[$. Then we have

$$\cup_{n\in N} Y_n = \{(x_i^{(k)})_{i\in N} : k \in N\}.$$

For $k \in N$, we set $X_k = [a_k, x_k^{(k)}[\cup[x_k^{(k)} + \frac{b_k - x_k^{(k)}}{2^{k+1}}, b_k[$, if $x_k^{(k)} \neq a_k$, and $X_k = [a_k + \frac{b_k - a_k}{2^{k+1}}, b_k[$, otherwise. Then it is clear, that

$$\lambda(\prod_{k\in N} X_k) > 0$$

and

$$\frac{\#(Y_n \cap \prod_{k\in N} X_k)}{\#(Y_n)} = 0$$

for $k \in N$, which follows

$$0 = \lim_{n\to\infty} \frac{\#(Y_n \cap \prod_{k\in N} X_k)}{\#(Y_n)} < \frac{\lambda(\prod_{k\in N} X_k)}{\lambda(\prod_{k\in N}[a_k, b_k[)}.$$

$\square$

Definition 3.4 Let $\prod_{k\in N}[a_k, b_k[\in \mathcal{R}$. A family of pairwise disjoint elementary rectangles $\tau = (U_k)_{1\leq k\leq n}$ of the $\prod_{k\in N}[a_k, b_k[$ is called Riemann partition of the $\prod_{k\in N}[a_k, b_k[$ if $\cup_{1\leq k\leq n} U_k = \prod_{k\in N}[a_k, b_k[$.

Definition 3.5 Let $\tau = (U_k)_{1\leq k\leq n}$ be Riemann partition of the $\prod_{k\in N}[a_k, b_k[$. Let $\ell(Pr_i(U_k))$ be a length of the i-th projection $Pr_i(U_k)$ of the U_k for $i \in N$. We set

$$d(U_k) = \sum_{i\in N} \frac{\ell(Pr_i(U_k))}{2^i(1 + \ell(Pr_i(U_k)))}.$$

It is obvious that $d(U_k)$ is a diameter of the elementary rectangle U_k for $k \in N$ with respect to Tychonoff metric ρ defined as follows

$$\rho((x_k)_{k\in N}, (y_k)_{k\in N}) = \sum_{k\in N} \frac{|x_k - y_k|}{2^k(1 + |x_k - y_k|)}$$

for $(x_k)_{k\in N}, (y_k)_{k\in N} \in \mathbf{R}^\infty$.

A number $d(\tau)$, defined by

$$d(\tau) = \max\{d(U_k) : 1 \le k \le n\}$$

is called mesh or norm of the Riemann partition τ.

Definition 3.6 Let $\tau_1 = (U_i^{(1)})_{1 \le i \le n}$ and $\tau_2 = (U_j^{(2)})_{1 \le j \le m}$ be Riemann partitions of the $\prod_{k \in N}[a_k, b_k[$. We say that $\tau_2 \le \tau_1$ iff

$$(\forall j)((1 \le j \le m) \to (\exists i_0)(1 \le i_0 \le n \ \& \ U_j^{(2)} \subseteq U_{i_0}^{(1)})).$$

Definition 3.7 Let f be a real-valued bounded function defined on the $\prod_{i \in N}[a_i, b_i[$. Let $\tau = (U_k)_{1 \le k \le n}$ be Riemann partition of the $\prod_{k \in N}[a_k, b_k[$ and $(t_k)_{1 \le k \le n}$ be a sample such that, for each k, $t_k \in U_k$. Then

   (i) a sum $\sum_{k=1}^{n} f(t_k)\lambda(U_k)$ is called Riemann sum of the f with respect to Riemann partition $\tau = (U_k)_{1 \le k \le n}$ together with sample $(t_k)_{1 \le k \le n}$;

   (ii) a sum $S_\tau = \sum_{k=1}^{n} M_k \lambda(U_k)$ is called the upper Darbu sum with respect to Riemann partition τ, where $M_k = \sup_{x \in U_k} f(x) (1 \le k \le n)$;

   (ii) a sum $s_\tau = \sum_{k=1}^{n} m_k \lambda(U_k)$ is called the lower Darbu sum with respect to Riemann partition τ, where $m_k = \inf_{x \in U_k} f(x) (1 \le k \le n)$.

Definition 3.8 Let f be a real-valued bounded function defined on $\prod_{i \in N}[a_i, b_i[$. We say that the f is Riemann-integrable on $\prod_{i \in N}[a_i, b_i[$ if there exists a real number s such that for every positive real number ϵ there exists a real number $\delta > 0$ such that, for every Riemann partition $(U_k)_{1 \le k \le n}$ of the $\prod_{k \in N}[a_k, b_k[$ with $d(\tau) < \delta$ and for every sample $(t_k)_{1 \le k \le n}$, we have

$$|\sum_{k=1}^{n} f(t_k)\lambda(U_k) - s| < \epsilon.$$

The number s is called Riemann integral and is denoted by $\int_{\prod_{k \in N}[a_k, b_k[} f(x)d\lambda(x)$.

Remark 3.1 In terms of measure λ, one can easily extend the concept of Riemann integrability for a bounded function defined on the union of countable family of elements of the $\mathcal{R}$.

Definition 3.9 A function f is called a step function on $\prod_{k \in N}[a_k, b_k[$ if it can be written as

$$f(x) = \sum_{k=1}^{n} c_k \mathcal{X}_{U_k}(x),$$

where $\tau = (U_k)_{1 \le k \le n}$ is any Riemann partition of the $\prod_{k \in N}[a_k, b_k[$, $c_k \in R$ for $1 \le k \le n$ and $\mathcal{X}_A$ is the indicator function of the A

Theorem 3.2 *Let f^* be a continuous function on $\prod_{k \in N}[a_k, b_k]$ with respect to Tychonoff metric ρ. Then the restriction f of the f^* to the $\prod_{k \in N}[a_k, b_k[$ is Reimann-integrable.*

Proof. It is obvious that, for every Riemann partition $\tau = (U_k)_{1 \le k \le n}$ of the $\prod_{k \in N}[a_k, b_k[$ and for every sample $(t_k)_{1 \le k \le n}$ with $t_k \in U_k (1 \le k \le n)$, we have

$$s_\tau \le \sum_{k=1}^{n} f(t_k)\lambda(U_k) \le S_\tau.$$

Note that if τ_1 and τ_2 are two Riemann partitions of the $\prod_{k \in N}[a_k, b_k[$ such that $\tau_2 \leq \tau_1$, then

$$s_{\tau_1} \leq s_{\tau_2} \leq \sum_{k=1}^{n} f(t_k)\lambda(U_k) \leq S_{\tau_2} \leq S_{\tau_1}.$$

Let us show the validity of the condition

$$(\forall \epsilon)(\epsilon > 0 \to (\exists r)(\forall \tau)(d(\tau) < r \to S_\tau - s_\tau < \epsilon)),$$

which yields that $\inf_\tau S_\tau = \sup_\tau s_\tau$.

Following Tychonoff theorem, the $\prod_{k \in N}[a_k, b_k]$ is compact set in the Polish group $\mathbf{R}^\infty$ equipped with Tychonoff metric ρ.

Following Cantor well known result, a function f^* is uniformly continuous on the $\prod_{k \in N}[a_k, b_k]$. Hence, for $\epsilon > 0$, there exists $r > 0$ such that

$$(\forall x, y)(x, y \in \prod_{k \in N}[a_k, b_k[\& \rho(x, y) < r \to |f(x) - f(y)| \leq \frac{\epsilon}{\lambda(\prod_{k \in N}[a_k, b_k[)}).$$

Thus, for every Riemann partition $\tau = (U_k)_{1 \leq k \leq n}$ with $d(\tau) < r$ we get

$$S_\tau - s_\tau \leq \frac{\epsilon}{\lambda(\prod_{k \in N}[a_k, b_k[)} \times \sum_{1 \leq k \leq n} \lambda(U_k) = \epsilon.$$

Thus, $\inf_\tau S_\tau = \sup_\tau s_\tau$.

Finally, setting $\delta = r$ and $s = \inf_\tau S_\tau$, we deduce that for every Riemann partition $(U_k)_{1 \leq k \leq n}$ of the $\prod_{k \in N}[a_k, b_k[$ with $d(\tau) < \delta$ and for every sample $(t_k)_{1 \leq k \leq n}$ with $t_k \in U_k (1 \leq k \leq n)$, we have

$$|\sum_{k=1}^{n} f(t_k)\lambda(U_k) - s| \leq S_\tau - s_\tau \leq \epsilon.$$

This ends the proof of Theorem 3.2

$\square$

We recall some important notions and well-known results from general topology and probability theory.

Definition 3.10 A topological Hausdorff space X is called normal if given any disjoint closed sets E and F, there are neighbourhoods U of E and V of F that are also disjoint.

Lemma 3.1 (Urysohn [14]) *A topological space is normal if and only if any two disjoint closed sets can be separated by a function. That is, given disjoint closed sets E and F, there is a continuous function f from X to $[0, 1]$ such that the preimages of 0 and 1 under f are E and F respectively.*

Remark 3.2 Since all compact Hausdorff spaces are normal, we deduce that $\prod_{k \in N}[a_k, b_k]$ equipped with Tychonoff topology, is normal. By Urysohn's lemma we deduce that any two disjoint closed sets in $\prod_{k \in N}[a_k, b_k]$ can be separated by a function.

Definition 3.11 A Borel measure μ, defined on a Hausdorff topological space X, is called Radon if

$$(\forall Y)(Y \in \mathcal{B}(X) \ \& \ 0 \leq \mu(Y) < +\infty \to \mu(Y) = \sup_{\substack{K \subseteq Y \\ K \text{ is compact in X}}} \mu(K)).$$

Lemma 3.2 (Ulam's [13]) *Every probability Borel measure defined on Polish metric space is Radon.*

Theorem 3.3 *Let $(Y_n)_{n \in N}$ be an increasing family of finite subsets of the $\prod_{i \in N}[a_i, b_i[$. Then $(Y_n)_{n \in N}$ is uniformly distributed in the $\prod_{k \in N}[a_k, b_k[$ if and only if for every f, being a restriction of any continuous (with respect to Tychonoff metric) function on $\prod_{k \in N}[a_k, b_k]$ to the $\prod_{k \in N}[a_k, b_k[$, the following equality*

$$\lim_{n \to \infty} \frac{\sum_{y \in Y_n} f(y)}{\#(Y_n)} = \frac{\int_{\prod_{k \in N}[a_k, b_k[} f(x) d\lambda(x)}{\lambda(\prod_{i \in N}[a_i, b_i[)}.$$

holds.

Proof. Necessity. Let $(Y_n)_{n \in N}$ be a uniformly distributed in the $\prod_{k \in N}[a_k, b_k[$ and let $f(x) := \sum_{k=1}^{m} c_k \mathcal{X}_{U_k}(x)$ be a step function. Then we have

$$\lim_{n \to \infty} \frac{\sum_{y \in Y_n} f(y)}{\#(Y_n)} = \lim_{n \to \infty} \frac{\sum_{y \in Y_n} \sum_{k=1}^{m} c_k \mathcal{X}_{U_k}(y)}{\#(Y_n)} =$$

$$\lim_{n \to \infty} \frac{\sum_{k=1}^{m} c_k \#(U_k \cap Y_n)}{\#(Y_n)} = \sum_{k=1}^{m} c_k \lim_{n \to \infty} \frac{\#(U_k \cap Y_n)}{\#(Y_n)} =$$

$$\sum_{k=1}^{m} c_k \frac{\lambda(U_k)}{\lambda(\prod_{i \in N}[a_i, b_i[)} = \frac{\int_{\prod_{k \in N}[a_k, b_k[} f(x) d\lambda(x)}{\lambda(\prod_{i \in N}[a_i, b_i[)}.$$

Now, let f be a restriction of any continuous function f^* on $\prod_{k \in N}[a_k, b_k]$ to the $\prod_{k \in N}[a_k, b_k[$. By Theorem 3.2 we deduce that f is Riemann-integrable. From the definition of the Reimann integral we deduce that, for every positive ϵ, there exists two step functions f_1 and f_2 on $\prod_{i \in N}[a_i, b_i[$ such that

$$f_1(x) \leq f(x) \leq f_2(x)$$

and

$$\int_{\prod_{i \in N}[a_i, b_i[} (f_1(x) - f_2(x)) d\lambda(x) < \epsilon.$$

Then we have

$$\int_{\prod_{i \in N}[a_i, b_i[} f(x) d\lambda(x) - \epsilon \leq \int_{\prod_{i \in N}[a_i, b_i[} f_1(x) d\lambda(x) =$$

$$\lambda(\prod_{i \in N}[a_i, b_i[) \times \lim_{n \to \infty} \frac{\sum_{y \in Y_n} f_1(y)}{\#(Y_n)} \leq \lambda(\prod_{i \in N}[a_i, b_i[) \times \varliminf_{n \to \infty} \frac{\sum_{y \in Y_n} f(y)}{\#(Y_n)} \leq$$

$$\lambda(\prod_{i \in N}[a_i, b_i[) \times \overline{\lim}_{n \to \infty} \frac{\sum_{y \in Y_n} f(y)}{\#(Y_n)} \leq \lim_{n \to \infty} \frac{\sum_{y \in Y_n} \hat{f}_2(y)}{\#(Y_n)} \leq$$

$$\lambda(\prod_{i \in N}[a_i, b_i[) \times \int_{\prod_{i \in N}[a_i, b_i[} f_2(x)d\lambda(x) \leq \int_{\prod_{i \in N}[a_i, b_i[} f(x)d\lambda(x) + \epsilon.$$

The latter relation yields an existence of the limit $\lim_{n \to \infty} \frac{\sum_{y \in Y_n} f(y)}{\#(Y_n)}$ such that

$$\lim_{n \to \infty} \frac{\sum_{y \in Y_n} f(y)}{\#(Y_n)} = \frac{\int_{\prod_{k \in N}[a_k, b_k[} f(x)d\lambda(x)}{\lambda(\prod_{i \in N}[a_i, b_i[)}.$$

This ends the proof of Necessity.

Sufficiency. Assume that $(Y_n)_{n \in N}$ is an increasing sequence of subsets of the $\prod_{k \in N}[a_k, b_k[$ such that for every f, being the restriction of the continuous function f^* on $\prod_{k \in N}[a_k, b_k]$ to the $\prod_{k \in N}[a_k, b_k[$, the following equality

$$\lim_{n \to \infty} \frac{\sum_{y \in Y_n} f(y)}{\#(Y_n)} = \frac{\int_{\prod_{k \in N}[a_k, b_k[} f(x)d\lambda(x)}{\lambda(\prod_{i \in N}[a_i, b_i[)}.$$

holds.

Let U be any elementary rectangle in $\prod_{i \in N}[a_i, b_i[$.

For $\epsilon > 0$, by Ulam's lemma we can choose such a compact set

$$F \subset \prod_{k \in N}[a_k, b_k] \setminus [U]_T,$$

that $\lambda((\prod_{k \in N}[a_k, b_k] \setminus [U]_T) \setminus F) < \frac{\epsilon}{2}$, where $[U]_T$ denotes completion of the set U by Tychonoff topology in $\prod_{k \in N}[a_k, b_k]$. Then, by Urysohn's lemma we deduce that there is a continuous function g_2^* from $\prod_{k \in N}[a_k, b_k]$ to $[0, 1]$ such the preimages of 0 and 1 under g_2^* are F and $[U]_T$ respectively. Then, for $x \in \prod_{k \in N}[a_k, b_k]$, we have

$$\mathcal{X}_U^*(x) \leq g_2^*(x)$$

and

$$(L) \int_{\prod_{k \in N}[a_k, b_k]} (g_2^*(x) - \mathcal{X}_U^*(x))d\lambda(x) \leq \frac{\epsilon}{2},$$

where $\mathcal{X}_U^*$ is indicator of the U defined on the $\prod_{k \in N}[a_k, b_k]$ and the symbol (L) indicates that an integral is taken in the Lebesgue sense.

Now let consider a set $[\prod_{k \in N}[a_k, b_k] \setminus U]_T$. Using Ulam's lemma, we can choose such a compact set

$$F_1 \subset \prod_{k \in N}[a_k, b_k] \setminus [\prod_{k \in N}[a_k, b_k] \setminus U]_T$$

that

$$\lambda((\prod_{k \in N}[a_k, b_k] \setminus [\prod_{k \in N}[a_k, b_k] \setminus U]_T) \setminus F_1) < \frac{\epsilon}{2}.$$

Then, by Urysohn's lemma we deduce that there is a continuous function g_1^* from $\prod_{k \in N}[a_k, b_k]$ to $[0, 1]$ such the preimages of 0 and 1 under g_1^* are $[\prod_{k \in N}[a_k, b_k]] \setminus U]_T$ and F_1, respectively. Then, for $x \in \prod_{k \in N}[a_k, b_k]$, we have

$$g_1^*(x) \leq \mathcal{X}_U^*(x)$$

and

$$(L) \int_{\prod_{k \in N}[a_k, b_k]} (\mathcal{X}_U^*(x) - g_1^*(x)) d\lambda(x) \leq \frac{\epsilon}{2}.$$

Now, if we denote by g_1, g_2 and $\mathcal{X}_U$ restrictions of functions g_1^*, g_2^* and $\mathcal{X}_U^*$ to the set $\prod_{k \in N}[a_k, b_k[$ and take into account a trivial fact that, for $i = 1, 2$,

$$(L) \int_{\prod_{k \in N}[a_k, b_k]} g_i^*(x)) d\lambda(x) = \int_{\prod_{k \in N}[a_k, b_k]} g_i(x) d\lambda(x),$$

we deduce that for every elementary rectangle U in $\prod_{i \in N}[a_i, b_i[$ there exists two continuous functions g_1^* and g_2^* on the $\prod_{i \in N}[a_i, b_i]$ such that

$$g_1(x) \leq \mathcal{X}_U(x) \leq g_2(x)$$

and

$$\int_{\prod_{i \in N}[a_i, b_i[} (g_2(x) - g_1(x)) d\lambda(x) \leq \epsilon.$$

Then we have

$$\lambda(U) - \epsilon \leq \int_{\prod_{i \in N}[a_i, b_i[} g_2(x) d\lambda(x) - \epsilon \leq \int_{\prod_{i \in N}[a_i, b_i[} g_1(x) d\lambda(x) =$$

$$\lambda(\prod_{i \in N}[a_i, b_i[) \times \lim_{n \to \infty} \frac{\sum_{y \in Y_n} g_1(y)}{\#(Y_n)} \leq \lambda(\prod_{i \in N}[a_i, b_i[) \times \underline{\lim}_{n \to \infty} \frac{\#(Y_n \cap U)}{\#(Y_n)} \leq$$

$$\lambda(\prod_{i \in N}[a_i, b_i[) \times \overline{\lim}_{n \to \infty} \frac{\#(Y_n \cap U)}{\#(Y_n)} \leq \lambda(\prod_{i \in N}[a_i, b_i[) \times \lim_{n \to \infty} \frac{\sum_{y \in Y_n} g_2(y)}{\#(Y_n)} =$$

$$\int_{\prod_{i \in N}[a_i, b_i[} g_2(x) d\lambda(x) \leq \int_{\prod_{i \in N}[a_i, b_i[} g_1(x) d\lambda(x) + \epsilon \leq \lambda(U) + \epsilon.$$

Since ϵ was taken arbitrary, we deduce that

$$\lambda(\prod_{i \in N}[a_i, b_i[) \times \lim_{n \to \infty} \frac{\#(Y_n \cap U)}{\#(Y_n)} = \lambda(U).$$

This ends the proof of Theorem 3.3

$\square$

References

[1] Baker R., "Lebesgue measure" on R^∞, *Proc. Amer. Math. Soc.*, **113**(4) (1991), 1023–1029. MR1062827 (92c:46051),Zbl 0741.28009

[2] Baker R., "Lebesgue measure" on $\mathbb{R}^\infty$. II. *Proc. Amer. Math. Soc.*,**132**(9), (2004), 2577–2591 (electronic). MR2054783 (2005d:28012), Zbl 1064.28015 Christensen,

[3] Engelking R., *Outline of general topology*, PWN, Warsaw.: North-Holland, Amsterdam (1974).

[4] Halmos P.R., *Measure theory*, Princeton, Van Nostrand (1950).

[5] Kuratowski K., Mostowski A., *Set theory*, Nauka, Moscow (1980) (in Russian).

[6] Kuipers L., Niederreiter H., *Uniform distribution of sequences,* Pure and Applied Mathematics. Wiley-Interscience [John Wiley & Sons], New York-London-Sydney (1974). xiv+390 pp. MR0419394 (54 # 7415)

[7] Pantsulaia G.R., On ordinary and Standard Lebesgue Measures on R^∞ , *Bull. Polish Acad.Sci.,* **73**(3) (2009), 209-222.

[8] Bruckner A.M., Bruckner, J. B., Thomson B.S., *Real Analysis* (2007), http: www.classicalrealanalysis.com.

[9] Hardy, G. H.; Littlewood, J. E. Erratum: l'Some problems of diophantine approximation', *Acta Math.* **41** (1916), no. 1, 196.

[10] Kharazishvili A.B., *Topological aspects of measure theory*, Naukova Dumka, Kiev (1984) (in Russian).

[11] Hardy G. H., Littlewood J. E., Some problems of diophantine approximation. *Acta Math.* **37** (1914), no. 1, 193–239.

[12] Weyl H, *Über* ein Problem aus dem Gebiete der diophantischen Approximation. *Marchr. Ges. Wiss.* Götingen. Math-phys. K1., 1914, 234-244

[13] Ulam S., Zur Masstheorier in der allgemeinen Mengenlehre, *Fund.Math.,* V 16, p. 140-150 (1930).

[14] Urysohn P., Zum Metrisationsproblem. (German) *Math. Ann.* **94** (1925), no. 1, 309–315

[15] Vinogradow I., On fractional parts of certain functions. *Ann. of Math.* (**2**) 37 (1936), no. 2, 448–455

In: Focus on Science and Technology...
Editor: Sergo Gotsiridze, pp. 261-294

ISBN 978-1-61209-970-5
© 2011 Nova Science Publishers, Inc.

Chapter 26

ON STRICT STANDARD AND STRICT ORDINARY PRODUCTS OF MEASURES AND SOME OF THEIR APPLICATIONS

Gogi Pantsulaia[*]
Department of Mathematics, Georgian Technical University,
Kostava Street-77, 0175 Tbilisi-75, Georgia
I.Vekua Institute of Applied Mathematics, Tbilisi State University,
University Street - 2, 0143 Tbilisi-43, Georgia

Abstract

Our method of construction of measures is a concept of strict standard and strict ordinary products of an infinite family of (no only σ-finite) measures, which allows us to construct Mankiewicz and Preiss-Tišer generators on $\mathbf{R}^\infty$. We show that if $f : \mathbf{R}^\infty \to R$ is a Lipschitz function and $R^{(N)}$ is a group of all eventually zero sequences, then, in Solovay model, f is λ-almost everywhere $R^{(N)}$-Gâteaux differentiable on $\mathbf{R}^\infty$, where λ denotes the strict standard Lebesgue measure on $\mathbf{R}^\infty$.

2010 Mathematics Subject Classification: Primary 28xx, 03xx; Secondary 28C10, 03Exx

Keywords and phrases: Strict Ordinary and strict standard products of σ-finite measures, Solovay model, Rademacher theorem, Uniform measures.

1. Introduction

The purpose of the present paper is to discuss the concept of strict ordinary and strict standard products of quasi-finite measures which is rather different from concepts [20], [21], [23], which do not yield constructions of Mankiewicz and Preiss-Tišer generators on $\mathbf{R}^\infty$ [19](see, [20], Examples 1-2, pp. 220-221).

The manuscript is organized as follows.

[*]E-mail address: gogi_pantsulaia@hotmail.com. The designated project has been fulfilled by financial support of the Georgia National Science Foundation (Grants: # GNSF / ST 08/3 − 391, # GNSF / ST 09_144 − 3 − 105)).

In Section 2, we give an auxiliary lemma concerned with an existence of a measure defined by the consistent family of finite measures. This lemma differs from the Lemma 1 established in [20].

In Section 3, we consider a concept of strict ordinary and strict standard products of a countable family of (no only σ-finite) measures, prove their existence and study the connection between such measures.

In Section 4, we consider a concept of strict standard product of an arbitrary family of (no only σ-finite) Borel measures with domain in Polish spaces.

In Section 5, we consider some auxiliary propositions in Solovay model.

In Section 6, we give infinite version of Rademacher theorem in R^∞. In particular, we show that if $f : \mathbf{R}^\infty \to R$ is a Lipschitz function and $R^{(N)}$ is a group of all eventually zero sequences, then, in Solovay model, f is λ-almost everywhere $R^{(N)}$-Gâteaux differentiable on $\mathbf{R}^\infty$.

In Section 7, for an arbitrary infinite parameter set α, we consider the problem of the existence of uniform measures on a Banach space ℓ^α of all real-valued bounded functions. We show that such measures always exist on ℓ^α and they are not defined uniquely by their values on balls for an infinite parameter set α.

In Section 8 we give an affirmative solution of Question 6.4 [23] in Solovay model and demonstrate that this question is independent from the theory $ZF + DC$.

2.   An Auxiliary Lemma

Let (E, S) be a measurable space and let $\mathcal{R}$ be any subclass of the σ-algebra S. Let $(\mu_B)_{B \in \mathcal{R}}$ be such a family of σ-finite measures that for $B \in \mathcal{R}$ we have $\operatorname{dom}(\mu_B) = S \cap \mathcal{P}(B)$, where $\mathcal{P}(B)$ denotes the power set of the set B.

Definition 2.1 A family $(\mu_B)_{B \in \mathcal{R}}$ is called to be consistent if

$$(\forall X)(\forall B_1, B_2)(X \in S \ \& \ B_1, B_2 \in \mathcal{R} \to \mu_{B_1}(X \cap B_1 \cap B_2) = \mu_{B_2}(X \cap B_1 \cap B_2)).$$

For $X \in S$ and for a countable family $(A_n)_{n \geq -1}$ of elements of $\mathcal{R}$ with $A_{-1} = A_0 = \emptyset$, we put

$$\mu_{\cup_{n \in N} A_n}(X \cap \cup_{n \in N} A_n) = \sum_{n \in N} \mu_{A_n}((A_n \setminus \cup_{k=0}^{n-1} A_k) \cap X).$$

If $(A_n)_{n \geq 1}$ and $(B_n)_{n \geq 1}$ are two countable families of elements of $\mathcal{R}$ with $A_{-1} = A_0 = B_{-1} = B_0 = \emptyset$, then for $X \in S \cap \cup_{n \in N} A_n \cap \cup_{n \in N} B_n$ we have

$$\mu_{\cup_{n \in N} A_n}(X) = \mu_{\cup_{n \in N} B_n}(X).$$

Indeed, we have

$$\mu_{\cup_{n \in N} A_n}(X) = \sum_{n \in N} \mu_{A_n}\left((A_n \setminus \cup_{k=0}^{n-1} A_k) \cap X\right) =$$

$$\sum_{n \in N} \mu_{A_n}\left((A_n \setminus \cup_{k=0}^{n-1} A_k) \cap \left(\cup_{m \in N} (B_m \setminus \cup_{l=0}^{m-1} B_l)\right) \cap X\right) =$$

$$\sum_{n \in N} \mu_{A_n} \left(\cup_{m \in N} \left(\left(A_n \setminus \cup_{k=0}^{n-1} A_k \right) \cap \left(B_m \setminus \cup_{l=0}^{m-1} B_l \right) \right) \cap X \right) =$$

$$\sum_{n \in N} \sum_{m \in N} \mu_{A_n} \left(\left(A_n \setminus \cup_{k=0}^{n-1} A_k \right) \cap \left(B_m \setminus \cup_{l=0}^{m-1} B_l \right) \cap X \right) =$$

$$\sum_{m \in N} \sum_{n \in N} \mu_{A_n} \left(\left(A_n \setminus \cup_{k=0}^{n-1} A_k \right) \cap \left(B_m \setminus \cup_{l=0}^{m-1} B_l \right) \cap X \right) =$$

$$\sum_{m \in N} \sum_{n \in N} \mu_{B_m} \left(\left(A_n \setminus \cup_{k=0}^{n-1} A_k \right) \cap \left(B_m \setminus \cup_{l=0}^{m-1} B_l \right) \cap X \right) =$$

$$\sum_{m \in N} \mu_{B_m} \left(\cup_{n \in N} \left(A_n \setminus \cup_{k=0}^{n-1} A_k \right) \cap \left(B_m \setminus \cup_{l=0}^{m-1} B_l \right) \cap X \right) =$$

$$\sum_{m \in N} \mu_{B_m} \left(\left(B_m \setminus \cup_{l=0}^{m-1} B_l \right) \cap X \right) = \mu_{\cup_{n \in N} B_n}(X).$$

Lemma 2.1 ([20], Lemma 1) *Let* $(\mu_B)_{B \in \mathcal{R}}$ *be a consistent family of σ-finite measures on* (E, S). *If* $X \in S$ *is covered by a countable family* $(A_n)_{n \in N}$ *of elements of the $\mathcal{R}$ such that* $A_{-1} = A_0 = \emptyset$, *then we put*

$$\lambda_{\mathcal{R}}^{(1)}(X) = \mu_{\cup_{n \in N} A_n}(X),$$

and $\lambda_{\mathcal{R}}^{(1)}(X) = +\infty$, *otherwise.*

Then $\lambda_{\mathcal{R}}^{(1)}$ *is such a measure on* (E, S) *that*

(i) $\lambda_{\mathcal{R}}^{(1)}(B) = \mu_B(B)$ *for every* $B \in \mathcal{R}$;

(ii) if there exists a non-countable family of pairwise disjoint sets $\{B_i \ : \ i \in I\} \subseteq \mathcal{R}$ *such that* $0 < \mu_{B_i}(B_i) < \infty$, *then the measure* $\lambda_{\mathcal{R}}^{(1)}$ *is non-σ-finite;*

(iii) if G is a group of measurable transformations of E such that $G(\mathcal{R}) = \mathcal{R}$ *and*

$$(\forall B)(\forall X)(\forall g)\left(\left(B \in \mathcal{R} \,\& X \in S \cap \mathcal{P}(B) \,\& \, g \in G \right) \rightarrow \mu_{g(B)}(g(X)) = \mu_B(X) \right),$$

where $\mathcal{P}(B)$ denotes a power set of the set B, then the measure $\lambda_{\mathcal{R}}^{(1)}$ *is G-invariant.*

Lemma 2.2 *Let* $(\mu_B)_{B \in \mathcal{R}}$ *be a consistent family of σ-finite measures on a measurable space* (E, S). *We set*

$$(\forall X)(X \in S \rightarrow \lambda_{\mathcal{R}}^{(2)}(X) = \sup\{\mu_{\cup_{n \in N} A_n}(X \cap \cup_{n \in N} A_n) :$$

$$(A_n)_{n \geq -1} \in \mathcal{R}^{\{-1\} \cup N} \,\& \, A_{-1} = A_0 = \emptyset\}.$$

Then the $\lambda_{\mathcal{R}}^{(2)}$ *is the measure on* (E, S) *such that*

(i) $\lambda_{\mathcal{R}}^{(2)}(B) = \mu_B(B)$ *for every* $B \in \mathcal{R}$;

(ii) if there exists a non-countable family of pairwise disjoint sets $\{B_i \ : \ i \in I\} \subseteq \mathcal{R}$ *such that* $0 < \mu_{B_i}(B_i) < \infty$, *then the measure* $\lambda_{\mathcal{R}}^{(2)}$ *is non-σ-finite;*

(iii) if G is a group of measurable transformations of E such that $G(\mathcal{R}) = \mathcal{R}$ *and*

$$(\forall B)(\forall X)(\forall g)\left(\left(B \in \mathcal{R} \,\& X \in S \cap \mathcal{P}(B) \,\& \, g \in G \right) \rightarrow \mu_{g(B)}(g(X)) = \mu_B(X) \right),$$

where $\mathcal{P}(B)$ denotes a power set of the set B, then the measure $\lambda_{\mathcal{R}}^{(2)}$ is G-invariant.

(iv) for $X \in S$, $\lambda_{\mathcal{R}}^{(2)}(X) = 0$ iff $(\forall B)(B \in \mathcal{R} \to \mu_B(X \cap B) = 0)$.

Proof.

Let us prove that the functional $\lambda_{\mathcal{R}}^{(2)}$ is σ-additive.

Let $(X_k)_{k \in N}$ be a countable family of pairwise disjoint elements of S.

Case I. $\lambda_{\mathcal{R}}^{(2)}(\cup_{k \in N} X_k) = A < +\infty$.

Then, for $i \in N$, there exists a family of elements $(A_n^{(i)})_{n \geq -1} \in \mathcal{R}^{\{-1\} \cup N}$ with $A_{-1}^{(i)} = A_1^{(i)} = \emptyset$ such that

$$A - \frac{1}{1+i} < \mu_{\cup_{n \in N} A_n^{(i)}}(\cup_{k \in N} X_k \cap \cup_{n \in N} A_n^{(i)}) < A.$$

Then we have that

$$\mu_{\cup_{i,n \in N} A_n^{(i)}}(\cup_{k \in N} X_k \cap \cup_{i,n \in N} A_n^{(i)}) = A.$$

It is obvious that

$$(\forall B)(B \in \mathcal{R} \to \mu_{\cup_{i,n \in N} A_n^{(i)}}((B \setminus \cup_{i,n \in N} A_n^{(i)}) \cap \cup_{k \in N} X_k) = 0).$$

The later relation follows from the definition of the $\lambda_{\mathcal{R}}^{(2)}(\cup_{k \in N} X_k)$. Indeed, if for any $B_0 \in \mathcal{R}$, we have

$$\mu_{\cup_{i,n \in N} A_n^{(i)}}((B_0 \setminus \cup_{n \in N} A_n^{(i)}) \cap \cup_{k \in N} X_k) = \delta > 0,$$

then

$$\mu_{\cup_{i,n \in N} A_n^{(i)} \cup B_0}(\cup_{k \in N} X_k \cap (B_0 \cup \cup_{i,n \in N} A_n^{(i)})) = \mu_{\cup_{i,n \in N} A_n^{(i)}}(\cup_{k \in N} X_k \cap \cup_{i,n \in N} A_n^{(i)}) +$$

$$\mu_{B_0 \setminus \cup_{i,n \in N} A_n^{(i)}}((B_0 \setminus \cup_{i,n \in N} A_n^{(i)}) \cap \cup_{k \in N} X_k) = A + \delta > A.$$

As a simple consequence, we get

$$\mu_{\cup_{i,n \in N} A_n^{(i)}}(X_k \cap \cup_{i,n \in N} A_n^{(i)}) =$$

$$\sup\{\mu_{\cup_{n \in N} A_n}(X_k \cap \cup_{n \in N} A_n) : (A_n)_{n \geq -1} \in \mathcal{R}^{\{-1\} \cup N} \text{ \& } A_{-1} = A_0 = \emptyset\}.$$

Hence, we have

$$A = \lambda_{\mathcal{R}}^{(2)}(\cup_{k \in N} X_k) = \mu_{\cup_{i,n \in N} A_n^{(i)}}(\cup_{k \in N} X_k \cap \cup_{n \in N} A_n^{(i)}) = \sum_{k \in N} \mu_{\cup_{i,n \in N} A_n^{(i)}}(X_k \cap \cup_{n \in N} A_n^{(i)}) =$$

$$\sum_{k \in N} \sup\{\mu_{\cup_{n \in N} A_n}(X_k \cap \cup_{n \in N} A_n) : (A_n)_{n \geq -1} \in \mathcal{R}^{\{-1\} \cup N} \text{ \& } A_{-1} = A_0 = \emptyset\} = \sum_{k \in N} \lambda_{\mathcal{R}}^{(2)}(X_k).$$

Case II. $\lambda_{\mathcal{R}}^{(2)}(\cup_{k \in N} X_k) = +\infty$.

Then, for $i \in N$, there exists a family of elements $(A_n^{(i)})_{n \geq -1} \in \mathcal{R}^{\{-1\} \cup N}$ with $A_{-1}^{(i)} = A_0^{(i)} = \emptyset$ such that

$$\mu_{\cup_{n\in N}^{(i)} A_n}(\cup_{k\in N} X_k \cap \cup_{n\in N} A_n^{(i)}) > i.$$

Then we have

$$\mu_{\cup_{i,n\in N} A_n^{(i)}}(\cup_{k\in N} X_k \cap \cup_{i,n\in N} A_n^{(i)}) = +\infty.$$

We also have

$$+\infty = \lambda_{\mathcal{R}}^{(2)}(\cup_{k\in N} X_k) = \mu_{\cup_{i,n\in N} A_n^{(i)}}(\cup_{k\in N} X_k \cap \cup_{n\in N} A_n^{(i)}) =$$

$$\sum_{k\in N} \mu_{\cup_{i,n\in N} A_n^{(i)}}(X_k \cap \cup_{n\in N} A_n^{(i)}) \le$$

$$\sum_{k\in N} \sup\{\mu_{\cup_{n\in N} A_n}(X_k \cap \cup_{n\in N} A_n)\ \&$$

$$(A_n)_{n\ge -1} \in \mathcal{R}^{\{-1\}\cup N}\ \&\ A_{-1} = A_0 = \emptyset\} = \sum_{k\in N} \lambda_{\mathcal{R}}^{(2)}(X_k).$$

Proof of the item (i)**.** We set $C_k = B$ for $k \ge 1$ and $C_{-1} = C_0 = \emptyset$. Then the family $(C_k)_{k\ge 1}$ covers B and by the definition of $\lambda_{\mathcal{R}}^{(2)}$ we have

$$\lambda_{\mathcal{R}}^{(2)}(B) = \sup\{\mu_{\cup_{n\in N} A_n}(X \cap \cup_{n\in N} A_n)\ \&$$

$$(A_n)_{n\ge -1} \in \mathcal{R}^{\{-1\}\cup N}\ \&\ A_{-1} = A_0 = \emptyset\} \ge$$

$$\mu_{\cup_{n\in N} C_n}(B \cap \cup_{n\in N} C_n) = \mu_B(B).$$

On the other hand, for $(A_n)_{n\ge -1} \in \mathcal{R}^{\{-1\}\cup N}$ with $A_{-1} = A_0 = \emptyset$, we have

$$\mu_{\cup_{n\in N} A_n}(B \cap \cup_{n\in N} A_n) = \sum_{n\in N} \mu_{A_n}((A_n \setminus \cup_{k=0}^{n-1} A_k) \cap B) =$$

$$\sum_{n\in N} \mu_B((A_n \setminus \cup_{k=0}^{n-1} A_k) \cap B) = \mu_B(\cup_{n\in N} A_n \cap B) \le \mu_B(B),$$

which implies that $\lambda_{\mathcal{R}}^{(2)}(B) \le \mu_B(B)$.

The proof of the item (ii) is obvious and we omit it.

Proof of the item (iii)**.** Let G be a group of measurable transformations of E such that $G(\mathcal{R}) = \mathcal{R}$ and

$$(\forall B)(\forall X)(\forall g)((B \in \mathcal{R}\ \&\ X \in B \cap S\ \&\ g \in G) \to \mu_{g(B)}(g(X)) = \mu_B(X)).$$

We are to show that the measure $\lambda_{\mathcal{R}}^{(2)}$ is G-invariant.

For $X \in S$, we have

$$\lambda_{\mathcal{R}}^{(2)}(g(X)) = \sup\{\mu_{\cup_{n\in N} A_n}(g(X) \cap \cup_{n\in N} A_n)\ \&$$

$$(A_n)_{n\ge -1} \in \mathcal{R}^{\{-1\}\cup N}\ \&\ A_{-1} = A_0 = \emptyset\} =$$

$$\sup\{\sum_{n\in N}\mu_{A_n}((A_n\setminus\cup_{k=0}^{n-1}A_k)\cap g(X))\ \&$$

$$(A_n)_{n\geq-1}\in\mathcal{R}^{\{-1\}\cup N}\ \&\ A_{-1}=A_0=\emptyset\}=$$

$$\sup\{\sum_{n\in N}\mu_{g^{-1}(A_n)}((g^{-1}(A_n)\setminus\cup_{k=0}^{n-1}g^{-1}(A_k))\cap X):$$

$$(g^{-1}(A_n))_{n\geq-1}\in\mathcal{R}^{\{-1\}\cup N}\ \&\ g^{-1}(A_{-1})=g^{-1}(A_0)=\emptyset\}=$$

$$\sup\{\sum_{n\in N}\mu_{B_n}((B_n\setminus\cup_{k=0}^{n-1}B_k)\cap X):(B_n)_{n\geq-1}\in\mathcal{R}^{\{-1\}\cup N}\ \&\ B_{-1}=B_0=\emptyset\}=$$

$$\sup\{\mu_{\cup_{n\in N}B_n}(X\cap\cup_{n\in N}B_n):(B_n)_{n\geq-1}\in\mathcal{R}^{\{-1\}\cup N}\ \&\ B_{-1}=B_0=\emptyset\}=\lambda_{\mathcal{R}}^{(2)}(X).$$

Proof of the item (iv). $(\Rightarrow)$. If we assume the contrary then for any $B_0\in\mathcal{R}$ we get $\mu_{B_0}(X\cap B_0)=\delta>0$.

Let $C_{-1}=C_0=\emptyset$ and $C_n=B_0$ for $n>0$. Then we get

$$\lambda_{\mathcal{R}}^{(2)}(X)=\sup\{\mu_{\cup_{n\in N}A_n}(X\cap\cup_{n\in N}A_n):$$

$$(A_n)_{n\geq-1}\in\mathcal{R}^{\{-1\}\cup N}\ \&\ A_{-1}=A_0=\emptyset\}\geq$$

$$\mu_{\cup_{n\in N}C_n}(X\cap\cup_{n\in N}C_n)=\delta>0,$$

which contradict to the condition $\lambda_{\mathcal{R}}^{(2)}(X)=0$.

$(\Leftarrow)$. The validity of the condition

$$(\forall B)(B\in\mathcal{R}\rightarrow\mu_B(X\cap B)=0)$$

implies that, for every $(A_n)_{n\geq-1}\in\mathcal{R}^{\{-1\}\cup N}$ with $A_{-1}=A_0=\emptyset$, we have

$$\mu_{\cup_{n\in N}A_n}(X\cap\cup_{n\in N}A_n)=0.$$

Hence,

$$\lambda_{\mathcal{R}}^{(2)}(X)=\sup\{\mu_{\cup_{n\in N}A_n}(X\cap\cup_{n\in N}A_n):$$

$$(A_n)_{n\geq-1}\in\mathcal{R}^{\{-1\}\cup N}\ \&\ A_{-1}=A_0=\emptyset\}=0.$$

$\square$

Remark 2.1 Let $E=R^2$ and $S=\mathcal{B}(R^2)$. We set $B_y=R\times\{y\}$ for $y\in R$ and

$$\mathcal{R}=\{B_y:y\in R\}.$$

Let μ_{B_y} be a linear Borel measure defined on B_y. Then the measure $\lambda_{\mathcal{R}}^{(1)}$ defined by Lemma 2.1 does not satisfy the condition of the Lemma 2.3. Indeed, for a set $X=\{0\}\times R$ we have $\mu_{B_y}(X \cap B_y) = 0$ for $y \in R$, but $\lambda_{\mathcal{R}}^{(1)}(X) = +\infty$, because the X is not covered by countable family of elements of $\mathcal{R}$. On the other hand, for the measure $\lambda_{\mathcal{R}}^{(2)}$ defined by Lemma 2.2 we have $\lambda_{\mathcal{R}}^{(2)}(X) = 0$. This example shows us that measures $\lambda_{\mathcal{R}}^{(1)}$ and $\lambda_{\mathcal{R}}^{(2)}$ are different.

Remark 2.2 Let $(\mu_B)_{B \in \mathcal{R}}$ be a consistent family of σ-finite measures on a measurable space (E, S). Let $\overline{\mu}_B$ be a completion of the μ_B for $B \in \mathcal{R}$. We set $S_B = \mathrm{dom}(\overline{\mu}_B) \cup \mathcal{P}(E \setminus B)$. Further, we set

$$\tilde{S} = \cap_{B \in \mathcal{R}} S_B$$

and $\tilde{\mu}_B$ be a measure defined by

$$\tilde{\mu}_B(X) = \overline{\mu}_B(X)$$

for $X \in \tilde{S}$.

Then it is not hard to show that a family of measures $(\tilde{\mu}_B)_{B \in \mathcal{R}}$ is a consistent family of σ-finite measures on a measurable space $(E, \tilde{S})$. We denote by $\tilde{\lambda}_{\mathcal{R}}^{(1)}$ and $\tilde{\lambda}_{\mathcal{R}}^{(2)}$ the measures defined Lemmas 2.1 and 2.2, respectively. Note that both measures $\tilde{\lambda}_{\mathcal{R}}^{(1)}$ and $\tilde{\lambda}_{\mathcal{R}}^{(2)}$ are complete, but we do not know whether the following equalities

$$\tilde{\lambda}_{\mathcal{R}}^{(1)} = \overline{\lambda_{\mathcal{R}}^{(1)}}$$

and

$$\tilde{\lambda}_{\mathcal{R}}^{(2)} = \overline{\lambda_{\mathcal{R}}^{(2)}}$$

hold.

In this context we consider the following example: *Let $(\mu_{B_y})_{y \in R}$ be a family of measures on R^2 defined in Remark 2.1. Then, under Continuum hypothesis (CH), the $\tilde{S}$ strict contains the class of all Lebesgue measurable subsets of the R^2. Indeed, if (A, B) is Sierpinski partition of the R^2 such that $card(A \cap B_y) \leq \omega$ for every $y \in R$, then $A \in \tilde{S}$ and $\tilde{\lambda}_{\mathcal{R}}(A) = 0$.*

On the other hand, in Solovay model, the $\tilde{S}$ coincides with the class of Lebesgue measurable subsets of the R^2.

Hence, a question asking whether the $\tilde{S}$ coincides with the class of Lebesgue measurable subsets of the R^2 is not solvable within the theory $ZF + DC$.

The later fact is equivalent to the assertion that a question asking whether the following equality

$$\sum_{y \in R} \overline{b_y} = \overline{\sum_{y \in R} b_y}$$

holds, where, for $y \in R$, the b_y denotes a Borel measure on R^2 defined by

$$b_y(X) = \mu_{B_y}(X \cap B_y),$$

is not solvable within the theory $ZF + DC$.

Example 2.1 Let $\mathcal{R}$ be the class of all infinite dimensional rectangles $R \in \mathcal{B}(\mathbf{R}^\infty)$ of the form

$$R = \prod_{i=1}^{\infty}(a_i, b_i), \quad -\infty < a_i < b_i < +\infty$$

with $0 \le \prod_{i=1}^{\infty}(b_i - a_i) < +\infty$, where

$$\prod_{i=1}^{\infty}(b_i - a_i) := \lim_{n \to \infty} \prod_{i=1}^{n}(b_i - a_i).$$

Let λ_R be defined by $\lambda_R(X) = \prod_{i=1}^{\infty} m(X_i)$ for every elementary rectangular set $X = \prod_{i \in N} X_i \in \mathcal{B}(R)$. By Charatheodory theorem, there exists a unique extension μ_R of the λ_R to whole $\mathcal{B}(R)$.

Since the family of measures $(\mu_R)_{R \in \mathcal{R}}$ is consistent and $\mathcal{R}$ is translation-invariant, by Lemma 2.1 , we claim that the $\lambda_{\mathcal{R}}^{(1)}$ is "Lebesgue measure" [1]. Analogously, by Lemma 2.2, we claim that the $\lambda_{\mathcal{R}}^{(2)}$ coincides with Baker generator λ(cf. [18], Theorem 5.1, p. 249).

Example 2.2 Let $\mathcal{R}$ be the class of all infinite dimensional rectangles $R \in \mathcal{B}(\mathbf{R}^{\infty})$ of the form

$$R = \prod_{i=1}^{\infty} R_i, \quad R_i \in \mathcal{B}(\mathbf{R})$$

such that

$$0 \le \prod_{i=1}^{\infty} m(R_i) := \lim_{n \to \infty} \prod_{i=1}^{n} m(R_i) < \infty,$$

where m denotes a one-dimensional classical Borel measure on $\mathbf{R}$.

Let λ_R be defined by $\lambda_R(X) = \prod_{i=1}^{\infty} m(X_i)$ for every elementary rectangular set $X = \prod_{i \in N} X_i \in \mathcal{B}(R)$. By Charatheodory theorem, there exists a unique extension μ_R of the λ_R to whole $\mathcal{B}(R)$.

Using same argument as in Example 2.1, by Lemma 2.1 we claim that $\lambda_{\mathcal{R}}^{(1)}$ is "Lebesgue measure" II [2]. Note here that $\lambda_{\mathcal{R}}^{(2)}$ also is "Lebesgue measure" on $\mathbf{R}^{\infty}$ which is different from the "Lebesgue measure" II.

Example 2.3 The Mankiewicz generator G_M [19] is a usual completion of the functional μ defined by

$$\mu(X) = \sum_{a \in \ell_1^{\perp}} \mu_{[0,1]^N}(X - a \cap B_{[0,1]^N})$$

for every $X \in \mathcal{B}(\mathbf{R}^{\infty})$, where

(i) $\mu_{[0,1]^N}$ denotes Kharazishvili quasi-generator of shy sets on $\mathbf{R}^{\infty}$(see, [19]),

(ii) $B_{[0,1]^N} = \cup_{n \in N}\left(\mathbf{R}^n \times [0, 1]^{N \setminus \{1, \cdots, n\}}\right)$,

(iii) $\ell_1^{\perp}$ denotes a linear complement of the vector subspace ℓ_1 in $\mathbf{R}^{\infty}$.

This measure G_M is an $\mathcal{G}$-invariant and has a property that X is a standard cube null set iff X is of G_M-measure zero for every $X \subset \mathbf{R}^{\infty}$.

We set

$$\mathcal{R} = \{R : R = [0, 1]^N + a \text{ for any } a \in \mathbf{R}^{\infty}\}$$

and

$$\mu_R(X) = \lambda(X - a \cap [0,1]^N)$$

for every $X \in \mathcal{B}(\mathbf{R}^\infty)$, where $\lambda = \mu^N$ and μ is a linear probability Lebesgue measure on $[0,1]$.

Then the class $\mathcal{R}$ and the family $(\mu_R)_{R \in \mathbf{R}}$ are invariant under the group $\mathcal{G}$, where $\mathcal{G}$ denotes a group generated by all shifts and canonical permutations of the $\mathbf{R}^\infty$.

Note that the family $(\mu_R)_{R \in \mathbf{R}}$ satisfies all conditions of Lemmas 2.1-2.2. Following Lemmas 2.1-2.2, the measures $\lambda_{\mathcal{R}}^{(1)}$ and $\lambda_{\mathcal{R}}^{(2)}$, are $\mathcal{G}$-invariant. Note here that the measure $\lambda_{\mathcal{R}}^{(2)}$, unlike the measure $\lambda_{\mathcal{R}}^{(1)}$, coincides with Mankiewicz's generator G_M.

Example 2.4 Let $(L_i)_{i \in I}$ be a family of all different n-dimensional vector subspaces of $\mathbf{R}^\infty$ and let $\ell_n^{(i)}$ be an n-dimensional Lebesgue measure on L_i. For $i \in I$, let denote by $L_i^\perp$ a linear compliment of L_i. Then a functional $G_{P\&T}$, defined by

$$G_{P\&T}^{(n)}(X) = \sum_{i \in I} \sum_{a \in L_i^\perp} \ell_n^{(i)}(X - a \cap L_i)$$

for $X \in \mathcal{B}(\mathbf{R}^\infty)$, is an $\mathcal{G}$-invariant (see, Example 2.3) Borel measure and $G_{P\&T}(Y) = 0$ iff Y is n-dimensional null in the sense of Preiss and Tišer for every $Y \subset \mathbf{R}^\infty$.

We set

$$\mathcal{R} = \{L_i + a : a \in \mathbf{R}^\infty, i \in I\}$$

and

$$\mu_{L_i+a}(X) = \ell_n^{(i)}(X - a \cap L_i)$$

for every $X \in \mathcal{B}(\mathbf{R}^\infty)$.

Then the class $\mathcal{R}$ and the family $(\mu_R)_{R \in \mathbf{R}}$, being $\mathcal{G}$-invariant, satisfy all conditions of Lemmas 2.1-2.2. Following Lemmas 2.1-2.2, the measures $\lambda_{\mathcal{R}}^{(1)}$ and $\lambda_{\mathcal{R}}^{(2)}$, are $\mathcal{G}$-invariant.

Note here that the measure $\lambda_{\mathcal{R}}^{(2)}$, unlike the measure $\lambda_{\mathcal{R}}^{(1)}$, coincides with Preiss and Tišer generator $G_{P\&T}$.

3. On Strict Ordinary and Standard Products of Infinite Family of σ-finite Measures

Suppose that $(\beta_i)_{i \in \mathbf{N}} \in [0, +\infty]^{\mathbf{N}}$.

Let us recall definitions of ordinary and standard products of non-negative real numbers $(\beta_i)_{i \in \mathbf{N}}$.

Definition 3.1 We say that a number $\beta \in [0, +\infty]$ is an ordinary product of numbers $(\beta_j)_{j \in \mathbf{N}}$ if

$$\beta = \lim_{n \to \infty} \prod_{i=1}^n \beta_i.$$

An ordinary product of numbers $(\beta_j)_{j \in \mathbf{N}}$ is denoted by $(\mathbf{O}) \prod_{i \in \mathbf{N}} \beta_i$.

Definition 3.2 A standard product of the family of numbers $(\beta_i)_{i \in \mathbf{N}}$ is denoted by $(\mathbf{S}) \prod_{i \in \mathbf{N}} \beta_i$ and defined as follows:

$(\mathbf{S}) \prod_{i \in \mathbf{N}} \beta_i = 0$ if $\sum_{i \in \mathbf{N}^-} \ln(\beta_i) = -\infty$, where $\mathbf{N}^- = \{i : ln(\beta_i) < 0\}$ [1], and $(\mathbf{S}) \prod_{i \in \mathbf{N}} \beta_i = e^{\sum_{i \in \mathbf{N}^-} \ln(\beta_i)}$ if $\sum_{i \in \mathbf{N}^-} \ln(\beta_i) \neq -\infty$.

Let $(E_i, S_i)_{i \in \mathbf{N}}$ be a sequence of measure spaces. Let $(\nu_i)_{i \in \mathbf{N}}$ be a sequence of finite measures with $\mathrm{dom}(\nu_i) = S_i$ for $i \in \mathbf{N}$. We put $\mathcal{C}(\prod_{i \in \mathbf{N}} E_i) = \prod_{i \in \mathbf{N}} S_i$.

Definition 3.3 Suppose that

$$0 \leq (\mathbf{O}) \prod_{i \in \mathbf{N}} \nu_i(E_i) < +\infty.$$

For $X \in \mathcal{C}(\prod_{i \in \mathbf{N}} E_i)$, we set

$$((\mathbf{O}) \prod_{i \in \mathbf{N}} \nu_i)(X) = ((\mathbf{O}) \prod_{i \in \mathbf{N}} \nu_i(E_i))[\prod_{i \in \mathbf{N}} \frac{\nu_i}{\nu_i(E_i)}](X),$$

if

$$0 < (\mathbf{O}) \prod_{i \in \mathbf{N}} \nu_i(E_i) < +\infty,$$

and

$$((\mathbf{O}) \prod_{i \in \mathbf{N}} \nu_i)(X) = 0$$

if

$$(\mathbf{O}) \prod_{i \in \mathbf{N}} \nu_i(E_i) = 0.$$

A measure $(\mathbf{O}) \prod_{i \in \mathbf{N}} \nu_i$ is called an ordinary product of the family of finite measures $(\nu_i)_{i \in \mathbf{N}}$.

Definition 3.4 Suppose that

$$0 \leq (\mathbf{S}) \prod_{i \in \mathbf{N}} \nu_i(E_i) < +\infty.$$

For $X \in \mathcal{C}(\prod_{i \in \mathbf{N}} E_i)$, we set

$$((\mathbf{S}) \prod_{i \in \mathbf{N}} \nu_i)(X) = ((\mathbf{S}) \prod_{i \in \mathbf{N}} \nu_i(E_i))[\prod_{i \in \mathbf{N}} \frac{\nu_i}{\nu_i(E_i)}](X),$$

if

$$0 < (\mathbf{S}) \prod_{i \in \mathbf{N}} \nu_i(E_i) < +\infty,$$

and

$$((\mathbf{S}) \prod_{i \in \mathbf{N}} \nu_i)(X) = 0$$

---

[1] We set $ln(0) = -\infty$

if

$$(\mathbf{S}) \prod_{i \in \mathbf{N}} \nu_i(E_i) = 0.$$

A measure $(\mathbf{S}) \prod_{i \in \mathbf{N}} \nu_i$ is called a standard product of the family of finite measures $(\nu_i)_{i \in \mathbf{N}}$.

Remark 3.1 If $\nu_i(E_i) > 0$ for $i \in \mathbf{N}$, then by Anderson well known results there exists a product of probability measures $\prod_{i \in \mathbf{N}} \frac{\nu_i}{\nu_i(E_i)}$. Coefficient $(\mathbf{O}) \prod_{i \in \mathbf{N}} \nu_i(E_i)$ and $(\mathbf{S}) \prod_{i \in \mathbf{N}} \nu_i(E_i)$ make only some correction. If $\nu_i(E_i) = e^{\frac{(-1)^i}{i+1}}$ for $i \in \mathbf{N}$, then $(\mathbf{S}) \prod_{i \in \mathbf{N}} \nu_i$ is identically a zero measure, whenever $(\mathbf{O}) \prod_{i \in \mathbf{N}} \nu_i$ is a non-zero measure. This example shows us that notions of ordinary and standard products are different.

Now let $(\mu_i)_{i \in \mathbf{N}}$ be a sequence of non-zero σ-finite measures with $\mathrm{dom}(\mu_i) = \mathcal{S}_i$ for $i \in \mathbf{N}$.

We denote by $\mathcal{R}(\mathcal{O})$ the family of all measurable rectangles $R \subseteq \prod_{i \in \mathbf{N}} E_i$ of the form $\prod_{i \in \mathbf{N}} Y_i$ with the property

$$0 \leq (\mathbf{O}) \prod_{i \in \mathbf{N}} \mu_i(Y_i) < \infty.$$

Analogously, we denote by $\mathcal{R}(\mathcal{S})$ the family of all measurable rectangles $R \subseteq \prod_{i \in \mathbf{N}} E_i$ of the form $\prod_{i \in \mathbf{N}} Y_i$ with the property

$$0 \leq (\mathbf{S}) \prod_{i \in \mathbf{N}} \mu_i(Y_i) < \infty.$$

We set

$$(\mu_i|_{Y_i})(Y) = \mu_i(Y_i \cap Y)$$

for every $Y \in Y_i \cap \mathcal{S}_i$.

Let $\mathcal{C}(\prod_{i \in \mathbf{N}} E_i)$ denotes the product of σ-algebras $(S_i)_{i \in \mathbf{N}}$.

Case 1. Let $R = \prod_{i \in \mathbf{N}} Y_i \in \mathcal{R}(\mathcal{O})$.
For $X \in \mathcal{C}(\prod_{i \in \mathbf{N}} E_j) \cap R$ we set

$$\lambda_R^{\mathrm{ordinary}}(X) = (\mathbf{O}) \prod_{i \in \mathbf{N}} (\mu_i|_{Y_i}))(X \cap R),$$

if $0 < (\mathbf{O}) \prod_{i \in \mathbf{N}} \mu_i(Y_i) < +\infty$, and

$$\lambda_R^{\mathrm{ordinary}}(X) = 0,$$

otherwise.

Case 2. Let $R = \prod_{i \in \mathbf{N}} Y_i \in \mathcal{R}(\mathcal{S})$.
For $X \in \mathcal{C}(\prod_{i \in \mathbf{N}} E_j) \cap R$ we set

$$\lambda_R^{\mathrm{standard}}(X) = (\mathbf{S}) \prod_{i \in \mathbf{N}} (\mu_i|_{Y_i}))(X \cap R),$$

if $0 < (\mathbf{S}) \prod_{i \in \mathbf{N}} \mu_i(Y_i) < +\infty$, and

$$\lambda_R^{\text{standard}}(X) = 0,$$

otherwise.

Let $\mathcal{R}(\mathcal{O})^+$ denotes the family of all measurable rectangles $R \subseteq \prod_{i \in \mathbf{N}} E_i$ of the form $\prod_{i \in \mathbf{N}} Y_i$ with the property

$$0 < (\mathbf{O}) \prod_{i \in \mathbf{N}} \mu_i(Y_i) < \infty.$$

Let $\mathcal{R}(\mathcal{O})^0$ denotes the family of all measurable rectangles $R \subseteq \prod_{i \in \mathbf{N}} E_i$ of the form $\prod_{i \in \mathbf{N}} Y_i$ with the property

$$0 = (\mathbf{O}) \prod_{i \in \mathbf{N}} \mu_i(Y_i).$$

Let $\mathcal{R}(\mathcal{S})^+$ denotes the family of all measurable rectangles $R \subseteq \prod_{i \in \mathbf{N}} E_i$ of the form $\prod_{i \in \mathbf{N}} Y_i$ with the property

$$0 < (\mathbf{S}) \prod_{i \in \mathbf{N}} \mu_i(Y_i) < \infty.$$

Let $\mathcal{R}(\mathcal{S})^0$ denotes the family of all measurable rectangles $R \subseteq \prod_{i \in \mathbf{N}} E_i$ of the form $\prod_{i \in \mathbf{N}} Y_i$ with the property

$$0 = (\mathbf{S}) \prod_{i \in \mathbf{N}} \mu_i(Y_i).$$

For $R \in \mathcal{R}(\mathcal{O})$, let us denote by $\overline{\lambda}_R^{\text{ordinary}}$ the completion of the $\lambda_R^{\text{ordinary}}$. We set

$$S^{\text{ordinary}} = \cap_{R \in \mathcal{R}(\mathcal{O})} \left(\text{dom} \overline{\lambda}_R^{\text{ordinary}} \cup \mathcal{P}(\prod_{i \in \mathbf{N}} E_i \setminus R) \right).$$

Analogously, for $R \in \mathcal{R}(\mathcal{S})$, let us denote by $\overline{\lambda}_R^{\text{standard}}$ the completion of the $\lambda_R^{\text{standard}}$. We set

$$S^{\text{standard}} = \cap_{R \in \mathcal{R}(\mathcal{O})} \left(\text{dom} \overline{\lambda}_R^{\text{standard}} \cup \mathcal{P}(\prod_{i \in \mathbf{N}} E_i \setminus R) \right).$$

Definition 3.5 *The measure μ is said to be a strict ordinary product of the family of measures $(\mu_i)_{i \in \mathbf{N}}$ if the following three conditions hold:*
(i) for every $R = \prod_{i \in \mathbf{N}} Y_i \in \mathcal{R}(\mathcal{O})$ we have

$$\mu(\prod_{i \in \mathbf{N}} Y_i) = (\mathbf{O}) \prod_{i \in \mathbf{N}} \mu_i(Y_i);$$

(ii)

$$(\forall X)(X \in dom(\mu) \to ((\mu(X) = 0) \Leftrightarrow (\forall R)(R \in \mathcal{R}(\mathcal{O})^+ \to \mu_R(R \cap X) = 0)));$$

(iii) $S^{\text{ordinary}} \subseteq dom(\mu)$.

Definition 3.6 *The measure μ is said to be a strict standard product of the family of measures $(\mu_i)_{i \in \mathbf{N}}$ if the following three conditions hold:*

(i) for every $R = \prod_{i \in \mathbf{N}} Y_i \in \mathcal{R}(\mathcal{S})$ we have

$$\mu(\prod_{i \in \mathbf{N}} Y_i) = (\mathbf{S}) \prod_{i \in \mathbf{N}} \mu_i(Y_i);$$

(ii)

$$(\forall X)(X \in dom(\mu) \to ((\mu(X) = 0) \Leftrightarrow (\forall R)(R \in \mathcal{R}(\mathcal{S})^+ \to \mu_R(R \cap X) = 0)));$$

(iii) $\mathcal{S}^{standard} \subseteq dom(\mu)$.

Lemma 3.1 *The family of measures $(\overline{\lambda}_R^{ordinary})_{R \in \mathcal{R}(\mathcal{O})}$, defined on the σ-algebra $\mathcal{S}^{ordinary}$ is consistent, i.e., for every $R_1, R_2 \in \mathcal{R}(\mathcal{O})$ the following equality*

$$(\forall X)(X \in \mathcal{S}^{ordinary} \to \overline{\lambda}_{R_1}^{ordinary}(X \cap R_1 \cap R_2) = \overline{\lambda}_{R_2}^{ordinary}(X \cap R_1 \cap R_2))$$

holds.

Proof. Let $R_1 = \prod_{i \in \mathbf{N}} R_i^{(1)}$ and $R_2 = \prod_{i \in \mathbf{N}} R_i^{(2)}$ be two elements of the class $\mathcal{R}(\mathcal{O})$.

Without loss of generality it can be assumed that $0 < (\mathbf{O}) \prod_{i \in \mathbf{N}} m_i(R_i^{(1)}) < \infty$ and $0 < (\mathbf{O}) \prod_{i \in \mathbf{N}} m_i(R_i^{(2)}) < \infty$.

Firstly, let us show that $\overline{\lambda}_{R_1}^{ordinary}(X) = \overline{\lambda}_{R_2}^{ordinary}(X)$ for every $X \in \mathcal{C}(R_1 \cap R_2)$. In this case it is sufficient to show that $\lambda_{R_1}^{ordinary}(Y) = \lambda_{R_2}^{ordinary}(Y)$ for every elementary measurable rectangle $Y = \prod_{i \in \mathbf{N}} Y_i$ in $R_1 \cap R_2$. Note here that under an elementary measurable rectangle $Y = \prod_{i \in \mathbf{N}} Y_i$ in $R_1 \cap R_2$ we assume a subset of $R_1 \cap R_2$ such that $Y_i \in \mathcal{C}(R_i^{(1)} \cap R_i^{(2)})$ for every $i \in \mathbf{N}$ and, in addition, there exists a finite subset $I_0 \subset \mathbf{N}$ such that $Y_i = R_i^{(1)} \cap R_i^{(2)}$ for $i \in I \setminus I_0$.

For $i \in \mathbf{N}$ and $Y_i \in \mathcal{C}(R_i^{(1)} \cap R_i^{(2)})$, we have

$$m_i(Y_i \cap R_i^{(1)} \cap R_i^{(2)}) = m_i(Y_i \cap R_i^{(1)}) = m_i(Y_i \cap R_i^{(2)}).$$

The latter relation implies that

$$(\mathbf{O}) \prod_{i \in \mathbf{N}} m_i(Y_i \cap R_i^{(1)} \cap R_i^{(2)}) = (\mathbf{O}) \prod_{i \in \mathbf{N}} m_i(Y_i \cap R_i^{(1)}) =$$

$$(\mathbf{O}) \prod_{i \in \mathbf{N}} m_i(Y_i \cap R_i^{(1)}).$$

Analogously, we have

$$(\mathbf{O}) \prod_{i \in \mathbf{N}} m_i(Y_i \cap R_i^{(1)} \cap R_i^{(2)}) = (\mathbf{O}) \prod_{i \in \mathbf{N}} m_i(Y_i \cap R_i^{(2)}) =$$

$$(\mathbf{O}) \prod_{i \in \mathbf{N}} m_i(Y_i \cap R_i^{(2)}).$$

Thus, we get

$$\lambda_{R_1}^{\text{ordinary}}(\prod_{i\in\mathbf{N}} Y_i) = (\mathbf{O})\prod_{i\in\mathbf{N}} m_i(Y_i \cap R_i^{(1)}) = (\mathbf{O})\prod_{i\in\mathbf{N}} m_i(Y_i \cap R_i^{(1)} \cap R_i^{(1)}) =$$

$$(\mathbf{O})\prod_{i\in\mathbf{N}} m_i(Y_i \cap R_i^{(2)}) = \lambda_{R_2}^{\text{ordinary}}(\prod_{i\in I} Y_i).$$

Since a class $\mathcal{A}(R_1 \cap R_2)$ of all finite disjoint unions of elementary measurable rectangles in $R_1 \cap R_2$ is a ring, and since, by definition, the class $\mathcal{C}(R_1 \cap R_2)$ of cylindrical sets of $R_1 \cap R_2$ is a minimal σ-ring generated by the ring $\mathcal{A}(R_1 \cap R_2)$, we claim (cf. [6], Theorem B, p. 27) that the class of all those sets of $R_1 \cap R_2$ for which this equality holds coincides with the class $\mathcal{C}(R_1 \cap R_2)$.

Hence, for $X \in \mathcal{C}(R_1 \cap R_2)$ we have

$$\lambda_{R_1}^{\text{ordinary}}(X) = \lambda_{R_2}^{\text{ordinary}}(X) = \lambda_{R_1 \cap R_2}^{\text{ordinary}}(X).$$

Now let $X \in S^{\text{ordinary}}$. Then $R_1 \cap R_2 \cap X$ also belongs to the S^{ordinary} because S^{ordinary} is a σ-algebra and it contains R_1, R_2, X, respectively.

Also,

$$R_1 \cap R_2 \cap X \in \text{dom}(\overline{\lambda}_{R_1}^{\text{ordinary}}) \cap \text{dom}(\overline{\lambda}_{R_2}^{\text{ordinary}}) \cap \text{dom}(\overline{\lambda}_{R_1 \cap R_2}^{\text{ordinary}}).$$

By using properties of completions of $\lambda_{R_1}^{\text{ordinary}}, \lambda_{R_2}^{\text{ordinary}}$ and $\lambda_{R_1 \cap R_2}^{\text{ordinary}}$, respectively, we claim that there exists cylindrical subsets B_1 and B_2 such that

(i) $\overline{\lambda}_{R_1}^{\text{ordinary}}(B_1 \Delta (R_1 \cap R_2 \cap X)) = 0,$

(ii) $\overline{\lambda}_{R_2}^{\text{ordinary}}(B_2 \Delta (R_1 \cap R_2 \cap X)) = 0,$

It is clear

(iii) $\overline{\lambda}_{R_1}^{\text{ordinary}}(B_1 \setminus (R_1 \cap R_2)) = 0,$

(iv) $\overline{\lambda}_{R_2}^{\text{ordinary}}(B_2 \setminus (R_1 \cap R_2)) = 0.$

Hence, without loss of generality we can assume that B_1 and B_2 are cylindrical subsets of the $R_1 \cap R_2$.

We have

$$R_1 \cap R_2 \cap X = (B_1 \setminus X_1) \cup X_2$$

and

$$R_1 \cap R_2 \cap X = (B_2 \setminus Y_1) \cup Y_2,$$

where

$$X_1 \subseteq Z_1 \subseteq R_1 \cap R_2, X_2 \subseteq Z_2 \subseteq R_1 \cap R_2,$$

$$\lambda_{R_1}^{\text{ordinary}}(Z_1) = \lambda_{R_1}^{\text{ordinary}}(Z_2) = 0$$

and

$$Y_1 \subseteq W_1 \subseteq R_1 \cap R_2, Y_2 \subseteq W_2 \subseteq R_1 \cap R_2, \lambda_{R_2}^{\text{ordinary}}(W_1) = \lambda_{R_1}^{\text{ordinary}}(W_2) = 0.$$

Assume the contrary and let

$$\overline{\lambda}_{R_1}^{\text{ordinary}}(B_1 \setminus B_2) > 0.$$

This means that

$$\emptyset = (R_1 \cap R_2 \cap X) \setminus (R_1 \cap R_2 \cap X) = ((B_1 \setminus X_1) \cup X_2) \setminus ((B_1 \setminus X_1) \cup X_2) \neq \emptyset$$

because

$$\overline{\lambda}_{R_1}^{\text{ordinary}}((B_1 \setminus X_1) \cup X_2) \setminus ((B_1 \setminus X_1) \cup X_2) = \overline{\lambda}_{R_1}^{\text{ordinary}}(B_1 \setminus B_2) > 0.$$

The consistency of the family of measures $(\overline{\lambda}_R^{\text{ordinary}})_{R \in \mathcal{R}(\mathcal{O})}$ is proved.

$\square$

By scheme used in the proof of Lemma 3.1, we can prove

Lemma 3.2 *The family of measures* $(\overline{\lambda}_R^{\text{standard}})_{R \in \mathcal{R}(\mathcal{S})}$, *defined on the σ-algebra* S^{standard} *is consistent, i.e., for every $R_1, R_2 \in \mathcal{R}(\mathcal{S})$ the following equality*

$$(\forall X)(X \in S^{\text{standard}} \to \overline{\lambda}_{R_1}^{\text{standard}}(X \cap R_1 \cap R_2) = \overline{\lambda}_{R_2}^{\text{standard}}(X \cap R_1 \cap R_2))$$

holds.

Theorem 3.1 *For $i \in \mathbf{N}$, let μ_i be a measure defined on a measure space (E_i, S_i). Assume, that there exists $R_0 = \prod_{i \in \mathbf{N}} Z_i \in \mathcal{R}(\mathcal{O})$ with*

$$0 < (\mathbf{O}) \prod_{i \in \mathbf{N}} \mu_i(Z_i) < +\infty.$$

Then there exists a complete measure $\overline{\nu}^{\text{ordinary}}$ on S^{ordinary} such that :
   (a) for every $R = \prod_{i \in \mathbf{N}} Y_i \in \mathcal{R}(\mathcal{O})$ we have

$$\overline{\nu}^{\text{ordinary}}\left(\prod_{i \in \mathbf{N}} Y_i\right) = (\mathbf{O}) \prod_{i \in \mathbf{N}} \mu_i(Y_i);$$

   (b) if each μ_i is a probability measure on E_i, then $\overline{\nu}^{\text{ordinary}}$ coincides with the completion of the $\prod_{i \in \mathbf{N}} \mu_i$;
   (c) if cardinality of indices i, for which E_i contains disjoint measurable subsets A_i, B_i such that $\mu_i(A_i) = \mu_i(B_i) = 1$, is not finite then the measure $\overline{\nu}^{\text{ordinary}}$ is not σ-finite;
   (d) the domain of the $\overline{\nu}^{\text{ordinary}}$ contains the class of all universally measurable subsets of the $\prod_{i \in \mathbf{N}} E_i$.

Proof. By Lemma 3.1 the family of measures $(\overline{\lambda}_R^{ordinary})_{R\in\mathcal{R}(\mathcal{O})}$ is consistent. Application of Lemma 2.1 yields that the measure $\overline{\nu}^{ordinary} = \mu_{\mathcal{R}(\mathcal{O})}$ satisfies conditions (a) and (c).

Now, let (E_i, S_i, μ_i) be a probability space for $i \in \mathbf{N}$. Then $\prod_{i\in\mathbf{N}} E_i$ belongs to the class $\mathcal{R}(\mathcal{O})$ and we have $\overline{\nu}^{ordinary}(A) = \mu_{\mathcal{R}(\mathcal{O})}(A) = \prod_{i\in\mathbf{N}} \mu_i(\prod_{i\in\mathbf{N}} A_i)$ for every elementary rectangular set $\prod_{i\in\mathbf{N}} A_i$ in $\prod_{i\in\mathbf{N}} E_i$. By Charatheodory theorem and the latter relation we claim that the measures $\overline{\nu}^{ordinary}$ and $\prod_{i\in\mathbf{N}} \mu_i$ coincide on the cylindrical σ-algebra $\mathcal{C}(\prod_{i\in\mathbf{N}} E_i)$. Since $\overline{\lambda}_{\prod_{i\in\mathbf{N}} E_i}^{ordinary}$ coincides with the completion of $\prod_{i\in\mathbf{N}} \mu_i$ and same times, with the $\overline{\lambda}^{ordinary}$, we claim that $\overline{\lambda}^{ordinary} = \overline{\prod_{i\in\mathbf{N}} \mu_i}$. This ends the proof of the validity of the Item (b).

For $R \in \mathcal{R}(\mathcal{O})^+$, we denote by μ_R a probability cylindrical measure defined by

$$(\forall X)(X \in \mathcal{C}(\prod_{i\in\mathbf{N}} E_i)) \to \mu_R(X) = \frac{\lambda^{ordinary}(X)}{\lambda^{ordinary}(R)}.$$

It is obvious that, for $R \in \mathcal{R}(\mathcal{O})^0$, the following equality

$$\mathrm{dom}(\overline{\lambda_R^{ordinary}}) \cup \mathcal{P}(\prod_{i\in\mathbf{N}} E_i \setminus R) = \mathcal{P}(\prod_{i\in\mathbf{N}} E_i)$$

holds.

It is clear that, for $R \in \mathcal{R}(\mathcal{O})^+$ the following equality

$$\mathrm{dom}(\overline{\mu}_R) = \mathrm{dom}(\overline{\lambda_R^{ordinary}}) \cup \mathcal{P}(\prod_{i\in\mathbf{N}} E_i \setminus R)$$

holds. Hence, we have

$$U(\prod_{i\in\mathbf{N}} E_i, \prod_{i\in\mathbf{N}} S_i) = \cap_{\mu\in K}(\prod_{i\in\mathbf{N}} S_i)^{\mu} \subseteq \cap_{R\in\mathcal{R}(\mathcal{O})}(\mathrm{dom}(\overline{\mu}_R) \cup \mathcal{P}(\prod_{i\in\mathbf{N}} E_i \setminus R)) =$$

$$\cap_{R\in\mathcal{R}(\mathcal{O})}(\mathrm{dom}(\overline{\lambda_R^{ordinary}}) \cup \mathcal{P}(\prod_{i\in\mathbf{N}} E_i \setminus R)) = S^{ordinary}.$$

This ends the proof of the validity of the Item (d).

$\square$

Definition 3.7 The measure $\overline{\nu}^{ordinary}$ is called a strict ordinary product of the family of measures $(\mu_i)_{i\in\mathbf{N}}$ and is denoted by $(sO)\prod_{i\in\mathbf{N}} \mu_i$.

Remark 3.1 In [21] has been constructed an ordinary product $(O)\prod_{i\in N} \mu_i$ of measures $(\mu_i)_{i\in\mathbf{N}}$. We do not know whether the following equality

$$\mathrm{dom}(\overline{(O)\prod_{i\in N} \mu_i}) = \mathrm{dom}((sO)\prod_{i\in N} \mu_i)$$

holds. In other words, is the completion of the ordinary product of measures a strict ordinary product?

Theorem 3.2 *Let* $D \in U(\prod_{i \in \mathbf{N}} E_i, \prod_{i \in \mathbf{N}} S_i)$. *For every* $(a_{n+1}, a_{n+2}, \cdots) \in R^{N \setminus \{1, \cdots, n\}}$, *we denote by* $D_{(a_{n+1}, a_{n+2}, \cdots)}$ $(a_{n+1}, a_{n+2}, \cdots)$-*section of the* D, *i.e.,*

$$D_{(a_{n+1}, a_{n+2}, \cdots)} = \{(x_1, \cdots, x_n) \in R^{\{1, \cdots, n\}} : (x_1, \cdots, x_n, a_{n+1}, a_{n+2}, \cdots) \in D\}.$$

Analogously, for $(a_1, a_2, \cdots, a_n) \in R^{\{1, \cdots, n\}}$, *we denote by* $D_{(a_1, a_2, \cdots, a_n)}$ $((a_1, a_2, \cdots, a_n))$-*section of the* D, *i.e.,*

$$D_{(a_1, a_2, \cdots, a_n)} = \{(x_{n+1}, x_{n+2}, \cdots) \in R^{N \setminus \{1, \cdots, n\}} : (a_1, \cdots, a_n, x_{n+1}, x_{n+2}, \cdots) \in D\}.$$

If D *is covered by countable family of elements* $(R_k)_{k \in N}$ *of the class* $\mathcal{R}(\mathcal{O})$, *then the following formula*

$$\overline{\nu}^{ordinary}(D) = \int_{R^{N \setminus \{1, \cdots, n\}}} (\overline{\nu}^{ordinary}_{[1, \cdots, n]}(D_{(a_{n+1}, a_{n+2}, \cdots)})) d\overline{\nu}^{ordinary}_{[n+1, \cdots[}(a_{n+1}, a_{n+2}, \cdots) =$$

$$\int_{R^{\{1, \cdots, n\}}} (\overline{\nu}^{ordinary}_{[n+1, \cdots[}(D_{(a_1, a_2, \cdots, a_n)})) d\overline{\nu}^{ordinary}_{[1, \cdots, n]}(a_1, a_2, \cdots, a_n)$$

is valid, where $\overline{\nu}^{ordinary}_{[1, \cdots, n]}$ *denotes an ordinary complete product of measures* $(\mu_i)_{1 \leq i \leq n}$ *and* $\overline{\nu}^{ordinary}_{[n+1, \cdots]}$ *denotes an ordinary complete product of measures* $(\mu_i)_{i > n}$.

Proof. D is covered by countable family of elements $(R_k)_{k \in N}$ of the class $\mathcal{R}(\mathcal{O})$. We have $R_k = \prod_{i \in N} Y_i^{(k)}$ for $k \in N$.

We set $A_0 = \prod_{i > n} (Y_i^{(0)})$. If we have constructed a family $(A_s)_{0 \leq s \leq m-1}$, then we set

$$A_m = \prod_{i > m} (Y_i^{(m)}) \setminus \cup_{0 \leq s \leq m-1} A_s$$

Thus a disjoint family $(A_s)_{s \in N}$ is constructed. We set $A_{-1} = \prod_{i > n} E_i \setminus \cup_{s \in N} A_s$. We have

$$D = \left(\sum_{s \in N} \prod_{0 \leq i \leq n} E_i \times A_s \right) \cap D = \sum_{s \in N \cup \{-1\}} \left(\prod_{0 \leq i \leq n} E_i \times A_s \cap D \right) = \sum_{s \in N \cup \{-1\}} D^{(s)},$$

where $D^{(s)} = \left(\prod_{0 \leq i \leq n} E_i \times A_s \cap D \right)$ for $s \in N \cup \{-1\}$.

Note, that for $s \in N \cup \{-1\}$, we have

$$\overline{\nu}^{ordinary}(D^{(s)}) = \int_{R^{N \setminus \{1, \cdots, n\}}} (\overline{\nu}^{ordinary}_{[1, \cdots, n]}(D^{(s)}_{(a_{n+1}, a_{n+2}, \cdots)})) d\overline{\nu}^{ordinary}_{[n+1, \cdots[}(a_{n+1}, a_{n+2}, \cdots)$$

Since $\overline{\nu}^{ordinary}(D) < +\infty$, we have

$$\overline{\nu}^{ordinary}(D) = \sum_{s \in N \cup \{-1\}} \overline{\nu}^{ordinary}(D^{(s)}) =$$

$$\sum_{s \in N \cup \{-1\}} \int_{R^{N \setminus \{1, \cdots, n\}}} (\overline{\nu}^{ordinary}_{[1, \cdots, n]}(D^{(s)}_{(a_{n+1}, a_{n+2}, \cdots)})) d\overline{\nu}^{ordinary}_{[n+1, \cdots[}(a_{n+1}, a_{n+2}, \cdots) =$$

$$\int_{R^{N\setminus\{1,\cdots,n\}}} \left(\sum_{s\in N\cup\{-1\}} \overline{\nu}_{[1,\cdots,n]}^{ordinary}(D_{(a_{n+1},a_{n+2},\cdots)}^{(s)}) \right) d\overline{\nu}_{[n+1,\cdots[}^{ordinary}(a_{n+1},a_{n+2},\cdots) =$$

$$\int_{R^{N\setminus\{1,\cdots,n\}}} (\overline{\nu}_{[1,\cdots,n]}^{ordinary}(D_{(a_{n+1},a_{n+2},\cdots)})) d\overline{\nu}_{[n+1,\cdots[}^{ordinary}(a_{n+1},a_{n+2},\cdots).$$

On the other hand we have

$$\overline{\nu}^{ordinary}(D) = \sum_{s\in N\cup\{-1\}} \overline{\nu}^{ordinary}(D^{(s)}) =$$

$$\sum_{s\in N\cup\{-1\}} \int_{R^{\{1,\cdots,n\}}} (\overline{\nu}_{[n+1,\cdots[}^{ordinary}(D_{(a_1,a_2,\cdots,a_n)}^{(s)})) d\overline{\nu}_{[1,\cdots,n]}^{ordinary}(a_1,a_2,\cdots,a_n) =$$

$$\int_{R^{\{1,\cdots,n\}}} \left(\sum_{s\in N\cup\{-1\}} \overline{\nu}_{[n+1,\cdots[}^{ordinary}(D_{(a_1,a_2,\cdots,a_n)}^{(s)}) \right) d\overline{\nu}_{[1,\cdots,n]}^{ordinary}(a_1,a_2,\cdots,a_n) =$$

$$\int_{R^{\{1,\cdots,n\}}} (\overline{\nu}_{[n+1,\cdots[}^{ordinary}(D_{(a_1,a_2,\cdots,a_n)})) d\overline{\nu}_{[1,\cdots,n]}^{ordinary}(a_1,a_2,\cdots,a_n).$$

This ends the proof of Theorem 2.3.

$\square$

Corollary 3.1(A construction of the Loev-Ross strict ordinary measure) *Let* $(X_i; \mathcal{B}_i; \mu_i)(i \in \mathbf{N})$ *be a sequence of regular Borel measure spaces, where X_i is a Hausdorff topological space. Let $\mathcal{K}(\mathcal{O})$ be a class of rectangles of the form $\prod_{i\in\mathbf{N}} K_i \in \mathcal{R}(\mathcal{O})$, such that K_i is compact in X_i for $i \in \mathbf{N}$. Then, by Lemma 3.1, the family of measures $(\lambda_R^{ordinary})_{R\in\mathcal{K}(\mathcal{O})}$ is consistent as a subfamily of the consistent family of measures $(\lambda_R^{ordinary})_{R\in\mathcal{R}(\mathcal{O})}$. Application of Lemma 2.1 yields that the measure $\mu = \mu_{\mathcal{K}(\mathcal{O})}$ satisfies the following conditions:*

(a) for every $R = \prod_{i\in\mathbf{N}} Y_i \in \mathcal{K}(\mathcal{O})$ we have

$$\mu(\prod_{i\in\mathbf{N}} Y_i) = (\mathbf{O}) \prod_{i\in\mathbf{N}} \mu_i(Y_i);$$

(b) if each μ_i is a probability measure on X_i, then $\overline{\nu}^{ordinary}$ coincides with the completion of the $\prod_{i\in\mathbf{N}} \mu_i$;

(c) if cardinality of indices i, for which X_i contains disjoint measurable subsets A_i, B_i such that $\mu_i(A_i) = \mu_i(B_i) = 1$, is not finite then the measure $\overline{\nu}^{ordinary}$ is not σ-finite;

(d) the domain of the $\overline{\nu}^{ordinary}$ contains the class of all universally measurable subsets of the $\prod_{i\in\mathbf{N}} X_i$.

Corollary 3.2(A construction of the Baker's strict ordinary measure) *Let $(X_i; \mathcal{B}_i; \mu_i)(i \in \mathbf{N})$ be a sequence of regular Borel measure spaces, where X_i is a Hausdorff topological space. Then, by Lemma 3.1, the family of measures $(\lambda_R^{ordinary})_{R\in\mathcal{R}(\mathcal{O})}$ is consistent. Application of Lemma 2.1 yields that the measure $\lambda = \mu_{\mathcal{R}(\mathcal{O})}$ satisfies the following conditions:*

(a) for every $R = \prod_{i \in \mathbf{N}} Y_i \in \mathcal{R}(\mathcal{O})$ we have

$$\lambda(\prod_{i \in \mathbf{N}} Y_i) = (\mathbf{O}) \prod_{i \in \mathbf{N}} \mu_i(Y_i);$$

(b) if each μ_i is a probability measure on X_i, then $\overline{\nu}^{ordinary}$ coincides with the completion of the $\prod_{i \in \mathbf{N}} \mu_i$;

(c) if cardinality of indices i, for which X_i contains disjoint measurable subsets A_i, B_i such that $\mu_i(A_i) = \mu_i(B_i) = 1$, is not finite then the measure $\overline{\nu}^{ordinary}$ is not σ-finite;

(d) the domain of the $\overline{\nu}^{ordinary}$ contains the class of all universally measurable subsets of the $\prod_{i \in \mathbf{N}} X_i$.

By the scheme used in the proof of Theorem 3.1, one can prove the validity of the following assertion.

Theorem 3.2 *For $i \in \mathbf{N}$, let μ_i be a measure defined on a measure space (E_i, S_i). Assume, that there exists $R_0 = \prod_{i \in \mathbf{N}} Z_i \in \mathcal{R}(\mathcal{S})$ with*

$$0 < (\mathbf{S}) \prod_{i \in \mathbf{N}} \mu_i(Z_i) < +\infty.$$

Then there exists a measure $\overline{\nu}^{standard}$ on $\mathcal{C}(\prod_{i \in \mathbf{N}} E_i)$ such that :
(a) for every $R = \prod_{i \in \mathbf{N}} Y_i \in \mathcal{R}(\mathcal{S})$ we have

$$\overline{\nu}^{standard}(\prod_{i \in \mathbf{N}} Y_i) = (\mathbf{S}) \prod_{i \in \mathbf{N}} \mu_i(Y_i);$$

(b) if each μ_i is a probability measure on X_i, then $\overline{\nu}^{standard}$ coincides with the completion of the $\prod_{i \in \mathbf{N}} \mu_i$;

(c) if cardinality of indices i, for which X_i contains disjoint measurable subsets A_i, B_i such that $\mu_i(A_i) = \mu_i(B_i) = 1$, is not finite then the measure $\overline{\nu}^{standard}$ is not σ-finite;

(d) the domain of the $\overline{\nu}^{standard}$ contains the class of all universally measurable subsets of the $\prod_{i \in \mathbf{N}} X_i$.

Definition 3.8 The measure $\overline{\nu}^{standard}$ is called a strict standard product of the family of measures $(\mu_i)_{i \in \mathbf{N}}$ and is denoted by $(sS) \prod_{i \in \mathbf{N}} \mu_i$.

Corollary 3.3(A construction of the Loev-Ross strict standard measure) *Let $(X_i; \mathcal{B}_i; \mu_i)(i \in \mathbf{N})$ be a sequence of regular Borel measure spaces, where X_i is a Hausdorff topological space. Let $\mathcal{K}(\mathcal{S})$ be a class of rectangles of the form $\prod_{i \in \mathbf{N}} K_i \in \mathcal{R}(\mathcal{S})$, such that K_i is compact in X_i for $i \in \mathbf{N}$. Then, by Lemma 3.2, the family of measures $(\lambda_R^{standard})_{R \in \mathcal{K}(\mathcal{S})}$ is consistent as a subfamily of the consistent family of measures $(\lambda_R^{standard})_{R \in \mathcal{R}(\mathcal{S})}$. Application of Lemma 2.1 yields that the measure $\mu = \mu_{\mathcal{K}(\mathcal{S})}$ satisfies the following conditions:*
(a) for every $R = \prod_{i \in \mathbf{N}} Y_i \in \mathcal{K}(\mathcal{S})$ we have

$$\mu(\prod_{i \in \mathbf{N}} Y_i) = (\mathbf{S}) \prod_{i \in \mathbf{N}} \mu_i(Y_i);$$

(b) *if each μ_i is a probability measure on E_i, then μ coincides with the completion of the $\prod_{i \in \mathbf{N}} \mu_i$;*

(c) *if cardinality of indices i, for which E_i contains disjoint measurable subsets A_i, B_i such that $\mu_i(A_i) = \mu_i(B_i) = 1$, is not finite then the measure μ is not σ-finite;*

(d) the domain of the μ contains the class of all universally measurable subsets of the $\prod_{i \in \mathbf{N}} X_i$.

Corollary 3.4(A construction of the Baker's strict standard measure) *Let $(X_i; \mathcal{B}_i; \mu_i)(i \in \mathbf{N})$ be a sequence of regular Borel measure spaces, where X_i is a Hausdorff topological space. Then, by Lemma 3.2, the family of measures $(\lambda_R^{standard})_{R \in \mathcal{R}(S)}$ is consistent. Application of Lemma 2.1 yields that the measure $\lambda = \mu_{\mathcal{R}(S)}$ satisfies the following conditions:*

(a) *for every $R = \prod_{i \in \mathbf{N}} Y_i \in \mathcal{R}(S)$ we have*

$$\lambda(\prod_{i \in \mathbf{N}} Y_i) = (\mathbf{S}) \prod_{i \in \mathbf{N}} \mu_i(Y_i);$$

(b) *if each μ_i is a probability measure on E_i, then λ coincides with the completion of the $\prod_{i \in \mathbf{N}} \mu_i$;*

(c) *if cardinality of indices i, for which E_i contains disjoint measurable subsets A_i, B_i such that $\mu_i(A_i) = \mu_i(B_i) = 1$, is not finite then the measure λ is not σ-finite;*

(d) the domain of the λ contains the class of all universally measurable subsets of the $\prod_{i \in \mathbf{N}} X_i$.

The next statements are immediate consequences of Theorems 3.1–3.2.

Theorem 3.3. *Let G_i be a group of measurable transformations of E_i for $i \in \mathbf{N}$. Under assumptions of Theorem 3.1, if each measure μ_i is G_i-invariant for $i \in \mathbf{N}$, then the measure $(\mathbf{sO}) \prod_{i \in \mathbf{N}} \mu_i$ is $\prod_{i \in \mathbf{N}} G_i$-invariant.*

Proof. We set $G = \prod_{i \in \mathbf{N}} G_i$. Let us show that the measure $(\mathbf{sO}) \prod_{i \in \mathbf{N}} \mu_i$ is G-invariant. This is a simple consequence of Lemma 2.1 (see, condition (iii)).

$\square$

By the scheme of the proof of Theorem 3.3 we can prove the following

Theorem 3.4. *Let G_i be a group of measurable transformations of E_i for $i \in \mathbf{N}$. Under assumptions of Theorem 3.2, if each measure μ_i is G_i-invariant for $i \in \mathbf{N}$, then the measure $(\mathbf{sS}) \prod_{i \in \mathbf{N}} \mu_i$ is $\prod_{i \in \mathbf{N}} G_i$-invariant.*

Remark 3.2 For $i \in \mathbf{N}$, we set $E_i = \mathbf{R}$ and $\mu_i = l_i$, where l_i denotes a linear Lebesgue measure on $\mathbf{R}$.

The measure $(\mathbf{sO}) \prod_{i \in \mathbf{N}} l_i$ is called a "strict ordinary Lebesgue measure" on $\mathbf{R}^\infty$.

The measure $(\mathbf{sS}) \prod_{i \in \mathbf{N}} l_i$ is called a "strict standard Lebesgue measure" on $\mathbf{R}^\infty$.

Note that strict ordinary and strict standard Lebesgue measures on $\mathbf{R}^\infty$, like ordinary and standard Lebesgue measures on the same space, are different.

4.   On a Strict Standard Product of an Arbitrary Family of σ-finite Borel Measures with Domain in Polish Spaces

The purpose of the present section is consider a new modification of a standard product of an arbitrary family of σ-finite Borel measures with domain in Polish spaces constructed [21].

Suppose that X is a topological space. The Borel sets $\mathcal{B}(X)$ are the σ-algebra generated by the open sets of a topological space X, and the Baire sets $\mathcal{B}_0(X)$ are the smallest σ-algebra making all real-valued continuous functions measurable. In 1957 (see [14]), Mařík proved that all normal, countably paracompact spaces have the following property: Every Baire measure extends to a regular Borel measure. Spaces which have this property have come to be known as Mařík spaces.

We recall some important notions and well-known results from general topology and probability theory.

X is a Hausdorff space iff distinct points in X have disjoint neighbourhoods. X is a regular space if and only if, given any closed set F and any point x that does not belong to F, there exists a neighbourhood U of x and a neighbourhood V of F that are disjoint. X is a normal space if and only if, given any disjoint closed sets E and F, there are neighbourhoods U of E and V of F that are also disjoint. X is a regular Hausdorff space if and only if it is both regular and Hausdorff. X is a completely regular space if and only if, given any closed set F and any point x that does not belong to F, there is a continuous function f from X to the real line $\mathbf{R}$ such that $f(x)$ is 0 and $f(y)$ is 1 for every y in F. X is a Tikhonov space, if and only if it is both completely regular and Hausdorff.

Recall that a Borel measure μ, defined on a Hausdorff topological space (X, τ), is called Radon if

$$(\forall Y)(Y \in \mathcal{B}(X) \ \& \ 0 \leq \mu(Y) < +\infty \to \mu(Y) = \sup_{\substack{K \subseteq Y \\ K \text{ is compact in } X}} \mu(K)) \ \Diamond$$

and called dense, if the condition $\Diamond$ holds for $Y = X$.

A family $(U_i)_{i \in I}$ of open subsets in (X, τ) is called a generalized sequence if

$$(\forall i_1)(\forall i_2)(i_1 \in I \ \& \ i_2 \in I \to (\exists i_3)(i_3 \in I \to (U_{i_1} \subset U_{i_3} \ \& \ U_{i_2} \subset U_{i_3}))).$$

A Borel probability measure μ defined on X is called τ-smooth if, for an arbitrary generalized sequence $(U_i)_{i \in I}$, the condition

$$\mu(\bigcup_{i \in I} U_i) = \sup_{i \in I} \mu(U_i)$$

is valid.

A Baire probability measure μ on X is called τ_0-smooth if, for an arbitrary generalized sequence $(U_i)_{i \in I}$ of open Baire subsets in X, for which $\bigcup_{i \in I} U_i$ is also a Baire subset, the condition

$$\mu(\bigcup_{i \in I} U_i) = \sup_{i \in I} \mu(U_i)$$

is valid .

Lemma 4.1 ([21], Lemma 4) *Let $(E_j, \rho_j)_{j \in \alpha}$ be a family of non-empty Polish metric spaces that at most $\aleph_0$ of them are noncompact and let μ_j be a Borel probability measure on E_j for $j \in \alpha$. Then the product measure $\prod_{j \in \alpha} \mu_j$ is τ_0-smooth and dense on $\prod_{j \in \alpha} E_j$.*

Lemma 4.2 ([21], Lemma 4) *Let $(E_j, \rho_j)_{j \in \alpha}$ be a family of non-empty Polish metric spaces that at most $\aleph_0$ of them are noncompact and let μ_j be a Borel probability measure on E_j for $j \in \alpha$. Then there exists a unique τ-smooth and Radon extension of the Baire measure $\prod_{j \in \alpha} \mu_j$ from the σ-algebra $\mathcal{B}_0(\prod_{j \in \alpha} E_j)$ to the σ-algebra $\mathcal{B}(\prod_{j \in \alpha} E_j)$.*

Let us recall a definition of the standard product of non-negative real numbers $(\beta_j)_{j \in \alpha} \in [0, +\infty]^\alpha$.

Definition 4.1 A standard product of the family of numbers $(\beta_j)_{j \in \alpha}$ is denoted by $(\mathbf{S}) \prod_{j \in \alpha} \beta_j$ and defined as follows:

$(\mathbf{S}) \prod_{j \in \alpha} \beta_j = 0$ if $\sum_{i \in \alpha^-} \ln(\beta_j) = -\infty$, where $\alpha^- = \{j : ln(\beta_j) < 0\}$ [2], and
$(\mathbf{S}) \prod_{j \in \alpha} \beta_j = e^{\sum_{j \in \alpha} \ln(\beta_j)}$ if $\sum_{j \in \alpha^-} \ln(\beta_j) \neq -\infty$.

Let (E, S) be a measurable space and let $\mathcal{R}$ be any subclass of the σ-algebra S. Let $(\mu_B)_{B \in \mathcal{R}}$ be such a family of σ-finite measures that for $B \in \mathcal{R}$ we have $\mathrm{dom}(\mu_B) = S \cap \mathcal{P}(B)$, where $\mathcal{P}(B)$ denotes the power set of the set B.

Remark 4.1 Let $(E_j, \rho_j)_{j \in \alpha}$ be again a sequence of non-empty Polish metric spaces that at most $\aleph_0$ of them are noncompact. Let $(\mu_j)_{j \in \alpha}$ be a sequence of Borel non-zero diffused finite measures with $\mathrm{dom}(\mu_j) = \mathcal{B}(E_j)$ for $j \in \alpha$ and

$$0 < (\mathbf{S}) \prod_{j \in \alpha} \mu_j(E_j) < +\infty.$$

By Lemmas 4.2, we claim that there exists a unique τ-smooth and Radon Borel extension λ of the Baire probability measure $\prod_{j \in \alpha} \frac{\mu_j}{\mu_j(E_j)}$. A Borel measure

$$(\mathbf{S}) \prod_{j \in \alpha} \mu_j(E_j) \times \lambda$$

is called a standard product of the family of finite Borel measures $(\mu_j)_{j \in \alpha}$ and is denoted by $(\mathbf{S}) \prod_{j \in \alpha} \mu_j$.

We put

$$\tau_i = \prod_{j \in \alpha_i} \mu_j.$$

Lemma 4.3 ([21], Lemma 7) *Let α be again an arbitrary infinite parameter set and let $(\alpha_i)_{i \in I}$ be its any partition such that α_i is a non-empty finite subset of the α for every $i \in I$. Let μ_j be a σ-finite diffused Borel measure defined on a Polish space (E_j, ρ_j) for $j \in \alpha$.*

We denote by $\mathcal{R}_{(\alpha_i)_{i \in I}}$ the family of all measurable rectangles $R \subseteq \prod_{j \in \alpha} E_j$ of the form $\prod_{i \in I} R_i$ with the property $0 \leq (\mathbf{S}) \prod_{i \in I} \tau_i(R_i) < \infty$ such that at most $\aleph_0$ of them are noncompact(i.e., $\mathrm{card}\{i : i \in I \ \& \ R_i$ is not compact in $\prod_{j \in \alpha_i} E_j\} \leq \aleph_0.)$

---

[2]We set $ln(0) = -\infty$

We suppose that there exists $R_0 = \prod_{i \in I} R_i^{(0)} \in \mathcal{R}_{(\alpha_i)_{i \in I}}$ such that

$$0 < (\mathbf{S}) \prod_{i \in I} \tau_i(R_i^{(0)}) < \infty.$$

For $X \in \mathcal{B}(R)$, we set $\mu_R(X) = 0$ if

$$(\mathbf{S}) \prod_{i \in I} \tau_i(R_i) = 0,$$

and

$$\mu_R(X) = (\mathbf{S}) \prod_{i \in I} \tau(R_i) \times \left(\prod_{i \in I} \frac{\tau_i R_i}{\tau_i(R_i)} \right)(X)$$

otherwise, where $\frac{\tau_i R_i}{\tau_i(R_i)}$ is a Borel probability measure defined on R_i as follows

$$(\forall X)(X \in \mathcal{B}(R_i) \rightarrow \frac{\tau_i R_i}{\tau_i(R_i)}(X) = \frac{\tau_i(Y \cap R_i)}{\tau_i(R_i)}).$$

Then the family of measures $(\mu_R)_{R \in \mathcal{R}}$ is consistent.

Let α be again an arbitrary infinite parameter set and let $(\alpha_i)_{i \in I}$ be its any partition such that α_i is a non-empty finite subset of the α for every $i \in I$. Let μ_j be a σ-finite continuous Borel measure defined on a Polish space (E_j, ρ_j) for $j \in \alpha$.

We denote by $\mathcal{R}_{(\alpha_i)_{i \in I}}$ the family of all measurable rectangles $R \subseteq \prod_{j \in \alpha} E_j$ of the form $\prod_{i \in I} R_i$ with the property $0 \le (\mathbf{S}) \prod_{i \in I} \tau_i(R_i) < \infty$ such that at most $\aleph_0$ of them are noncompact.

We suppose that there exists $R_0 = \prod_{i \in I} R_i^{(0)} \in \mathcal{R}_{(\alpha_i)_{i \in I}}$ such that

$$0 < (\mathbf{S}) \prod_{i \in I} \tau_i(R_i^{(0)}) < \infty.$$

We say that a Borel measure $\nu_{(\alpha_i)_{i \in I}}$ defined on $\mathcal{B}(\prod_{j \in \alpha} E_j)$ is a standard $(\alpha_i)_{i \in I}$-product of the family of σ-finite continuous Borel measures $(\mu_j)_{j \in \alpha}$ if for every

$$R = \prod_{i \in I} R_i \in \mathcal{R}_{(\alpha_i)_{i \in I}}$$

we have

$$\nu_{(\alpha_i)_{i \in I}}(R) = (\mathbf{S}) \prod_{i \in I} \tau_i(R_i),$$

where $\tau_i = \prod_{j \in \alpha_i} \mu_j$ for $i \in I$.

We say that a Borel measure $\nu_{(\alpha_i)_{i \in I}}$ defined on $\mathcal{B}(\prod_{j \in \alpha} E_j)$ is a strict standard $(\alpha_i)_{i \in I}$-product of the family of σ-finite continuous Borel measures $(\mu_j)_{j \in \alpha}$ if :
(i) for every

$$R = \prod_{i \in I} R_i \in \mathcal{R}_{(\alpha_i)_{i \in I}}$$

we have

$$\nu_{(\alpha_i)_{i \in I}}(R) = (\mathbf{S}) \prod_{i \in I} \tau_i(R_i),$$

where $\tau_i = \prod_{j \in \alpha_i} \mu_j$ for $i \in I$.

(ii) for every $X \in \mathcal{B}(\prod_{i \in I} E_i)$ we have

$$(\nu_{(\alpha_i)_{i \in I}}(X) = 0 \Leftrightarrow (\forall R)(R \in \mathcal{R}_{(\alpha_i)_{i \in I}} \rightarrow \nu_{(\alpha_i)_{i \in I}}(X \cap R) = 0)).$$

Theorem 4.1 ([21], Theorem 1) *Let μ_j be a quasi-finite [3] Borel measure defined on a Polish space (E_j, ρ_j) for $j \in \alpha$. Let α be again an arbitrary infinite parameter set and let $(\alpha_i)_{i \in I}$ be its any partition such that α_i is a non-empty finite subset of the α for every $i \in I$, and let us suppose that there exists $R_0 = \prod_{i \in I} R_i^{(0)} \in \mathcal{R}_{(\alpha_i)_{i \in I}}$ such that*

$$0 < (\mathbf{S}) \prod_{i \in I} \tau_i(R_i^{(0)}) < \infty.$$

Then there exists a standard $(\alpha_i)_{i \in I}$-product of the family $(\mu_j)_{j \in \alpha}$.

Theorem 4.2 *Let μ_j be a quasi-finite Borel measure defined on a Polish space (E_j, ρ_j) for $j \in \alpha$. Let α be again an arbitrary infinite parameter set and let $(\alpha_i)_{i \in I}$ be its any partition such that α_i is a non-empty finite subset of the α for every $i \in I$, and let us suppose that there exists $R_0 = \prod_{i \in I} R_i^{(0)} \in \mathcal{R}_{(\alpha_i)_{i \in I}}$ such that*

$$0 < (\mathbf{S}) \prod_{i \in I} \tau_i(R_i^{(0)}) < \infty.$$

Then there exists a strict standard $(\alpha_i)_{i \in I}$-product of the family $(\mu_j)_{j \in \alpha}$.

Proof. We denote by $\mathcal{R}_{(\alpha_i)_{i \in I}}$ the family of all measurable rectangles $R \subseteq \prod_{j \in \alpha} E_j$ of the form $\prod_{i \in I} R_i$ with the property $0 \leq (\mathbf{S}) \prod_{i \in I} \tau_i(R_i) < \infty$ such that at most $\aleph_0$ of them are noncompact.

For $X \in \mathcal{B}(R)$, we set $\mu_R(X) = 0$ if

$$(\mathbf{S}) \prod_{i \in I} \tau_i(R_i) = 0,$$

and

$$\mu_R(X) = (\mathbf{S}) \prod_{i \in I} \tau(R_i) \times \Big(\prod_{i \in I} \frac{\tau_i R_i}{\tau_i(R_i)} \Big)(X)$$

otherwise, where $\frac{\tau_i R_i}{\tau_i(R_i)}$ is a Borel probability measure defined on R_i as follows

$$(\forall X)(X \in \mathcal{B}(R_i) \rightarrow \frac{\tau_i R_i}{\tau_i(R_i)}(X) = \frac{\tau_i(Y \cap R_i)}{\tau_i(R_i)}).$$

---

[3]A measure μ is called quasi-finite if there exists $A \in \mathrm{dom}(\mu)$ such that $0 < \mu(A) < \infty$.

By Lemma 4.3, the family of measures $(\mu_R)_{R \in \mathcal{R}}$ is consistent. We set

$$\nu_{(\alpha_i)_{i \in I}} = \lambda^{(2)}_{\mathcal{R}_{(\alpha_i)_{i \in I}}},$$

where the measure $\lambda^{(2)}_{\mathcal{R}_{(\alpha_i)_{i \in I}}}$ is defined by Lemma 2.2. Application of Lemma 2.3 ends the proof of Theorem 4.2.

$\square$

Following [21], a standard $(\alpha_i)_{i \in I}$-product of the family $(\mu_j)_{j \in \alpha}$ is denoted by

$$(\mathbf{S}, (\alpha_i)_{i \in I}) \prod_{j \in \alpha} \mu_j.$$

We denote a strict standard $(\alpha_i)_{i \in I}$-product of the family $(\mu_j)_{j \in \alpha}$ by

$$(\mathbf{S}, (\alpha_i)_{i \in I}) \prod_{j \in \alpha} \mu_j.$$

Here we present a certain example of the family of σ-finite continuous Borel measures $(\mu_j)_{j \in N}$ defined on the real axis $\mathbf{R}$ and of two different partitions $(\alpha_i)_{i \in N}$ and $(\beta_i)_{i \in N}$ of N for which

$$(\mathbf{sS}, (\alpha_i)_{i \in N}) \prod_{j \in N} \mu_j \neq (\mathbf{sS}, (\beta_i)_{i \in N}) \prod_{j \in N} \mu_j.$$

Example 4.1 We set $\alpha = N$. For $j \in N$, let l_j be a linear Lebesgue measure on $\mathbf{R}$. Let $\alpha_i = \{i\}$ and $\beta_i = \{2i+1, 2(i+1)\}$ for $i \in N$.
We set

$$Y_i = [0, \frac{1}{2}] \times [0, 2].$$

It is obvious that

$$\left((\mathbf{sS}, (\beta_i)_{i \in N}) \prod_{j \in N} l_j\right)\left(\prod_{i \in N} Y_i\right) = 1$$

and

$$\left((\mathbf{sS}, (\alpha_i)_{i \in N}) \prod_{j \in N} l_j\right)\left(\prod_{i \in N} Y_i\right) = 0.$$

In connection with Theorems 4.2 and Example 4.1 we state the following

Problem 4.1 Under assumptions of Theorem 4.2, describe all pairs of partitions $(\alpha_i)_{i \in I}$ and $(\beta_i)_{i \in I}$ of the α for which $(\mathbf{sS}, (\alpha_i)_{i \in I}) \prod_{j \in \alpha} \mu_j = (\mathbf{sS}, (\beta_i)_{i \in I}) \prod_{j \in \alpha} \mu_j$.

The next two statements are immediate consequences of Theorem 4.2 and Lemma 2.2.

Theorem 4.3. *Under assumptions of Theorem 4.2, if each measure μ_j is G_j-left-and-right-invariant, where G_j denotes a group of Borel transformations of the E_j for $j \in \alpha$, then the measure $(\mathbf{sS}, (\alpha_i)_{i \in I}) \prod_{j \in \alpha} \mu_j$ is $\prod_{j \in \alpha} G_j$-left-and-right-invariant.*

Theorem 4.4 *Under assumption of Theorem 4.2, if each measure μ_j is G_j-left-invariant, where G_j denotes a group of Borel transformations of the E_j for $j \in \alpha$, then the measure $(\mathbf{sS}, (\alpha_i)_{i \in I}) \prod_{j \in \alpha} \mu_j$ is a $\prod_{j \in \alpha} G_j$-left-invariant.*

Observation 4.1. *Under conditions of Theorem 4.2, the measure $(\mathbf{sS}, (\alpha_i)_{i \in I}) \prod_{j \in \alpha} \mu_j$ is Radon.*

Proof. Let $0 < \left((\mathbf{sS}, (\alpha_i)_{i \in I}) \prod_{j \in \alpha} \mu_j\right)(X) < \infty$. It means that $X \in \mathcal{B}(\prod_{j \in \alpha} E_j)$ is presented as follows

$$X = (\cup_{n \in \{0\} \cup N} A_n \cap X) \cup (X \setminus \cup_{n \in \{0\} \cup N} A_n),$$

where $(A_n)_{n \in \{0\} \cup N}$ is a countable family of elements of the $\mathcal{R}_{(\alpha_i)_{i \in I}}$ with $A_{-1} = A_0 = \emptyset$ and

$$(\mathbf{sS}, (\alpha_i)_{i \in I}) \prod_{j \in \alpha} \mu_j (X \setminus \cup_{n \in \{0\} \cup N} A_n) = 0.$$

Since the measure μ_{A_n} is Radon, we can choose a compact set

$$F_n \subseteq (A_n \setminus \cup_{k=1}^{n-1} A_k) \cap X$$

such that

$$\mu_{A_n}\left(((A_n \setminus \cup_{k=1}^{n-1} A_k) \cap X) \setminus F_n\right) < \frac{\epsilon}{2^{n+1}}$$

for $n \in N$.

Also, we can choose a natural number n_ϵ such that

$$\sum_{n=n_\epsilon+1}^{\infty} \lambda_{A_n}((A_n \setminus \cup_{k=1}^{n-1} A_k) \cap X) < \frac{\epsilon}{2}.$$

Finally, we get

$$\left((\mathbf{S}, (\alpha_i)_{i \in I}) \prod_{j \in \alpha} \mu_j\right)(X \setminus \cup_{s=1}^{n_\epsilon} F_s) =$$

$$\sum_{n=1}^{\infty} \lambda_{A_n}((A_n \setminus \cup_{k=1}^{n-1} A_k) \cap (X \setminus \cup_{s=1}^{n_\epsilon} F_s)) =$$

$$\sum_{n=1}^{n_\epsilon} \lambda_{A_n}((A_n \setminus \cup_{k=1}^{n-1} A_k) \cap (X \setminus \cup_{s=1}^{n_\epsilon} F_s)) +$$

$$\sum_{n=n_\epsilon+1}^{\infty} \lambda_{A_n}((A_n \setminus \cup_{k=1}^{n-1} A_k) \cap (X \setminus \cup_{s=1}^{n_\epsilon} F_s)) \leq$$

$$\sum_{n=1}^{n_\epsilon} \lambda_{A_n}(((A_n \setminus \cup_{k=1}^{n-1} A_k) \cap X) \setminus F_n) +$$

$$\sum_{n=n_\epsilon+1}^{\infty} \lambda_{A_n}((A_n \setminus \cup_{k=1}^{n-1} A_k) \cap X) \leq \frac{\epsilon}{2} + \frac{\epsilon}{2} = \epsilon.$$

$\square$

Remark 4.3 For $j \in \alpha$, we set $E_j = \mathbf{R}$ and $\mu_j = m$, where m denotes a linear Lebesgue measure on $\mathbf{R}$.

Let $(\alpha_i)_{i \in I}$ be any partition of α such that α_i is a non-empty finite for every $i \in I$.

It is clear that $\prod_{j \in \alpha}[a_j, b_j] \in \mathcal{R}_{(\alpha_i)_{i \in I}}$ if $0 \leq (\mathbf{S}) \prod_{i \in I} m^{\alpha_i}(\prod_{j \in \alpha_i}[a_j, b_j]) < \infty$, where m^{α_i} is a Lebesgue measure on $\mathbf{R}^{\alpha_i}$.

Then the measure $(\mathbf{sS}, (\alpha_i)_{i \in I}) \prod_{j \in \alpha} \mu_j$ has the following properties:

$$(i) \left((\mathbf{sS}, (\alpha_i)_{i \in I}) \prod_{j \in \alpha} \mu_j\right)(\prod_{j \in \alpha}[a_j, b_j]) = (\mathbf{S}) \prod_{j \in \alpha}(b_i - a_i);$$

$$(ii) \left((\mathbf{sS}, (\alpha_i)_{i \in I}) \prod_{j \in \alpha} \mu_j\right)(X) = 0 \Leftrightarrow (\forall R)(R \in \mathcal{R}_{(\alpha_i)_{i \in I}} \rightarrow (\mathbf{sS}, (\alpha_i)_{i \in I}) \prod_{j \in \alpha} \mu_j)(X \cap R) = 0)$$

for every $X \in \mathcal{B}(\mathbf{R}^\alpha)$.

The measure $(\mathbf{sS}, (\alpha_i)_{i \in I}) \prod_{j \in \alpha} \mu_i$ is called a strict standard "$(\alpha_i)_{i \in I}$-Lebesgue measure" on $\mathbf{R}^\alpha$.

When $\operatorname{card}(\alpha_i) = 1$ for every $i \in I$, then $(\mathbf{sS}, (\alpha_i)_{i \in I}) \prod_{j \in \alpha} \mu_i$ is called a strict standard Lebesgue measure on $\mathbf{R}^\alpha$ and is denoted by ℓ_α.

Let f be any permutation of α. A mapping $A_f : \mathbf{R}^\alpha \rightarrow \mathbf{R}^\alpha$, defined by $A_f((x_i)_{i \in \alpha}) = (x_{f(i)})_{i \in \alpha}$ for $(x_i)_{i \in \alpha} \in \mathbf{R}^\alpha$, is called a canonical permutation of $\mathbf{R}^\alpha$.

Note that in our situation $\mathcal{R}_{(\alpha_i)_{i \in I}}$ is the family of all measurable rectangles $R \subseteq \mathcal{B}(\mathbf{R}^\alpha)$ of the form $\prod_{i \in \alpha} Y_i$ with the property $0 \leq (\mathbf{S}) \prod_{i \in \alpha} m(Y_i) < \infty$ such that at most $\aleph_0$ of them are noncompact(i.e., $\operatorname{card} \{i : i \in I \ \& \ Y_i \text{ is not compact in } \prod_{i \in \alpha_i} E_i\} \leq \aleph_0.$). It is obvious to see that a measure ℓ_α is invariant under a group $\mathcal{P}(\mathbf{R}^\alpha)$ generated by shifts and canonical permutations of $\mathbf{R}^\alpha$ and

$$\ell_\alpha(\prod_{i \in \alpha} Y_i) = (\mathbf{S}) \prod_{i \in \alpha} \ell_1(Y_i).$$

Remark 4.4 It can be shown that a standard Lebesgue measure m^α [21] and a strict standard Lebesgue measure ℓ_α are different when $\operatorname{card}(\alpha) \geq \omega$. Such a difference between the measures m^α and ℓ_α is caused by the phenomena that every set of finite m^α-measure can be covered by countable elements of the family $\mathcal{R}_{(\alpha_i)_{i \in I}}$ which we can not say for the measure ℓ_α.

5. Some Auxiliary Propositions in Solovay Model

Solovay model (cf. [25]) is the following system of axioms:

$$(ZF)\&(DC)\&(\text{every subset of } \mathbf{R} \text{ is measurable in the Lebesgue sense}),$$

where (ZF) denotes the Zermelo-Fraenkel set theory and (DC) denotes the axiom of Dependent Choices (cf. [3]).

Now, let (E_1, S_1, μ_1) and (E_2, S_2, μ_2) be two measure spaces. The measures μ_1 and μ_2 are called isomorphic if there exists a measurable isomorphism from E_1 onto E_2 such that

$$(\forall X)(X \in S_1 \rightarrow \mu_1(X) = \mu_2(f(X))).$$

Lemma 5.1 *Let E_1 and E_2 be any two Polish topological spaces. Let μ_1 be a probability diffused Borel measure on E_1 and let μ_2 be a probability diffused Borel measure on E_2. Then there exists a Borel isomorphism $\varphi : (E_1, B(E_1)) \to (E_2, B(E_2))$ such that*

$$\mu_1(X) = \mu_2(\varphi(X))$$

for every $X \in B(E_1)$.

The proof of Lemma 5.1 can be found in [3].

Lemma 5.2 ([18], Lemma 7.2, p. 118) *Let E be a Polish space and let μ be a probability diffused Borel measure on E. Then, in Solovay's model the completion $\bar{\mu}$ of μ is defined on the power set of E.*

We have the following simple consequence of Lemma 5.2.

Corollary 5.1 Let E be a Polish space. Then the class of of universally measurable subsets of E coincides with the power set of the E in Solovay model.

6. On Infinite Versions of Rademacher Theorem for R^∞ in Solovay Model

Let $(R^\infty, +)$ be an abelian Polish group with metric ρ defined by

$$\rho((x_k)_{k\in N}, (y_k)_{k\in N}) = \sum_{k\in N} \frac{|x_k - y_k|}{2^k(1 + |x_k - y_k|)}$$

for $(x_k)_{k\in N}, (y_k)_{k\in N} \in R^\infty$.

We say that a function $f : R^\infty \to R$ is differentiable at the point $a \in R^\infty$ if there exists a continuous strict linear mapping $L : \mathbf{R}^\infty \to R$ such that the following condition

$$\lim_{t\to 0} \frac{f(a + tv) - f(a)}{t} = L(v)$$

for every $v \in \mathbf{R}^\infty$.

Under a continuous linear mapping $L : \mathbf{R}^\infty \to R$ we understood such a mapping for which the following two conditions holds:

(i) if $\sum_{k\in N} c_k$ tends to the c in the metric ρ then $\sum_{k\in N} L(c_k)$ tends to the $L(c)$ in the usual sense on R.

(ii) for every $x, y \in \mathbf{R}^\infty$ and $\alpha, \beta \in R$ we have $L(\alpha x + \beta y) = \alpha L(x) + \beta L(y)$.

It is obvious that if $L : \mathbf{R}^\infty \to R$ is a continuous linear mapping then

$$(\forall (x_k)_{k\in N})((x_k)_{k\in N} \in \mathbf{R}^\infty \to L(\sum_{k\in N} x_k e_k) = \sum_{k\in N} x_k L(e_k)).$$

Hence, a continuous linear mapping is uniquely defined by its values on the family $\{e_k : k \in N\}$. The latter relation means that $card(\{L(e_k) : L(e_k) \neq 0\}) < \omega$ because for every infinite sequence of non-zero real numbers $(L(e_k))_{k \in N}$ we can found such a sequence of real numbers $(x_k)_{k \in N} \in R^\infty$ that the sum $\sum_{k \in N} x_k L(e_k)$ will be divergent.

Let G be any subset of R^N. We say that a function $f : R^\infty \to R$ is called an G-Gâteaux differentiable at the point $a \in \mathbf{R}^\infty$ if there exists a continuous linear operator $A_a : R^N \to R$ such that for each $v \in G$ one has

$$f(a + tv) - f(a) - tA_a(v) = r(v, t), \lim_{t \to 0} |t^{-1} r(v, t)| = 0.$$

We say that a function $f : R^\infty \to R$ is called an G-Fréchet differentiable at the point $a \in \mathbf{R}^\infty$ if there exists a continuous linear operator $A_a : R^N \to R$ such that for each $v \in G$ one has

$$f(a + v) - f(a) - \rho(0, v) A_a(v) = r(v), \quad \lim_{\rho(0,v) \to 0} \frac{r(v)}{\rho(0, v)} = 0.$$

We say that f is Lipschitz function on $\mathbf{R}^\infty$ if there exists a positive number L such that

$$(\forall x, y)(x, y \in \mathbf{R}^\infty \to |f(x) - f(y)| < L\rho(x, y)).$$

Lemma 6.1 *Let* $f : R^n \to R$ *be a Lipschitz function. Then* f *is* ℓ_n-*almost everywhere* R^n-*Gâteaux differentiable on* R^n.

Lemma 6.2. *Let* $f : \mathbf{R}^\infty \to R$ *be a Lipschitz function. For* $n \in N$, *we set* $G_n = R^n \times \{0\}^{N \setminus \{1, \cdots, n\}}$.

Let $a \in R^{(N)}$. *Then, for* $n \in N$, *we have*

$$\ell_n(\{(x_1, \cdots, x_n) : f \text{ is not } G_n - \text{ Gâteaux differentiable at the point}$$

$$(x_1, \cdots, x_n, a_{n+1}, a_{n+2}, \cdots) \}) = 0.$$

Proof. Let us define $g : R^n \to R$ by

$$g(x_1, \cdots, x_n) = f(x_1, \cdots, x_n, a_{n+1}, a_{n+2}, \cdots).$$

Let $(v_1, \cdots, v_n, 0, \cdots) \in G_n$.
We have

$$\lim_{t \to 0} \frac{f((x_1, \cdots, x_n, a_{n+1}, a_{n+2}, \cdots) + t(v_1, \cdots, v_n, 0, \cdots)) - f((x_1, \cdots, x_n, a_{n+1}, a_{n+2}, \cdots))}{t} =$$

$$\lim_{t \to 0} \frac{g((x_1, \cdots, x_n) + t(v_1, \cdots, v_n)) - g(x_1, \cdots, x_n)}{t}.$$

Let us show that g is Lipschitz function on R^n. Indeed, we have

$$|g(x_1, \cdots, x_n) - g(y_1, \cdots, y_n)| = |f(x_1, \cdots, x_n, a_{n+1}, a_{n+2}, \cdots) - f(y_1, \cdots, y_n, a_{n+1}, a_{n+2}, \cdots)|$$

$$L \sum_{k=1}^{n} \frac{|x_k - y_k|}{2^k(1 + |x_k - y_k|)} \le L \sum_{k=1}^{n} |x_k - y_k| \le nL \sup_{1 \le k \le n} |x_k - y_k|.$$

By Lemma 6.1 we claim that g is ℓ_n-almost everywhere R^n-Gâteaux differentiable on R^n. This ends the proof of Lemma 6.2.

$\square$

Lemma 6.3 *Let $f : R^\infty \to R$ be a Lipschitz function. For $n \in N$, we set $G_n = R^n \times \{0\}^{N\setminus\{1,\cdots,n\}}$. Then, in Solovay model f is λ-almost everywhere G_n-Gâteaux differentiable on R^N, where the λ denotes the strict ordinary Lebesgue measure on R^N.*

Proof. Let λ_n be a Lebesgue measure on $R^{\{1,\cdots,n\}}$ and $\lambda_{N\setminus\{1,\cdots,n\}}$ be a strict standard Lebesgue measure on $R^{N\setminus\{1,\cdots,n\}}$. It is clear that λ is a strict standard product of measures λ_n and $\lambda_{N\setminus\{1,\cdots,n\}}$.

We denote by D a collection of points in R^N for which f is not G_n-Gâteaux differentiable.

For every $(a_{n+1}, a_{n+2}, \cdots) \in R^{N\setminus\{1,\cdots,n\}}$, we denote by $D_{(a_{n+1},a_{n+2},\cdots)}$ a $(a_{n+1}, a_{n+2}, \cdots)$-section of the D, i.e.,

$$D_{(a_{n+1},a_{n+2},\cdots)} = \{(x_1, \cdots, x_n) \in R^{\{1,\cdots,n\}} : (x_1, \cdots, x_n, a_{n+1}, a_{n+2}, \cdots) \in D\}.$$

By Lemma 6.2, for every $(a_{n+1}, a_{n+2}, \cdots) \in R^{N\setminus\{1,\cdots,n\}}$ we have that

$$\ell_n(\{(x_1, \cdots, x_n) : f \text{ is not } G_n - \text{ differentiable at the point}$$

$$(x_1, \cdots, x_n, a_{n+1}, a_{n+2}, \cdots) \}) = 0.$$

For every $E = \prod_{k \in N} E_k \in \mathcal{R}(\mathcal{O})^+$, by Fubini theorem we get

$$\mu_R(D \cap E) = \int_{\prod_{k \in N\setminus\{1,\cdots,n\}} E_k} \lambda_n((D \cap E)_{(a_{n+1},a_{n+2},\cdots)}) d\lambda_{N\setminus\{1,\cdots,n\}}(a_{n+1}, a_{n+2}, \cdots) \le$$

$$\int_{\prod_{k \in N\setminus\{1,\cdots,n\}} E_k} \lambda_n(D_{(a_{n+1},a_{n+2},\cdots)}) d\lambda_{N\setminus\{1,\cdots,n\}}(a_{n+1}, a_{n+2}, \cdots) = 0.$$

By Lemma 6.2, we claim that $\lambda(D) = 0$. This ends the proof of Lemma 6.3.

$\square$

Theorem 6.1 *Let $f : R^\infty \to R$ be a Lipschitz function and $R^{(N)}$ be a group of all eventually zero sequences. Then, in Solovay model, f is λ-almost everywhere $R^{(N)}$-Gâteaux differentiable on R^N.*

Proof. Let D_n be a set of points in R^N where f is not G_n-Gâteaux differentiable. Let D be a set of points in R^N where f is not $R^{(N)}$-Gâteaux differentiable. Since $R^{(N)} = \cup_{n \in N} G_n$ we claim that $D = \cup_{n \in N} D_n$. By Lemma 6.3 we have that $\lambda(D_n) = 0$. Hence, $\lambda(D) \le \sum_{n \in N} \lambda(D_n) = 0$. This ends the proof of Theorem 6.1.

$\square$

7. On Uniform Measures on ℓ^α

We begin this section by the following definition.

Definition 7.1 Let S be a translation-invariant σ-algebra of subsets of B such that

$$\mathcal{F}(B) \subseteq S,$$

where $\mathcal{F}(B)$ denotes a minimal σ-algebra of subsets of B generated by all open balls.

We say that a measure u defined on S is uniformly distributed (or, shortly is uniform measure) if the u measure of a ball depends only of the radius and not of the center.

It is clear that an arbitrary translation-invariant measure on B with $\mathcal{F}(B) \subseteq \text{dom}(\mu)$ is uniformly distributed.

In [22], has been investigated the following

Problem 7.1 *Let μ_1 and μ_2 are two uniform measures on Banach space B such that*

$$\mathcal{F}(B) \subseteq (dom)(\mu_1) = (dom)(\mu_2)$$

and the values of measures μ_1 and μ_2 coincide on an arbitrary ball.
Can we assert that the measures μ_1 and μ_2 coincide?

In [22], has been obtained the following results.

Theorem 7.1 ([22], Theorem 4.1) *There exist two translation-invariant uniform measures μ_1 and μ_2 defined on the σ-algebra $\mathcal{B}(\ell^\infty) \cap \mathcal{B}(\mathcal{R}^\infty)$ such that*
(i) for every closed ball $B(x,r)$ with center at x and radius r we have $\mu_1(B(x,r)) = \mu_2(B(x,r))$;
(ii) the measure μ_1 is absolutely continuous with respect to μ_2, and μ_1 and μ_2 are not equivalent.

Theorem 7.2 ([22], Theorem 4.2)   *For an arbitrary non-empty parameter set α, we denote by the ℓ^α a Banach space of all real-valued bounded functions defined on the α and equipped with a norm $||\cdot||_\alpha$ defined by*

$$||(x_i)_{i\in\alpha}||_\alpha = \sup_{i\in\alpha} |x_i|$$

for $(x_i)_{i\in\alpha} \in \ell^\alpha$.
Let $\mathbb{F}(\ell^\alpha)$ be a minimal σ-algebra of subsets of the ℓ^α generated by the family of all open balls in ℓ^α.
Further, let B_0 be a closed unite ball defined by

$$B_0 = \{(x_i)_{i\in\alpha} : (x_i)_{i\in\alpha} \in \ell^\alpha \ \& \ ||(x_i)_{i\in\alpha}||_\alpha \leq 1\}.$$

Then there exists a translation-invariant measure μ defined on the measure space $(\ell^\infty, \mathbb{F}(\ell^\alpha))$ such that $\mu(B_0) = 1$.

Theorem 7.3 ([22], Theorem 4.3) *For an arbitrary infinite parameter set α, there exist two uniform measures λ_1 and λ_2 defined on the Banach space ℓ^α such that:*

(i) $dom(\lambda_1) = dom(\lambda_2) = \mathcal{F}(\ell^\alpha)$;

(ii) the values of measures λ_1 and λ_2 coincide on an arbitrary ball such that $\lambda_1(B_0) = \lambda_2(B_0) = 1$;

(iii) the measures λ_1 and λ_2 are not equivalent.

The validity of Theorems 7.1-7.3 has been obtained in [22] by the technique of ordinary and standard products of linear Lebesgue measures. Note here that same results can be obtained by the technique of strict ordinary and strict standard products of linear Lebesgue measures.

8.   On a Completion of the $\mathcal{B}(R^\infty)$ by the Baker Measure

Suppose $\mathcal{R}_1$ is the class of all infinite dimensional rectangles $R \in B(\mathbb{R}^\mathbb{N})$ of a form

$$R = \prod_{k=1}^\infty (a_i, b_i), \quad -\infty < a_i \le b_i < \infty$$

such that $0 \le \prod_{k=1}^\infty (b_k - a_k) < \infty$.

Following [18](see, Definition 15.4.1), a translation-invariant Borel measure λ in $\mathbb{R}^\mathbb{N}$ is called Baker measure if

$$\lambda(R) = \prod_{k=1}^\infty (b_k - a_k)$$

for $R \in \mathcal{R}_1$.

In [23] has been considered the following

Question 8.1 Is it valid that

$$\mathcal{B}_{||\cdot||_\infty}(\ell^\infty) \subseteq \overline{\mathcal{B}}_\lambda(\mathbf{R}^\infty),$$

where $\overline{\mathcal{B}}_\lambda(\mathbf{R}^\infty)$ denotes the completion of the $\mathcal{B}(\mathbf{R}^\infty)$ by the Baker's measure λ?

This question has been solved negatively by Petr Holický and Jaroslav Tišer for Baker's measures [1] and [2] in the theory ZFC (see, [23]).

The purpose of the final section is to show that Question 8.1 is independent from the theory $ZF + DC$. In this direction it is sufficient to construct such a consistent extension of the theory $ZF + DC$ that Question 8.1 is solved negatively. Indeed, we have the following

Theorem 8.1 *In Solovay model there exists a Baker measure on $\mathbf{R}^\infty$ for which Question 8.1 is solved positively.*

Proof. Let λ be Baker's strict ordinary measure. Following condition (a) of Corollary 3.3, the λ is Baker's measure. Following condition (d) of Lemma 3.2, the domain of the λ contains the class of all universally measurable subsets of the $\mathbf{R}^\infty$. Following Corollary 5.1, the class of universally measurable subsets of the $\mathbf{R}^\infty$ coincides with the power set $\mathcal{P}(\mathbf{R}^\infty)$ of the $\mathbf{R}^\infty$ in Solovay model. Thus, in Solovay model we have

$$\mathcal{B}_{||\cdot||_\infty}(\ell^\infty) \subseteq \mathcal{P}(\mathbf{R}^\infty) = \overline{\mathcal{B}}_\lambda(\mathbf{R}^\infty).$$

This ends the proof of Theorem 8.1.

$\square$

References

[1] Baker R., "Lebesgue measure" on R^∞, *Proc. Amer. Math. Soc.*, **113**(**4**) (1991), 1023–1029. MR1062827 (92c:46051),Zbl 0741.28009

[2] Baker R., "Lebesgue measure" on $\mathbb{R}^\infty$. II. *Proc. Amer. Math. Soc.*,**132**(**9**), (2004), 2577–2591 (electronic). MR2054783 (2005d:28012), Zbl 1064.28015 Christensen,

[3] Cichon J., Kharazishvili A., Weglorz B., *Subsets of the real line*, Wydawnictwo Uniwersytetu Lodzkiego, Lodz (1995).

[4] Christensen J.P.R. Uniform Measures, *Proc. of the functional analysis week,* March (1969), 3–7,

[5] Fremlin D.H.,[p98*] *Measure theory,* in preparation (draft available by anonymous ftp.at ftp. essex. ac. uk /pub / measure theory).

[6] Halmos P.R., *Measure theory*, Princeton, Van Nostrand (1950).

[7] Kakutani S., Oxtoby J., Construction of non–separable invariant extension of the Lebesgue measure space, *Ann Math.*, **52**, 1950.

[8] Kharazishvili A.B., *Elements of combinatorical theory of infinite sets*, Tbilisi University Press., Tbilisi, (1981) (in Russian).

[9] Kharazishvili A.B.,*Invariant extensions of the Lebesgue measure*, Tbilisi (1983) (in Russian).

[10] Kharazishvili A.B., On cardinalities of isodyne topological spaces, *Bull. Acad. Sci. Georgia,* **137**, 1 (1990), 33-36.

[11] Kodaira K., Kakutani S., A nonseparable tranlation–invariant extension of the Lebesgue measure space, *Ann. Math.,* vol. 52, 1950, pp. 574–579.

[12] Kuratowski K., Mostowski A., *Set theory*, Nauka, Moscow (1980) (in Russian).

[13] Loeb Peter A.; Ross David A. Infinite products of infinite measures. *Illinois J. Math.* **49** (2005), no. 1, 153–158 (electronic). MR2157373 (2006j:28012)

[14] Mařik J., The Baire and Borel measure, *Czechoslovak Math. J.* **7** (1957), 248. 253

[15] Oxtoby J.C., Measure and Category, A survey of the analogies between topological and measure spaces. Second edition. *Graduate Texts in Mathematics,* **2**. Springer-Verlag, New York-Berlin (1980), x+106 pp.

[16] Pantsulaia G.R., An application of independent families of sets to the measure extension problem, *Georgian Math. J.,* **11** (2004), no. 2, 379–390.

[17] Pantsulaia G.R., Relations between shy sets and sets of ν_p-measure zero in Solovay's Model, *Bull. Polish Acad.Sci.*, vol.52, No.1, (2004).

[18] Pantsulaia G.R., *Invariant and Quasiinvariant Measures in Infinite-Dimensional Topological Vector Spaces,* Nova Science Publishers, 2007, xii+231 pp.

[19] Pantsulaia G.R., On generators of shy sets on Polish topological vector spaces, *New York Journal of Mathematics,* Volume 14, 2008 , 235-261 .

[20] Pantsulaia G.R., On ordinary and Standard Lebesgue Measures on R^∞ , *Bull. Polish Acad.Sci.,* **73**(3) (2009), 209-222.

[21] Pantsulaia G.R., On a standard product of an arbitrary family of σ-finite Borel measures with domain in Polish spaces, *Theory Stoch. Process* (Accepted).

[22] Pantsulaia G.R., On uniform measures in Banach spaces, *Georgian International journal of Science and Technology, Nova Publishers.Inc* (Accepted).

[23] Pantsulaia G.R., On ordinary and standard products of infinite family of σ-finite measures and some of their applications, *Acta Mathematica Sinica, English Series* (Accepted)

[24] Rogers C.A., *Hausdorff Measures*, Cambridge Univ. Press, 1970.

[25] Solovay R.M., A model of set theory in which every set of reals is Lebesgue measurable,*Ann.Math.,***92** (1970), 1–56.

[26] Szpilrajn E., Ensembles indépendants et mesures non séparables. *Comptes Rendus,* **207**(1938), 768–770.

[27] Stone A. H., Paracompactness and product spaces. *Bull. Amer. Math. Soc.* **54**, (1948). 977–982. MR0026802 (10,204c)

[28] Stephen Willard , *General Topology,* (1970) Addison-Wesley Publishing Company, Reading Massachusetts

[29] Bruckner A.M., Bruckner, J. B., Thomson B.S., *Real Analysis* (2007), http: www.classicalrealanalysis.com.

[30] Wen Sheng You, A certain regular property of the method I construction and packing measure. *Acta Math. Sin. (Engl. Ser.)* **23** (2007), no. 10, 1769–1776. MR2352292

[31] Vakhanya N.N., Tarieladze V.I., Chobanyan S.A. , *Probability distributions in Banach spaces*, Nauka, Moscow (1985) (in Russian).

In: Focus on Science and Technology...
Editor: Sergo Gotsiridze, pp. 295-301

ISBN 978-1-61209-970-5
© 2011 Nova Science Publishers, Inc.

Chapter 27

LEBESGUE CONSTANTS FOR (C, α)-SUMMATION OF FOURIER SERIES OVER GENERALIZED SPHERICAL FUNCTIONS

S.B. Topuria * and N.D. Macharashvili †
Department of Mathematics, Georgian Technical University,
77 Kostava St. 0175 Tbilisi 75, Georgia

Abstract

It is shown how Lebesgue constants for Cesáro (C, α)-means of Fourier series in terms of generalized spherical functions depend on the summation index α.

2010 Mathematics Subject Classification: Primary 42Cxx, 42Axx; Secondary 42C05, 42A38, 42C10.

Key words and phrases: Lebesgue constants, Cesáro means, summation, Fourier series, generalized spherical functions, Legendre polynomial.

1. Notation, Definitions and Auxiliary Statements

We use the following notations:

R^k is the k-dimensional Euclidean space ($k = 1, 2, \ldots$);
$x = (x_1, x_2, x_3)$, $y = (y_1, y_2, y_3)$ are points (vectors) of the space R^3;
(ρ, θ, φ) are the spherical coordinates of the point $x = (x_1, x_2, x_3)$;
$(x, y) = \sum_{k=1}^{3} x_i y_i$ is the scalar product of vectors from R^3;
$|x| = \sqrt{(x, x)}$ is the length of the vector x;
$S^2 = \{x : x \in R^3; |x| = 1\}$ is the unit sphere and $|S^2| = 4\pi$ is its area.
$f \in L(S^2)$ means that f is summable on S^2;
dS^2 is an element of the surface area S^2, defined by $dS^2 = \sin \theta d\theta d\varphi$.

---

*E-mail address: topur@list.ru
†E-mail address: n_macharashvili@mail.ru

Assume that $f \in L(S^2)$. Its Fourier series in terms of generalized spherical functions is the series

$$S(f; x) = \sum_{v=|m|}^{\infty} I_v^{(m)}(f; \theta, \varphi), \qquad (1.1)$$

where $m = 0, \pm 1$,

$$I_v^{(m)}(f; \theta, \varphi) = \frac{(-1)^m}{4\pi}(2v+1) \int_{S^2} f(y) e^{-im(\varphi_1+\varphi_2)} P_{m,m}^v(\cos \gamma) dS^2(y) =$$

$$= \frac{(-1)^m}{4\pi}(2v+1) \int_0^{2\pi} \int_0^{\pi} f(\theta', \varphi') e^{-im(\varphi_1+\varphi_2)} P_{m,m}^v(\cos \gamma) \sin \theta' d\varphi' d\theta'.$$

$$\cos \gamma = (x, y) = \cos \theta \cos \theta' - \sin \theta \sin \theta' \cos \beta,$$
$$\beta = \pi + \varphi' - \varphi,$$
$$\operatorname{tg} \varphi_1 = \frac{\sin \beta \sin \theta'}{\cos \theta \sin \theta' \cos \beta + \cos \theta' \sin \theta};$$
$$\operatorname{tg} \varphi_2 = \frac{\sin \beta \sin \theta'}{\cos \theta' \sin \theta \cos \beta + \cos \theta \sin \theta'};$$

$$P_{m,n}^v(\mu) = \frac{(-1)^{v-m} i^{m-n}}{2^v (v-m)!} \sqrt{\frac{(v+m)!(v-m)!}{(v+n)!(v-n)!}} (1-\mu)^{\frac{n-m}{2}} (1+\mu)^{\frac{m+n}{2}} \times$$

$$\times \frac{d^{v-m}}{d\mu^{v-m}} [(1-\mu)^{v-n}(1+\mu)^{v+n}], \quad \mu = \cos \theta.$$

For $m = 0$, a system of generalized spherical functions coincides with a system of ordinary spherical functions [1], while the series (1.1) coincides with a Fourier-Laplace series on a sphere, i.e., it can be rewritten in the form

$$S(f; \theta, \varphi) = \frac{1}{4\pi} \sum_{v=0}^{\infty}(2v+1) \int_0^{\pi} \int_0^{2\pi} f(\theta', \varphi') P_v(\cos \gamma) \sin \theta' d\theta' d\varphi', \qquad (1.2)$$

where $P_v(\cos \gamma)$ is a Legendre polynomial, $P_v(\mu) = P_{0,0}^v(\mu)$.

The relation [2]

$$P_{1,1}^v(\mu) = P_{-1,-1}^v(\mu) = \frac{1}{2v+1}\frac{d}{d\mu}[P_{v+1}(\mu) - P_{v-1}(\mu)] -$$

$$- \frac{1}{2v+1}\frac{1}{1+\mu}[P_{v+1}(\mu) - P_{v-1}(\mu)]$$

with the following virtue of ([3], p. 37)

$$\frac{d}{d\mu}[P_{v+1}(\mu) - P_{v-1}(\mu)] = (2v+1)P_v(\mu)$$

implies the validity of the following equality

$$P_{1,1}^v(\mu) = P_v(\mu) - \frac{1}{1+\mu}\int_{-1}^{\mu} P_v(t)dt,$$

i.e.,

$$P_{1,1}^v(\cos\gamma) = P_v(\cos\gamma) - \frac{1}{1+\cos\gamma}\int_\gamma^\pi P_v(\cos t)\sin t\, dt. \tag{1.3}$$

From (1.3) we have $P_{1,1}^0(\cos\gamma) = 0$.

Therefore for $m = \pm 1$ the series (1) can rewritten as follows

$$S(f; x) = \sum_{v=0}^\infty I_v^{(m)}(f; \theta, \varphi), \tag{1.4}$$

where

$$I_v^{(m)}(f; \theta, \varphi) = -\frac{1}{4\pi}(2v+1)\int_0^{2\pi}\int_0^\pi f(\theta', \varphi')e^{-im(\varphi_1+\varphi_2)}[P_v(\cos\gamma)-$$
$$-\frac{1}{1+\cos\gamma}\int_\gamma^\pi P_v(\cos t)\sin t\, dt]\sin\theta'\, d\varphi'\, d\theta'.$$

If we take the point $x(\theta, \varphi)$ as the initial one and introduce the new spherical coordinates $(\gamma, \overline{\varphi})$, then we obtain

$$I_v^{(m)}(f; \theta, \varphi) = -\frac{2v+1}{4\pi}\int_0^{2\pi}\int_0^\pi F(\gamma, \overline{\varphi})e^{-im(\overline{\varphi}_1+\overline{\varphi}_2)}[P_v(\cos\gamma)-$$
$$-\frac{1}{2\cos^2\frac{\gamma}{2}}\int_\gamma^\pi P_v(\cos t)\sin t\, dt]\sin\gamma\, d\gamma\, d\overline{\varphi},$$

where $F(\gamma, \overline{\gamma}) = f(\theta', \varphi')$.

Denote by $\sigma_n^\alpha(f; x)$ the Cesáro means (C, α), $\alpha > -1$, of the series (1.4), i.e.,

$$\sigma_n^\alpha(f; \theta, \varphi) = -\frac{1}{4\pi}\int_0^{2\pi}\int_0^\pi F(\gamma, \overline{\varphi})\phi_n^\alpha(\cos\gamma)e^{im(\overline{\varphi}_1+\overline{\varphi}_2)}\sin\gamma\, d\gamma\, d\overline{\varphi},$$

where

$$\phi_n^\alpha(\cos\gamma) = K_n^\alpha(\cos\gamma) - N_n^\alpha(\cos\gamma), \tag{1.5}$$

$$K_n^\alpha(\cos\gamma) = \frac{1}{A_n^\alpha}\sum_{v=0}^n\left(v+\frac{1}{2}\right)A_{n-v}^\alpha P_v(\cos\gamma),$$

$$N_n^\alpha(\cos\gamma) = \frac{1}{2\cos^2\frac{\gamma}{2}}\int_\gamma^\pi K_n^\alpha(\cos t)\sin t\, dt,$$

$K_n^\alpha(\cos\gamma)$ is the kernel of the (C, α)-means of the Fourier-Laplace series (1.2) ([4], p.20).

When considering problems of convergence and summation of Fourier series, an important role is played by the so-called Lebesgue constants which, for the Cesáro means (C, α), $\alpha > -1$, of Fourier series in terms of generalized spherical functions have the form

$$L_n^\alpha(x) = \frac{1}{2\pi}\int_{S^2}|\phi_n^\alpha(\cos\gamma)|dS^2(y) = \frac{1}{2\pi}\int_0^\pi d\gamma\int_{(x,y)=\cos\gamma}|\phi_n^\alpha(\cos\gamma)|dS(y) =$$
$$= \frac{1}{2\pi}\int_0^\pi|\phi_n^\alpha(\cos\gamma)|2\pi\sin\gamma\, d\gamma = \int_0^\pi|\phi_n^\alpha(\cos\gamma)|\sin\gamma\, d\gamma. \tag{1.6}$$

2.   Main Result

Generally speaking, in what follows we will denote by C various positive constants.
The following statement is true.

Theorem. *For a Lebesgue constant of Fourier series in terms of generalized spherical functions the following estimates are valid:*

$$\left.\begin{array}{rl} C_1 n^{\frac{1}{2}-\alpha} < L_n^\alpha < C_2 n^{\frac{1}{2}-\alpha} & \text{for}\ \ \alpha < \dfrac{1}{2}, \\[2mm] C_1 \ln n < L_n^\alpha < C_2 \ln n & \text{for}\ \ \alpha = \dfrac{1}{2}, \\[2mm] 0 < L_n^\alpha < C & \text{for}\ \ \alpha > \dfrac{1}{2}. \end{array}\right\} \tag{2.7}$$

Proof 2.1 *From (1.5) and (1.6) we have*

$$L_n^\alpha < \int_0^\pi |K_n^\alpha(\cos\gamma)| \sin\gamma d\gamma + \int_0^\pi |N_n^\alpha(\cos\gamma)| \sin\gamma d\gamma = I_1 + I_2.$$

The estimate (2.7) holds for I_1 ([4], p. 21; [5]).
Let us estimate I_2. We obtain

$$I_2 = \int_0^\pi \frac{1}{2\cos^2\frac{\gamma}{2}} \left| \int_\gamma^\pi K_n^\alpha(\cos t) \sin t dt \right| \sin\gamma d\gamma \le$$

$$\le \int_0^\pi \frac{\sin\gamma}{2\cos^2\frac{\gamma}{2}} \left(\int_\gamma^\pi |K_n^\alpha(\cos t)| \sin t dt \right) d\gamma =$$

$$= \int_0^\pi \left(\int_0^t \frac{\sin\gamma d\gamma}{2\cos^2\frac{\gamma}{2}} \right) |K_n^\alpha(\cos t)| \sin t dt = 2 \int_0^\pi |K_n^\alpha(\cos t)| |\ln\cos\frac{t}{2}| \sin t dt =$$

$$= 2\left(\int_0^{\frac{1}{n}} |K_n^\alpha(\cos t)| |\ln\cos\frac{t}{2}| \sin t dt + \int_{\frac{1}{n}}^{\pi-\frac{1}{n}} |K_n^\alpha(\cos t)| |\ln\cos\frac{t}{2}| \sin t dt + \right.$$

$$\left. \int_{\pi-\frac{1}{n}}^\pi |K_n^\alpha(\cos t)| |\ln\cos\frac{t}{2}| \sin t dt \right) = 2(A_1 + A_2 + A_3).$$

Since ([6], p. 134)

$$|K_n^\alpha(\cos\gamma)| \le Cn^2 \ \ \text{for}\ \ \alpha > -1, \ \ 0 \le \gamma \le \pi;$$

$$|K_n^\alpha(\cos\gamma)| \le Cn \ \ \text{for}\ \ \alpha > -1, \ \ \frac{\pi}{2} \le \gamma \le \pi,$$

we have

$$A_1 < Cn^2 \int_0^{\frac{1}{n}} \sin\gamma d\gamma < C; \tag{2.8}$$

$$A_3 < Cn \int_{\pi-\frac{1}{n}}^\pi |\ln\cos\frac{t}{2}| \sin t dt < C. \tag{2.9}$$

To estimate A_2, we apply an asymptotic expression of the kernel $K_n^\alpha(\cos\gamma)$ which, for $|\gamma - \frac{\pi}{2}| < \frac{n}{n+1}\frac{\pi}{2}$, has the form ([6], p. 133)

$$K_n^\alpha(\cos\gamma) = \frac{A_n^{\frac{1}{2}}}{A_n^\alpha} \frac{\sin\left[\left(n+1+\frac{\alpha}{2}\right)\gamma - \frac{1+2\alpha}{4}\pi\right]}{\sqrt{2\sin\gamma}\left(2\sin\frac{\gamma}{2}\right)^{\alpha+1}} +$$

$$+ \frac{(n+1)^{-\frac{1}{2}-\alpha}\eta_n^\alpha(\gamma)}{(\sin\gamma)^{\frac{3}{2}}\left(\sin\frac{\gamma}{2}\right)^{\alpha+1}} + \frac{\zeta_n^\alpha(\gamma)}{(n+1)\left(\sin\frac{\gamma}{2}\right)^3}, \qquad (2.10)$$

where $|\eta_n^\alpha(\gamma)| < C$ and $|\zeta_n^\alpha(\gamma)| < C$.

Let us represent A_2 as follows

$$A_2 = \int_{\frac{1}{n}}^{\frac{\pi}{2}} |K_n^\alpha(\cos t)| \left| \ln\cos\frac{t}{2}\right| \sin t\, dt + \int_{\frac{\pi}{2}}^{\pi-\frac{1}{n}} |K_n^\alpha(\cos t)| \left| \ln\cos\frac{t}{2}\right| \sin t\, dt = B_1 + B_2.$$

It is easy to show that the estimate

$$B_1 \leq \begin{cases} C_1 n^{\frac{1}{2}-\alpha} & for \ \ \alpha < \frac{1}{2}, \\ C \ln n & for \ \ \alpha = \frac{1}{2}, \\ C & for \ \ \alpha > \frac{1}{2}, \end{cases} \qquad (2.11)$$

holds for B_1. By virtue of (2.10), for B_2 we have

$$B_2 = \int_{\frac{\pi}{2}}^{\pi-\frac{1}{n}} |K_n^\alpha(\cos t)| \left| \ln\cos\frac{t}{2}\right| \sin t\, dt < Cn^{\frac{1}{2}-\alpha}\int_{\frac{\pi}{2}}^{\pi-\frac{1}{n}} |\ln\cos\frac{t}{2}| \frac{\sin t}{\sqrt{\sin t}} dt +$$

$$+ Cn^{-\frac{1}{2}-\alpha}\int_{\frac{\pi}{2}}^{\pi-\frac{1}{n}} \frac{1}{(\sin t)^{\frac{3}{2}}} |\ln\cos\frac{t}{2}| \sin t\, dt + \frac{C}{n}\int_{\frac{\pi}{2}}^{\pi-\frac{1}{n}} |\ln\cos\frac{t}{2}| \sin t\, dt <$$

$$< Cn^{\frac{1}{2}-\alpha} + \frac{C}{n^{\frac{1}{2}+\alpha}}\int_{\frac{\pi}{2}}^{\pi-\frac{1}{n}} \frac{dt}{\sqrt{\sin t}} + C < Cn^{\frac{1}{2}-\alpha} + C. \qquad (2.12)$$

From (2.8), (2.9), (2.11) and (2.12) it follows that

$$I_2 \leq \begin{cases} Cn^{\frac{1}{2}-\alpha} & for \ \ \alpha < \frac{1}{2}, \\ C \ln n & for \ \ \alpha = \frac{1}{2}, \\ C & for \ \ \alpha > \frac{1}{2}. \end{cases}$$

Let us now estimate L_n^α from below. Taking the inequality

$$L_n^\alpha \geq \left| \int_0^\pi |K_n^\alpha(\cos\alpha)| \sin\gamma\, d\gamma - \int_0^\pi |N_n^\alpha(\cos\gamma)| \sin\gamma\, d\gamma \right| = |I_1 - I_2|$$

into account, it suffices to estimate I_2 from below, but for this we need to estimate B_1 and B_2 from below. By virtue of (2.10), for B_1 we have

$$B_1 = \int_{\frac{1}{n}}^{\frac{\pi}{2}} |K_n^\alpha(\cos t)| \left| \ln \cos \frac{t}{2} \right| \sin t \, dt = D_1 + D_2 + D_3.$$

$$D_1 = \frac{A_n^{\frac{1}{2}}}{A_n^\alpha} \int_{\frac{1}{n}}^{\frac{\pi}{2}} \frac{\left| \sin \left[\left(n + 1 + \frac{\alpha}{2} \right) t - \frac{1+2\alpha}{4} \pi \right] \right|}{\sqrt{2 \sin t} \left(2 \sin \frac{t}{2} \right)^{\alpha+1}} \sin t \, dt =$$

$$= \frac{A_n^{\frac{1}{2}}}{2^{\alpha+1} A_n^\alpha} \int_{\frac{1}{n}}^{\frac{\pi}{2}} \frac{\left| \sin \left[\left(n + 1 + \frac{\alpha}{2} \right) t - \frac{1+2\alpha}{4} \pi \right] \right|}{\left(2 \sin \frac{t}{2} \right)^{\alpha+\frac{1}{2}}} \sqrt{\cos \frac{t}{2}} \, dt \geq$$

$$\geq C n^{\frac{1}{2}-\alpha} \int_{\frac{1}{n}}^{\frac{\pi}{2}} \frac{\left| \sin \left[\left(n + 1 + \frac{\alpha}{2} \right) t - \frac{1+2\alpha}{4} \pi \right] \right|}{t^{\alpha+\frac{1}{2}}} \, dt.$$

Let us introduce a new variable $\left(n + 1 + \frac{\alpha}{2} \right) t - \frac{1+2\alpha}{4} \pi = \beta$, *then*

$$D_1 \geq C n^{\frac{1}{2}-\alpha} \int_{(n+1+\frac{\alpha}{2})\frac{1}{n} - \frac{1+2\alpha}{4}\pi}^{(n+1+\frac{\alpha}{2})\frac{\pi}{2} - \frac{1+2\alpha}{4}\pi} \frac{|\sin \beta|}{\left(\frac{\beta + \frac{1+2\alpha}{4}\pi}{n+1+\frac{\alpha}{2}} \right)^{\alpha+\frac{1}{2}}} \frac{d\beta}{n+1+\frac{\alpha}{2}} \geq$$

$$\geq C n^{\frac{1}{2}-\alpha} \left(n + 1 + \frac{\alpha}{2} \right)^{\alpha-\frac{1}{2}} \int_{\pi}^{\frac{n}{2}\pi} \frac{|\sin \beta| d\beta}{\left(\beta + \frac{1+2\alpha}{4}\pi \right)^{\alpha+\frac{1}{2}}}.$$

Since $\left(n + 1 + \frac{\alpha}{2} \right) \frac{1}{n} - \frac{1+2\alpha}{4}\pi \leq \pi$ *and* $\left(n + 1 + \frac{\alpha}{2} \right) \frac{\pi}{2} - \frac{1+2\alpha}{4}\pi \geq \frac{n}{2}\pi$ *for* $\alpha \leq 1$, *we*
have

$$D_1 \geq C \int_{\pi}^{[\frac{n}{2}]\pi} \frac{|\sin \beta| d\beta}{\left(\beta + \frac{1+2\alpha}{4}\pi \right)^{\alpha+\frac{1}{2}}} = C \sum_{v=1}^{[\frac{n}{2}]-1} \int_{v\pi}^{(v+1)\pi} \frac{|\sin \beta| d\beta}{\left(\beta + \frac{1+2\alpha}{4}\pi \right)^{\alpha+\frac{1}{2}}}.$$

By the change of the variable $\beta = t + v\pi$ *we obtain*

$$D_1 \geq C \sum_{v=1}^{[\frac{n}{2}]-1} \int_0^{\pi} \frac{|\sin(t + v\pi)| dt}{\left(t + v\pi + \frac{1+2\alpha}{4}\pi \right)^{\alpha+\frac{1}{2}}} \geq C \sum_{v=1}^{[\frac{n}{2}]-1} \int_0^{\pi} \frac{\sin t \, dt}{v^{\alpha+\frac{1}{2}}} =$$

$$= C \sum_{v=1}^{[\frac{n}{2}]-1} \frac{1}{v^{\alpha+\frac{1}{2}}} \geq C \int_1^{[\frac{n}{2}]-1} \frac{dt}{t^{\alpha+\frac{1}{2}}}.$$

Let us consider the following cases:
I. If $\alpha < \frac{1}{2}$, *then*

$$D_1 \geq C \int_1^{[\frac{n}{2}]-1} t^{-\alpha-\frac{1}{2}} dt = \frac{C}{\frac{1}{2}-\alpha} \left\{ \left(\left[\frac{n}{2} \right] - 1 \right)^{\frac{1}{2}-\alpha} - 1 \right\} \geq C n^{\frac{1}{2}-\alpha}. \qquad (2.13)$$

II. If $\alpha = \frac{1}{2}$, then

$$D_1 \geq C \int_1^{[\frac{n}{2}]-1} \frac{dt}{t} = C \ln \left(\left[\frac{n}{2}\right] - 1 \right) \geq C \ln n. \tag{2.14}$$

III. If $\alpha > \frac{1}{2}$, then

$$D_1 \geq C \int_1^{[\frac{n}{2}]-1} t^{-\alpha-\frac{1}{2}} dt = \frac{C}{\alpha - \frac{1}{2}} \left\{ 1 - \left(\left[\frac{n}{2}\right] - 1 \right)^{\frac{1}{2}-\alpha} \right\} \geq C. \tag{2.15}$$

From (2.13), (2.14) and (2.15) we have

$$D_1 \geq \begin{cases} C n^{\frac{1}{2}-\alpha} & \text{for} \quad \alpha < \frac{1}{2}, \\ C \ln n & \text{for} \quad \alpha = \frac{1}{2}, \\ C & \text{for} \quad \alpha > \frac{1}{2}. \end{cases} \tag{2.16}$$

In an analogous manner it can be shown that an analogue of the estimate (2.16) holds for D_2, D_3 and B_2.

Applying the inequality $|a - b| \geq \left| |a| - |b| \right|$ it can be established that an analogue of the estimate (2.16) also holds for I_2.

From (1.5) it follows that $L_n^\alpha \geq \left| |I_1| - |I_2| \right|$. Therefore

$$L_n^\alpha \geq \begin{cases} C n^{\frac{1}{2}-\alpha} & \text{for} \quad \alpha < \frac{1}{2}, \\ C \ln n & \text{for} \quad \alpha = \frac{1}{2}, \\ C & \text{for} \quad \alpha > \frac{1}{2}. \end{cases}$$

The theorem is proved.

References

[1] Gelfand J. M., Shapiro Z. I., Representations of a group of rotations of a three-dimensional space and their applications. *Usp. Mat. Nauk,* **7**:1, 1952, 3–117 (in Russian).

[2] Litvinkov S. S., On the convergence of Fourier series with respect to generalized spherical functions, *Izv. Visš. Učhebn. Zaveden. Matematika*, **4**, 1962, 92–103 (in Russian).

[3] Hobson E. W., *The theory of spherical and ellipsoidal harmonics,* Moscow, 1952 (in Russian).

[4] Topuria S. B., *The Fourier-Laplace series on a sphere,* Tbilisi, Izdat. Tbil. Gos. Univ., 1987 (in Russian).

[5] Khocholava V. V., On Lebesgue constants of Fourier-Laplace series, *Trudy Gruz. Polytekhn. Inst.,* No. **5**(237), 1981, 24–34 (in Russian).

[6] Kogbetliantz E., Recherches sup la Summabilite des series Ultraspheriques par methode des moyennes arithmetiques. *J. Math. Pures appl.* **9**, No. 3 (1924), 107–187.

INDEX

E

I

J

K

L

T